Integrated Study of Viral Genomes

Integrated Study of Viral Genomes

Edited by **Liam Romano**

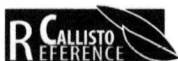

R Callisto
Reference

New York

Published by Callisto Reference,
106 Park Avenue, Suite 200,
New York, NY 10016, USA
www.callistoreference.com

Integrated Study of Viral Genomes
Edited by Liam Romano

International Standard Book Number: 978-1-63239-434-7 (Hardback)

Printed in the United States of America.

Contents

Preface

This book provides an overview of viral genomes. It is an all-inclusive book on the vast subject of viral genomes. Viruses are small agents which cause infection within the host by multiplying inside the functioning cells of the host. The comprehension of the molecular incidents underlying the infectious procedure has been of particular significance to advance strategies intended to combat viral disorders of medicinal, veterinary and agricultural significance. Few viruses cause horrible diseases, while others are also of interest due to various reasons. The writers of this book present the readers with a viewpoint on the broad range of virus-host systems. This book deals with a wide variety of topics such as genome diversity and development, virus-host interactions, counting technically significant features, regulation of replication and gene expression. This book also covers a huge range of scientific approaches to assess genome deviation or stability. It intends to provide useful data regarding many features related to viral genomes for both students and experts.

After months of intensive research and writing, this book is the end result of all who devoted their time and efforts in the initiation and progress of this book. It will surely be a source of reference in enhancing the required knowledge of the new developments in the area. During the course of developing this book, certain measures such as accuracy, authenticity and research focused analytical studies were given preference in order to produce a comprehensive book in the area of study.

This book would not have been possible without the efforts of the authors and the publisher. I extend my sincere thanks to them. Secondly, I express my gratitude to my family and well-wishers. And most importantly, I thank my students for constantly expressing their willingness and curiosity in enhancing their knowledge in the field, which encourages me to take up further research projects for the advancement of the area.

Editor

Part 1

Virus Genomes Organization and Functions

Nudivirus Genomics and Phylogeny

Yongjie Wang[1,*], Olaf R.P. Bininda-Emonds[2], and Johannes A. Jehle[3]
*[1]Laboratory of Marine and Food Microbiology,
College of Food Science and Technology,
Shanghai Ocean University, Shanghai,
[2] Institute for Biology and Environmental Sciences (IBU),
Carl von Ossietzky University Oldenburg, Oldenburg,
[3]Institute for Biological Control,
Federal Research Centre for Cultivated Plants,
Julius Kühn-Institut, Darmstadt,
[1]China
[2,3]Germany*

1. Introduction

The nudiviruses (NVs) are a diverse group of arthropod-specific large DNA viruses. They form rod-shaped, enveloped virions, and replicate in the nucleus of infected cells. Nudivirus genomes are covalently closed circles of double stranded DNA molecules. Some nudiviruses have been used as potential bio-control agent for management of economically important arthropod pests (Burand 1998, Huger 1966). A variety of non-occluded rod-shaped dsDNA viruses replicating in the host nucleus have been observed in various host species, belonging to Lepidoptera, Trichoptera, Diptera, Siphonaptera, Hymenoptera, Neuroptera, Coleoptera, Homoptera, Thysanura, Orthoptera, Acarina, Araneina, and Crustacea. They had been considered as "non-occluded baculoviruses" (Huger and Krieg 1991) or more recently as nudiviruses (Burand, 1998). Most of these viruses were identified solely based on morphological features and very limited biological data. Accordingly, it remains unclear whether they are evolutionarily monophyletic or polyphyletic lineages, and whether they are genetically related to each other, to the well-investigated baculoviruses, or to other large dsDNA viruses.

Thus far, only a few nudiviruses have somehow been studied in detail. The *Oryctes rhinoceros* nudivirus (OrNV), formerly known as the rhinoceros beetle virus or *Oryctes baculovirus*, was discovered in the 1960s and has been widely used to control rhinoceros beetle (*O. rhinoceros*) in coconut and oil palm in Southeast Asia and the Pacific (Huger 1966, Jackson et al. 2005). It has an enveloped rod-shaped virion and replicates in the nucleus of infected midgut and fat body cells (Huger 1966, Payne 1974, Payne et al. 1977). *Heliothis zea* nudivirus 1 (HzNV-1), formerly known as Hz-1 virus or the non-occluded baculovirus Hz-1, was originally described as a persistent viral infection in the IMC-Hz-1 cell line isolated

---

*Corresponding Author

from the adult ovarian tissues of the corn earworm *Heliothis zea* (Granados et al. 1978). It can also persistently infect several other lepidopterous cell lines, e.g. IPLB-1075 (*H. zea*), IPLB-SF-21 (*Spodoptera frugiperda*), IPLB-65Z (*Lymantria dispar*) and TN-368 (*Trichoplusia ni*) (Granados et al. 1978, Kelly et al. 1981, Lu and Burand 2001). In contrast, clear infections have not been observed when the virus was inoculated into larvae of *H. zea*, *H. armigera*, *Estigmene acrea*, *S. frugiperda*, and *S. littoralis* (Granados et al. 1978, Kelly et al. 1981). The potential molecular mechanisms associated with this defective host infection of HzNV-1 need to be explored, which will shed light on the viral evolution. *Gryllus bimaculatus* nudivirus (GbNV) infects nymphs and adults of several field crickets *G. bimaculatus*, *G. campestris*, *Teleogryllus oceanicus* and *T. commodus*, and replicates in the nuclei of the infected fat body cells (Huger 1985). *Heliothis zea* nudivirus 2 (HzNV-2), previously known as gonad-specific virus, *H. zea* reproductive virus or Hz-2V, was first observed in the gonads of adult corn earworm *H. zea*. Its infection brings about deformities of the reproductive organs of insect hosts, which in turn lead to sterility in both female and male moths (Burand and Rallis 2004, Raina et al. 2000). HzNV-2 is also able to infect other Noctuid species and to replicate in two lepidopteran insect cell lines of TN-368 and Ld652Y, derived from ovarian tissues (Burand and Lu 1997, Lu and Burand 2001, Raina and Lupiani 2006).

1.1 Infection cycles and gene expression

Only very limited data on the infection cycle of nudiviruses are available. Their life cycle in either cell culture or natural hosts is still poorly understood. HzNV-1 has a bi-phasic infection process of latency and productivity in its life cycle. In the latent phase of infection, viruses either exist as episomes or insert their DNA into the host genome (Lin et al. 1999), and keep latency for many passages in the infected insect cells (Chao et al. 1992, Lin et al. 1999, Wood and Burand 1986); virus particles are undetectable in most of these latently infected cells. Sometimes virions are released from as few as 0.2% of latently infected cells, resulting in the presence of low viral titers (around 10^3 PFU/ml) in the culture medium (Chao et al. 1998, Lin et al. 1999). During the productive infection cycle, in contrast, high titers of virus progeny are produced, resulting in the death of most cells. Often, however, a small proportion of the cells, usually less than 5%, are latently infected, and viruses stay in these cells for a prolonged period of time (Chao et al. 1992, Wood and Burand 1986). Upon in vitro infection, OrNV appears to attach to and subsequently internalizes into cultured cells by pinocytosis (Crawford and Sheehan 1985), a mechanism involving the formation of invaginations by the cell membrane, which close and break off to generate virus-containing vacuoles in the cytoplasm. While it remains unknown how the viral DNA is released into the cytoplasm and eventually enters the nucleus. During the later stage of replication, along with the cytopathic changes to the nucleus, the virogenic stroma is developed, where the viral envelopes and nucleocapsid shells are produced and subsequently packaged with viral DNA. At last, the matured virions enter the cytoplasm followed by budding through the cell membrane (Crawford and Sheehan 1985) .

In vitro sequential expression of viral genes encoding structural and intracellular proteins has been divided into early, intermediate and late stage in the replication cycle of OrNV (Crawford and Sheehan 1985). The temporal gene expression profiles of HzNV-1 during productive infection are divided into three stages: (i) the early stage, 0 to 2 h p.i.; (ii) the intermediate stage, 2 to 6 h p.i.; and (iii) the late stage, which includes all virus-specific events appearing after 6 h p.i. (Chao et al. 1992). Persistency-associated transcript 1 (PAT1), expressed by persistency-associated gene 1 (*pag1*), is the only detectable transcript during latent infection of HzNV-1 (Chao et al. 1998).

1.2 Taxonomy and nomenclature

Given that they share similar structural and replication aspects with baculoviruses of insects, nudiviruses were previously classified as the so-called "non-occluded baculoviruses" (NOBs) (Huger and Krieg 1991). NOBs were later removed from the family *Baculoviridae* because no genetic data were available which would have supported their relationship (Mayo 1995). Nudiviruses have been also referred to as intranuclear bacilliform viruses (IBVs). Notably, unlike baculoviruses, nudiviruses generally lack occlusion bodies (OBs). The genus name *Nudivirus* has been proposed to accommodate this group. Based on the currently available morphological and molecular data, the following demarcation criteria were proposed for classification of a candidate virus into the genus *Nudivirus*: (i) Viral genome is consist of large circular dsDNA molecule; (ii) A set of conserved core genes are shared among members and viruses propagate in the nuclei of infected host cells; (iii) Morphology of virion is rod-shaped and enveloped; (iv) Viruses are transmitted per oral and/or per parenteral route, and infect larvae and/or adults with diverse tissue and cell tropisms (Wang et al. 2007a). Obviously, these demarcation criteria need to be complemented with more biological properties, such as virion properties, infection and replication strategies, as well as host range and virus ecology, becoming available. To name a nudivirus species, it was suggested to follow the nomenclature for other large eukaryotic dsDNA viruses, host name with the suffix name of nudivirus (Wang et al. 2007c).

Presently, nudiviruses comprise five tentative species, OrNV, GbNV, HzNV-1, HzNV-2, and *Penaeus monodon* nudivirus (PmNV) (Wang and Jehle 2009). Considering their similarities to baculoviruses and, on the other hand, taking their distinct biological, ecological features and virion properties into account, the establishment of an independent family "Nudiviridae" within a new order "Baculovirales" along with the *Baculoviridae* seems most appropriate. The establishment of an order "Baculovirales" would allow subsequent flexible integration of other "baculovirus-related" but highly diverged viruses, such as the proposed "Hytrosaviridae" (Abd-Alla et al. 2009) or the *Nimaviridae*, without taxonomic re-definition of the family *Baculoviridae*.

2. Genome structure

2.1 Genome size

HzNV-1 was the first completely sequenced nudivirus (Cheng et al. 2002). Its genome is 228,089 bp in size, has a G+C content of 42%, and encodes 154 ORFs (Table 1). HzNV-1 ORFs are randomly distributed on both DNA strands with 45% clockwise orientation and 55% counterclockwise orientation. HzNV-2, the close relative to HzNV-1, has a genome of 231,621 bp, only slightly longer than that of HzNV-1, with a G+C content of 42% identical to HzNV-1 (Wang et al. 2007a). Later on, the genome of OrNV, the first discovered nudivirus, was partially sequenced (Wang et al. 2007c). Recently, the complete genome of OrNV was successfully achieved using DNA generated with multiple displacement amplification (MDA) (Wang et al. 2008). The OrNV genome is 127,615 bp in size with a G+C content of

42% and contains 139 ORFs (Table 1, Fig. 1) (Wang et al. 2008, Wang et al. 2011). Thus far, the smallest nudivirus genome sequenced is GbNV, which is 96,944 bp in length with a G+C content of 28% and contains 98 ORFs. Among them, 58% are in clockwise distribution and 42% are in reverse direction (Table 1, Fig. 2) (Wang et al. 2007b). Genome sequencing of other nudiviruses such as the *Tipula oleracea* nudivirus (ToNV) (E. Herniou, personal communication) is ongoing. Partial nucleotide sequences of the shrimp PmNP genome are already accessible in GenBank.

Virus	Size in bp	GC content (mol %)	No. of ORFs	Clockwise orientation* (No. of ORF/ ORF%)	Gene density (kbp per ORF)	No. of Rsr
HzNV-1	228,089	41.8	154	69/45	1.47	3
GbNV	96,944	28.0	98	57/58	0.93	17
OrNV	127,615	42.0	139	64/46	0.82	20

Table 1. Characteristics of nudivirus genomes. *Clockwise orientation means in the same orientation as the DNA polymerase B ORF. Rsr = Repetitive sequence regions.

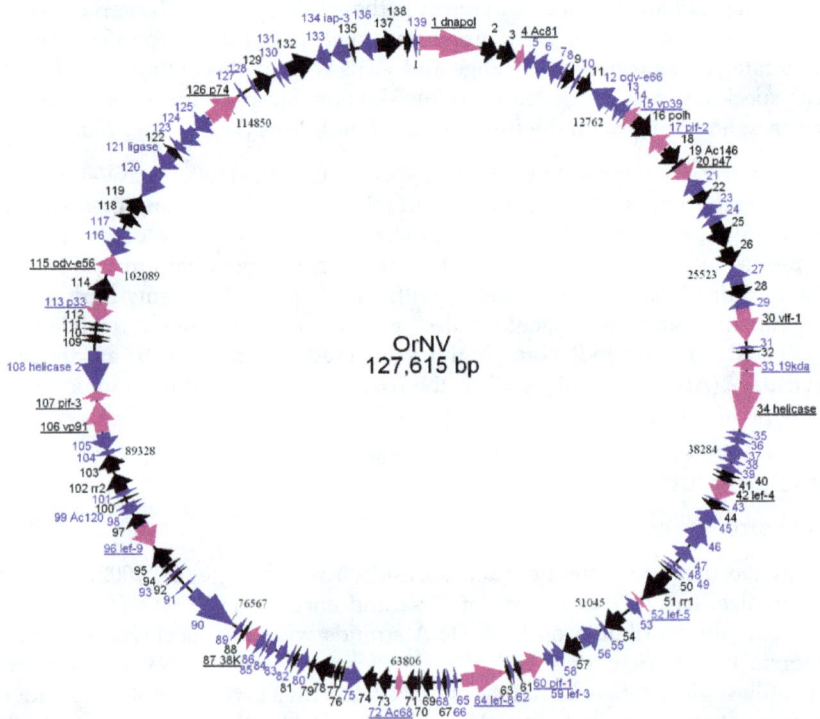

Fig. 1. The genome map of OrNV. ORFs and their transcriptional directions are indicated in arrows. Black color, clockwise coding; blue color, counterclockwise coding; pink color, the 20 baculovirus core gene homologues.

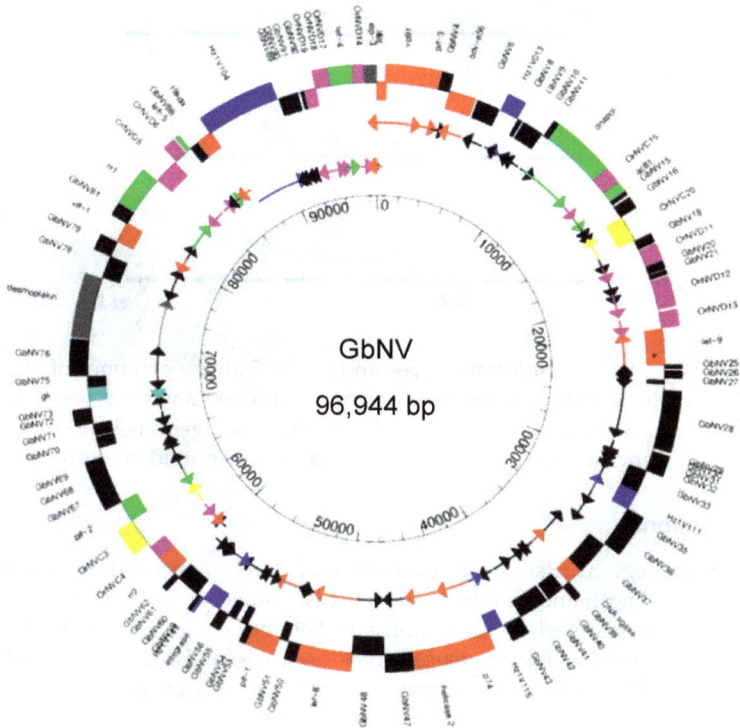

Fig. 2. Circular map of the GbNV genome. ORFs and their transcriptional directions are indicated in boxes and arrows, respectively. Red color, baculovirus and HzNV-1 homologues; black color, GbNV specific ORFs; blue color, HzNV-1 homologues; green color, baculovirus, HzNV-1 and OrNV homologues; pink color, OrNV homologues; yellow color, HzNV-1 and OrNV homologues; blue color, HzNV-1 homologues; light blue, cellular homologues; grey color, baculovirus homologues. Taken from Wang et al. (2007b) with permission from the American Society for Microbiology.

2.2 Gene order

Similar to what is observed in other viral families (e.g., the *Baculoviridae*), gene order is poorly conserved in nudivirus genomes as well. OrNV and GbNV share a number of gene clusters, comprising 2–7 collinearly arranged genes, distributed throughout their genomes (Wang and Jehle 2009). In contrast, only two gene clusters were detected between OrNV and HzNV-1 (Wang and Jehle 2009, Wang et al. 2011). However, a gene cluster of *helicase, pif-4/19 kda,* and/or *lef-5* is present in all three nudivirus genomes (Fig. 3), which is similar to the conserved core gene cluster of four genes of *helicase, pif-4/19 kda, 38K* and *lef-5* in all sequenced baculoviruses (Herniou et al. 2003, Jehle and Backhaus 1994). Hence, core gene clustering strongly supports the hypothesis of a common ancestor of nudiviruses and baculoviruses.

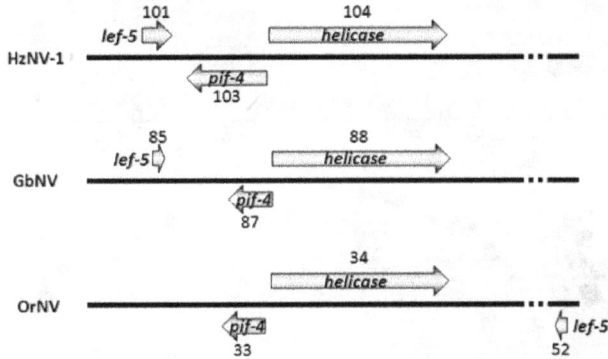

Fig. 3. Conserved gene cluster of *helicase*, *pif-4*, and/or *lef-5* on the genomes of HzNV-1, GbNV and OrNV. ORF is indicated by boxed arrows; number above and below the boxed arrows represents the sequential ordering of ORFs on the viral genomes; line in bold represents the viral genomes; dashed bold line indicates the omitted genomic ranges.

2.3 Repetitive regions

Repetitive sequence regions (Rsr) were detected in all three sequenced nudivirus genomes. They are variable in length and numbers and are distributed throughout the genome. They are homologous neither to each other within and between genomes, nor to those of other large dsDNA viruses, such as baculoviruses, hytrosaviruses and white spot syndrome virus (WSSV). Rsr appear to be a universal feature of all large dsDNA viruses.

3. Nudivirus gene structure

3.1 Promoter motifs

A promoter motif of TTATAGTAT was identified at the upstream regulatory regions of HzNV-1 late gene *p34* (ORF79) and *p51* (ORF64) (Guttieri and Burand 1996, Guttieri and Burand 2001). It was also found within 200 bp of the initiation codon of HzNV-1 ORF81 based on in silica sequence analysis (Cheng et al. 2002). Although consensus early and late promoter motif sequences similar to those of baculoviruses were predicted in nudivirus ORFs, convincing experimental data remain unavailable (Cheng et al. 2002, Wang et al. 2007c).

3.2 Untranslated regions

In the HzNV-1 transcripts, the early gene *hhi1* (HzNV-1 HindIII fragment 1 gene) contains 270 nucleotides (nts) of 5′ untranslated region (UTR) which, together with its upstream 62 bps, compose *hhi1* early promoter (Wu et al. 2008, Wu et al. 2010); the HzNV-1 late gene *p34* (ORF79) possesses 16 and 17 nts of 5′ UTR, respectively, differing by 1 nt, and both 5′ UTRs overlap with the identified 9 bp late promoter motif of the *p34* (Guttieri and Burand 1996); as for the HzNV-1 late gene *p51* (ORF64), the major late transcriptional initiation site is at −205 bp relative to the translational start codon and seven minor late start sites locate at various positions upstream of this primary site (Guttieri and Burand 2001). The putative polyadenylation signals (AATAAA) downstream of the stop codon of the *p34* and *p51* were found (Guttieri and Burand 1996, Guttieri and Burand 2001). Thus far, nothing is known on how UTR mediate the translational efficiency of nudivirus genes.

3.3 Open reading frames (ORFs)

Computer-assisted ORF prediction included all sequences starting with ATG followed by 50 or more amino acid (aa) codons and minimum overlap with other ORFs. ORFs with less than 50 aa are only considered as putative genes in cases of clear homology to ORFs in other dsDNA viruses.

4. Gene content and conserved gene functions

There are 66, 34, and 33 homologous genes shared by OrNV and GbNV, OrNV and HzNV-1, and GbNV and HzNV-1, respectively (Table 2), suggesting that OrNV and GbNV are more closely related to each other than to HzNV-1. OrNV, GbNV and HzNV-1 have 33 genes in common (Table 2). Strikingly, 20 out of them are homologues of baculovirus core genes, which are present in all 54 baculovirus genomes that have been deposited in GenBank as of July 2011. Baculovirus 31 core genes play crucial role in virus replication cycle and are the evolutionarily conserved marker genes in identification, classification and phylogeny of baculoviruses (Herniou et al. 2003, Herniou and Jehle 2007, Jehle et al. 2006a, Jehle et al. 2006b, van Oers and Vlak 2007). Nine other ORFs are likely involved in DNA replication, repair and recombination, and nucleotide metabolism; one is homologous to baculovirus *iap-3* gene; two others are nudivirus-specific ORFs of unknown function (Table 2). The presence of 20 baculovirus core genes in nudiviruses strongly indicates that nudiviruses and baculoviruses are the closest lineages among the viruses known so far.

Besides in nudiviruses, homologues to baculovirus core genes were also detected in two salivary gland hypertrophy viruses (SGHVs) MdSGHV infecting the house fly *Musca domestica* and GpSGHV infecting the tsetse fly *Glossina pallidipes* (Abd-Alla et al. 2008, Garcia-Maruniak et al. 2008). GpSGHV and MdSGHV share 37 homologous ORFs and are phylogenetically closely related (Garcia-Maruniak et al. 2009). Surprisingly, several core gene homologues of baculoviruses was identified in the marine WSSV as well (Wang et al. 2011), suggesting that WSSV, as suspected since it was observed, is evolutionarily related, albeit distantly, to baculoviruses. Most strikingly, nudiviruses, SGHVs and WSSV have the homologues to the genes encoding peroral infectivity factors (*p74*, *pif-1*, *pif-2* and *pif-3*) (Wang et al. 2011). These four *pif* genes are conserved among all sequenced baculoviruses and are absolutely crucial for successful peroral infection of insect hosts. As midgut infection is the essential first step in the invasion of baculoviruses, PIFs may be the key determinants of host range and virulence. Accordingly, it seems to be reasonable to hypothesize that a highly conserved interaction mode of viruses and hosts upon primary infection is present in nudiviruses, baculoviruses, SGHVs and WSSV. However, only limited data on the function of the PIF proteins has been delineated in baculoviruies (Slack and Arif 2007), let alone in nudiviruses, SGHVs and WSSV. Obviously, deeply exploring of the molecular mechanisms of the PIF proteins as well as their homologues is crucial for better understanding of host range, zoonotic behaviour, and epizootic or enzootic disease of these viruses. In addition, nudiviruses appear to share homologues of the transcription apparatus of baculoviruses, suggesting that a similar mode of late gene transcription is used in nudiviruses as well (Wang et al. 2011). Taken together, this finding provides crucial clues to the origin and evolution of arthropod specific large dsDNA viruses. The biochemical and biological function of the genes predicted in nudiviruses remains unknown. Only the occlusion body protein-encoding gene of PmNV has been molecularly characterised, revealing no homology to any other genes deposited in Genbank (Chaivisuthangkura et al. 2008).

Function	Name	OrNV	GbNV	HzNV-1	PmNV
DNA replication, repair, and recombination	dnapol	1	12	131	N.d.
	helicase	34	88	104	N.d.
	helicase 2	108	46	60	N.d.
	integrase	75	57	144	+
	ligase	121	38	36	N.d.
	lef-3	59	86	–	N.d.
Nucleotide metabolism	rr1	51	82	95	N.d.
	rr2	102	63	73	N.d.
	tk	58	74	115	+
	tk	117	34	111	+
	tk	125	44	71	+
	tk	137	17	51	+
Transcription	p47	20	69	75	N.d.
	lef-4	42	96	98	N.d.
	lef-8	64	49	90	N.d.
	lef-9	96	24	75	+
	lef-5	52	85	101	+
	vlf-1	30	80	121	+
Oral infectivity	p74	126	45	11	N.d.
	pif-1	60	52	55	N.d.
	pif-2	17	66	123	N.d.
	pif-3	107	3	88	N.d.
	19 kda/pif-4	33	87	103	N.d.
	odv-e56/pif-5	115	5	76	N.d.
Packaging, assembly, and morphogenesis	polh/gran	16	65	69	N.d.
	ac68	72	55	74	N.d.
	38 K	87	1	10	+
Inhibition of apoptosis	iap-3	134	98	138	N.d.
Unknown function	vp39	15	64	89	N.d.
	vp91	106	2	46	N.d.
	ac81	4	14	33	N.d.
	ac92	113	7	13	N.d.
		47	19	30	N.d.
		76	58	143	N.d.
		3	13	–	N.d.
		18	67	–	N.d.
		22	72	–	N.d.

Function	Name	OrNV	GbNV	HzNV-1	PmNV
		23	74	–	N.d.
		24	75	–	N.d.
		25	76	–	N.d.
		27	78	–	N.d.
		29	81	–	N.d.
		39	93	–	N.d.
		40	94	–	N.d.
		41	95	–	N.d.
		44	97	–	N.d.
		45	23	–	N.d.
		46	22	–	N.d.
		53	84	–	N.d.
		54	83	–	N.d.
		61	51	–	N.d.
		79	59	–	N.d.
		80	60	–	N.d.
		86	61	–	N.d.
		90	28	–	N.d.
		95	9	–	N.d.
		104	62	–	N.d.
		105	43	–	N.d.
		114	6	–	N.d.
		116	33	–	N.d.
		118	35	–	N.d.
		119	36	–	N.d.
		120	37	–	N.d.
		122	39	–	N.d.
		123	41	–	N.d.
		132	48	–	N.d.
		6	–	109	N.d.
		–	–	52	+
		–	–	64	+
		–	–	93	+
		–	–	118	+
		–	–	141	+

Table 2. Homologous genes conserved in nudiviruses. –: Absent; +: Present; N.d.: Not determined. The predicted ORFs in nudiviruses are presented in number. Homologues to baculovirus core genes are marked in bold face.

5. Nudivirus phylogeny

5.1 Phylogenetic analysis

Due to the poorness of information of other distinguishing features, single gene phylogeny and/or phylogenomics became the most important approach to delineate the relationship of concerned viruses on strain and species level. However, single gene phylogenies have fallen increasingly into disfavor given the recognition that gene trees can often differ substantially from the underlying species tree due to a variety of evolutionary events in addition to simply stochastic or analytical error. This is likely to be especially true in DNA viruses with the substantial evolutionary dynamics intrinsic to their genomes. Thus, (i) horizontal gene transfer (HGT), i.e., exchange with other viruses, symbiotic bacteria and hosts; (ii) homologous or non-homologous recombination with other viruses; (iii) gene/domain duplication and rearrangement; and (iv) lineage specific gene loss/expansion all impose significant complications on both the bioinformatic detection of orthologous genes and on the accuracy of the resulting gene trees with respect to the overall species tree (Shackelton and Holmes 2004).

To overcome these problems, a set of conserved genes were analysed using both the supertree and supermatrix approaches. Multiple sequence alignments of individual genes were performed using any of T-Coffee (Notredame, Higgins and Heringa 2000), MUSCLE (Edgar 2004), ClustalW/X (Chenna et al. 2003), MAFFT (Katoh et al. 2002) and Kalign (Lassmann and Sonnhammer 2005), and were manually refined as needed. Sequence alignment quality was assessed by using MUMSA (Lassmann and Sonnhammer 2005). In particular, 20 of the 30 baculovirus core genes (Table 2) were analysed, considering that they are evolutionarily more conserved than other nonessential genes and that homologues to all or most of them are present in NVs, SGHVs, and WSSV as well as, albeit more distantly, in other large eukaryotic dsDNA viruses such as NCLDVs (nucleocytoplasmic large DNA viruses) and herpesviruses (Table 3).

NALDV			WSSV	NCLDV						Herpes-virus	Homologue	
Baculo-virus	Nudi-virus	SGHV		ASFV	Pox-virus	Irido-virus	Asco-virus	Phy-codnavirus	Mimi-virus		Definition	Putative function
■	■	■	■	■	■	■	■	■	■	■	DNA polymerase B	DNA replication, recombination and repair
■	■	■	-	■	■	■	■	■	■	-	Helicase / Helicase III (D5R like helicase)	
■	■	■	-	■	■	■	■	□	■	-	LEF-8 / Rpb2	
■	■	■	-	■	■	■	■	□	■	-	LEF-9 / Rpb1	Transcription
■	■	■	-	■	■	■	□	■	■	-	LEF-5 / TFIIS-like transcription factor	
■	■	■	■	■	■	■	■	■	■	-	P33 / Thiol oxidoreductase	Virion packaging and morphogenesis

Table 3. Ancient core genes identified in NALDVs, WSSV, NCLDVs, and herpesviruses. Black squares, homologue detected in all available genomes; grey squares, in many but not in all available genomes; white squares, in few available genomes; –, not detected. Homologue definition, gene name in NALDVs / in NCLDVs. Ascovirus is considered to be a member of the NCLDVs because of its high sequence similarity to iridovirus.

The supertree and supermatrix framework represent alternative strategies to the issue of data combination. In the supermatrix approach, all the primary character data are combined into a single supermatrix that is analysed using standard phylogenetic methods (de Queiroz and Gatesy 2007). By contrast, the supertree approach combines phylogenetic trees derived from individual partitions of the full data set (here the individual gene trees) to likewise derive a single, joint phylogenetic estimate (Bininda-Emonds 2004a). Thus, the supertree approach addresses conflict and congruence at the level of the source trees rather than at the level of the primary data (Bininda-Emonds 2004b). Although this approach has been criticised because of the inherent loss of information (among others, see de Queiroz and Gatesy 2007), numerous simulation studies have demonstrated that this loss of information is not detrimental in practice (see Bininda-Emonds 2004a). Moreover, the contrasting approaches of the supertree and supermatrix frameworks form the basis of the global congruence framework (Bininda-Emonds 2004b), whereby increased confidence is placed in those clades common to both approaches and increased attention is demanded on conflicting solutions, particularly when each is strongly supported.

For the supertree analyses, phylogenetic analyses of the individual gene trees were performed under a maximum likelihood (ML) framework using RAxML 7.0.4 (Stamatakis, Hoover and Rougemont 2008). Optimal substitution matrices for each amino acid data were selected initially using the Perl script ProteinModelSelector (http://icwww.epfl.ch/~stamatak/index-Dateien/Page443.htm) as implemented in batchRAxML (http://www.molekulare systematik.uni-oldenburg.de/33997.html) and then applied for the full ML analysis of each gene tree. In all cases, rate heterogeneity between sites was accounted for using the CAT approximation of the gamma distribution (Stamatakis 2006). The former is an approximation of the latter that is both computationally more efficient in terms of its memory demands and overall speed, and provides equivalent results (Stamatakis 2006). However, all final likelihood values were obtained under a true gamma distribution. ML analysis used the new fast bootstrapping approach (Stamatakis et al. 2008) that simultaneously obtained the ML tree as well as estimates of nodal support based on a non-parametric bootstrap (Felsenstein 1985). Bootstrap values were based on 1000 replicates. Gene trees were rooted on the herpesviruses HHV-3, HHV-4, and HHV-5 (as a monophyletic group) because they share the minimum number of conserved ancestral genes with the other viruses (Table 3); trees lacking herpesviruses were treated as unrooted.

The supertree analysis used the method of matrix representation with parsimony (MRP) (Baum 1992, Ragan 1992), whereby the topology of each gene tree was then encoded using additive binary coding: for each node in turn, all taxa descended from that node are scored as "1", all taxa otherwise present on the tree are scored as "0", and all remaining taxa as "?". Semi-rooted coding was employed in that rooted gene trees included an all-zero fictitious outgroup taxon to root the supertree; for unrooted gene trees, this taxon was coded using "?" (Bininda-Emonds, Beck and Purvis 2005). The matrix representations of all source trees were then combined into a single matrix that was analyzed using maximum parsimony (MP). Individual pseudocharacters in the matrix were weighted according to the bootstrap support of their corresponding nodes, a procedure that improves the accuracy of the supertree analysis by helping account for differential support within the primary character matrices (Bininda-Emonds and Sanderson 2001). MP searches in PAUP* v4.0b10 (Swofford 2002) used a heuristic search strategy based on a random addition sequence (10000 replicates), TBR branch swapping, and with up to 50000 equally most parsimonious trees (MPTs) being saved. The supertree was taken to be the 50% majority-rule consensus of all MPTs. Support for the nodes in the supertree was estimated using the rQS index (Bininda-Emonds et al. 2003, Price, Bininda-Emonds and Gittleman 2005) restricted to informative gene trees only; analyses used the Perl script QualiTree (http://www.molekularesystematik.uni-oldenburg.de/33997.html).

The rQS index measures the number of gene trees that explicitly support or conflict with a given node on the supertree. Values of 1 and -1 indicate universal support or conflict, respectively, among the set of gene trees (Fig. 4).

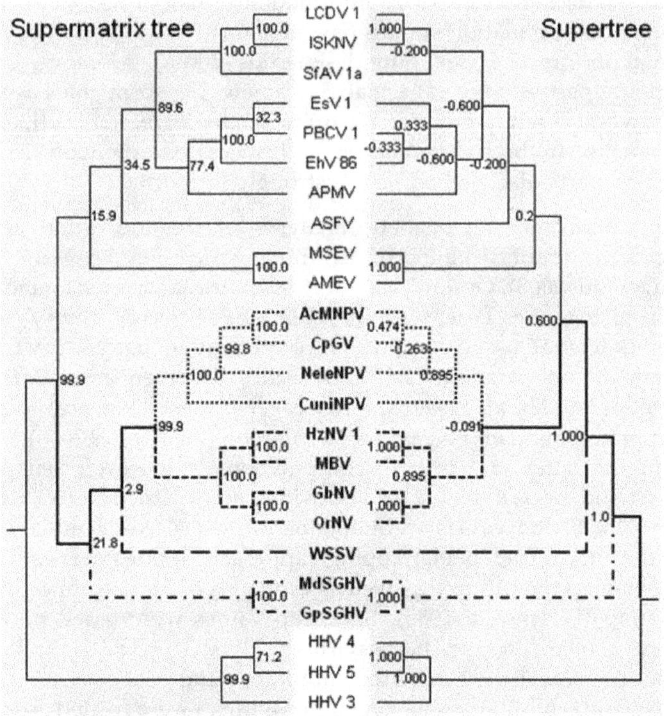

Fig. 4. Combined data trees based on the 20 conserved baculovirus core gene sequences: (A) ML supermatrix tree derived from a simultaneous analysis of the concatenated sequences and (B) weighted MRP supertree of the 20 gene trees in Wang et al, 2011. The latter represents the 50% majority-rule consensus of 71 equally most parsimonious solutions. Nodal support is given as non-parametric bootstrap frequencies (*n* = 1000) determined from the supermatrix data set / degree of support among the informative source trees for a given node as measured by the rQS index. Branch lengths in (A) are proportional to the average number of substitutions per site per unit time. GenBank accession numbers for these viral genomes and virus full names are listed as follows: NC_001623 (*Autographa californica* NPV, AcMNPV), NC_002816 (*Cydia pomonella* GV, CpGV), NC_005906 (*Neodiprion lecontei* NPV, NeleNPV), NC_003084 (*Culex nigripalpus* NPV, CuniNPV), NC_004156 (*Heliothis zea* NV 1, HzNV-1), NC_009240 (*Gryllus bimaculatus* NV, GbNV), EU747721 (*Oryctes rhinoceros* NV, OrNV), NC_010356 (*Glossina pallidipes* SGHV, GpSGHV), NC_010671 (*Musca domestica* SGHV, MdSGHV), NC_003225 (Shrimp white spot syndrome virus, WSSV), NC_001659 (African swine fever virus, ASFV), NC_002520 (*Amsacta moorei* EV, AMEV), NC_001993 (*Melanoplus sanguinipes* EV, MSEV), NC_008361 (*Spodoptera frugiperda* AV 1a, SfAV-1a), NC_001824 (Lymphocystis disease virus 1, LCDV-1), NC_003494 (Infectious spleen and kidney necrosis virus, ISKNV), NC_006450 (*Acanthamoeba polyphaga* mimivirus, APMV), NC_000852 (*Paramecium bursaria* Chlorella virus 1, PBCV-1), NC_007346 (*Emiliania huxleyi* virus 86, EhV-86), NC_002687 (*Ectocarpus siliculosus* virus 1, EsV-1), NC_001348 (Human herpesvirus 3, HHV-3), NC_001347 (Human herpesvirus 5, HHV-5), and NC_007605 (Human herpesvirus 4, HHV-4).

For the supermatrix analysis, all individual gene data sets were concatenated into a single, larger matrix that was analyzed using RAxML. Analysis used the same method as for the individual gene trees, except that a partitioned model was used whereby each gene partition was modeled individually according to the optimal model of evolution determined previously. Support values for each tree were also estimated using the support measure for the other technique. In other words, the rQS index was also applied to the supermatrix tree to estimate the support for its nodes across the gene trees and the bootstrap values for the nodes on the supertree were estimated using the 1000 bootstrap replicate trees derived from the supermatrix analysis (Fig. 4).

5.2 Phylogeny and evolution

5.2.1 Common ancestry of NVs, baculoviruses and SGHVs

In the light of gene content analysis, an evolutionary link among NVs, baculoviruses, SGHVs and WSSV is most plausible. Consequently, it should be possible to analyze their phylogenetic relationship on the basis of their shared conserved ancestral genes. When these 20 single gene trees were inferred, most of the nodes showed medium to high bootstrap values, with average values across an entire gene tree ranging from 57.6 ± 10.9 (*helicase*; n = 17 nodes) and 99.0 ± 0.8 (*p47*; n = 4 nodes), suggesting the trees are topologically reliable on the whole (Wang et al. 2011).

The supermatrix (on the basis of the 20 core genes indicated in Table 2) and the supertree using these 20 single core gene trees in (Wang et al. 2011) analyses were performed. Both the supermatrix tree and supertree were highly congruent (Fig. 4). In both cases, the monophyly of each of the NVs, baculoviruses, and SGHVs was strongly supported, the branching patterns within each of the baculovirus and NV clades were also in good agreement with the current picture of their phylogeny, and a common ancestor of baculoviruses and NVs was suggested (Fig. 4). Hence, we recognized from both the supertree and supermatrix tree that baculoviruses and NVs are monophyletic; they can be considered as the minimally forming group that we term the nuclear arthropod-specific large DNA viruses (NALDVs). Both methods conflict in positioning the SGHVs within (or at least as sister lineage to the NALDV; supermatrix tree) or outside (supertree) the NALDV group. For each tree, the preferred position enjoys better support than that from the other analysis based on the most appropriate support measure. For instance, the supertree placement of the SGHVs has an rQS index value of 0.455 compared to a value of 0.143 supporting the grouping of SGHVs with the baculoviruses and NVs. However, whereas the supermatrix placement of the SGHVs enjoys some rQS support (0.143 as mentioned), the supertree placement has no bootstrap support whatsoever (0.6 compared to 99.4). Thus, the supermatrix placement of the SGHVs as sister lineage to baculoviruses and NVs, and the SGHVs being members of the NALDVs seem to be justifiable.

5.2.2 The "Monodon baculovirus" represents a nudivirus

Blastp searches revealed that a number of nudivirus homologues are present in the partially sequenced genome of the so-called "Monodon baculovirus" of the shrimp *Penaeus monodon* (21,150 bp in total; GenBank accession no. EU246943, EU246944, EF458632, AY819785).

When using the annotated shrimp MBV ORFs as query in BLAST similarity search, best hits were frequently found with HzNV-1 (Wang et al. 2011). Different phylogenetic analyses, including single gene tree inference as well as both supermatrix and supertree analyses, of the homologues of baculovirus core genes *lef-9, vlf-1, lef-5, 38K*, revealed unequivocally an obvious relationship between MBV and the non-occluded HzNV-1 (Fig. 4) (Wang et al. 2011). Given that seven other MBV and HzNV-1 ORFs are also highly similar, it is strongly suggested to consider MBV as an occluded member of the NVs and to rename it to *Penaeus monodon* nudivirus (PmNV) (Wang et al. 2011).

5.2.3 WSSV might be related to the NALDVs

The position of WSSV differs between the two trees, however, being nested deep with the NALDVs in the supermatrix tree and as sister to the clade of NCLDVs plus NALDVs in the supertree (Fig. 4). Support for either position based on either the rQS index or the bootstrap is worse than that for other clades in the tree. The different positions for WSSV reflect how the two different methods used deal with the restricted, conflicting information that is available for this virus. Although WSSV shares six genes with the other viruses, only two of these (*DNA polymerase* and *p33*) are phylogenetically informative as to its potential placement with respect to the NALDV and NCLDV groups because homologue counterparts are available in both groups. The remaining four genes (*p74, pif-1, pif-2*, and *pif-3*) are restricted to baculoviruses, NVs, SGHVs, and WSSV only. The resulting trees are therefore essentially unrooted and it is not possible to determine if WSSV nests within NALDVs (contradicting the supermatrix placement) or is sister to them (consistent with both placements). Of the two informative genes, only *p33* has associated WSSV with the NALDVs; the *DNA polymerase* has grouped it within the NCLDVs (Wang et al. 2011). The supermatrix analysis is influenced largely by the relative number of amino acids (aa) supporting a given position. In the current context, *DNA polymerase* with ~3000 aa residues is clearly outweighing the ~1000 aa residues of *p33*, thereby favouring the placement of WSSV with the NCLDVs. By contrast, the supertree analysis is more sensitive to the number of trees supporting a given position and, importantly, the relative node support within those trees (in a weighted supertree analysis). Thus, although the *DNA polymerase* tree places WSSV within the NCLDVs, this position is very poorly supported and outweighed by its more robust placement within NALDVs in the *p33* tree (Wang et al. 2011). As a result, WSSV was excluded from the NCLDV group in the supermatrix tree.

Thus, the phylogenetic analyses are equivocal with respect to the evolutionary relationships of WSSV based on the current data set and more genes need to be sampled to resolve its placement. Nevertheless, other sources of evidence suggest that WSSV is more closely related to the NALDVs than to other DNA viruses. Notably, WSSV shares six conserved homologous genes with the NALDVs, but rarely possesses homologous genes with numerous other marine viruses colonising the same aquatic ecological niches. It therefore seems that WSSV is a very ancient virus that has undergone extremely divergent evolution, as witnessed by the branch lengths generally subtending this virus (Fig. 4). This fact, in turn, hampers identification of its gene homologues and reconstruction of its phylogenetic affinities using present-day alignment based methods. In contrast, when an alignment-free whole-proteome phylogenetic analysis was applied, WSSV clustered with SGHVs (Wu et al. 2009), which coincidently is in agreement with the presented hypothesis of WSSV`s evolutionary link to the NALDVs. However, in the study by Wu et al. (2009) the SGHV and WSSV were placed within the herpesviruses, although there is no evidence of relationship among these viruses, when considering structural, biological and other genome features.

5.2.4 A common ancestry of nudiviruses, baculoviruses, hytrosaviruses, and WSSV

Taking together, 20 baculovirus core gene homologues were identified in nudiviruses, 12 in SGHVs, and six in WSSV, respectively. Consequently, this shared gene content of baculoviruses, nudiviruses, SGHVs, and WSSV is an important evidence for a proposed common ancestry of these viruses. Any other explanation, e. g., horizontal gene transfer of these genes, seems to be less probable. Therefore it is proposed that baculoviruses, nudiviruses, hytrosaviruses, and WSSV most likely shared a common ancestor and form a highly diverse group of nuclear arthropod-specific large DNA viruses (Wang and Jehle 2009; Wang et al. 2011).

6. Acknowledgement

This work was funded by grants from Shanghai Municipal Education Commission (the Eastern Scholar Project and the Leading Academic Discipline Project) and Shanghai Municipal Science and Technology Commission (Project no. 10540503000).

7. References

Abd-Alla, A. M., F. Cousserans, A. G. Parker, J. A. Jehle, N. J. Parker, J. M. Vlak, A. S. Robinson & M. Bergoin (2008) Genome analysis of a Glossina pallidipes salivary gland hypertrophy virus reveals a novel, large, double-stranded circular DNA virus. *J Virol*, Vol. 82, No. 9, (May 2008), pp. 4595-4611, ISSN 1098-5514

Abd-Alla, A. M., J. M. Vlak, M. Bergoin, J. E. Maruniak, A. Parker, J. P. Burand, J. A. Jehle & D. G. Boucias (2009) Hytrosaviridae: a proposal for classification and nomenclature of a new insect virus family. *Arch Virol*, Vol. 154, No. 6, (2009), pp. 909-918, ISSN 1432-8798

Bézier, A., M. Annaheim, J. Herbinière, C. Wetterwald, G. Gyapay, S. Bernard-Samain, P. Wincker, I. Roditi, M. Heller, M. Belghazi, R. Pfister-Wilhem, G. Periquet, C. Dupuy, E. Huguet, A. N. Volkoff, B. Lanzrein & J. M. Drezen (2009) Polydnaviruses of braconid wasps derive from an ancestral nudivirus. *Science*, Vol. 323, No. 5916, (February 2009), pp. 926-930, ISSN 1095-9203

Baum, B. R. (1992) Combining trees as a way of combining data sets for phylogenetic inference, and the desirability of combining gene trees. *Taxon*, Vol. 41, No. 1, (1992), pp. 3-10.

Bininda-Emonds, O. R. (2004a) The evolution of supertrees. *Trends Ecol Evol*, Vol. 19, No. 6, (June 2004), pp. 315-322, ISSN 0169-5347

Bininda-Emonds, O. R. (2004b) Trees versus characters and the supertree/supermatrix "paradox". *Syst Biol*, Vol. 53, No. 2, (April 2004), pp. 356-359, ISSN 1063-5157

Bininda-Emonds, O. R., R. M. Beck & A. Purvis (2005) Getting to the roots of matrix representation. *Syst Biol,* Vol. 54, No. 4, (August 2005), pp. 668-672, ISSN 1063-5157

Bininda-Emonds, O. R., K. E. Jones, S. A. Price, R. Grenyer, M. Cardillo, M. Habib, A. Purvis & J. L. Gittleman (2003) Supertrees are a necessary not-so-evil: a comment on Gatesy et al. *Syst Biol,* Vol. 52, No. 5, (October 2003), pp. 724-729, ISSN 1063-5157

Bininda-Emonds, O. R. & M. J. Sanderson (2001) Assessment of the accuracy of matrix representation with parsimony analysis supertree construction. *Syst Biol,* Vol. 50, No. 4, (August 2001), pp. 565-579, ISSN 1063-5157

Burand, J. P. 1998. Nudiviruses. In *The Insect Viruses,* eds. L. K. Miller & L. A. Ball, 69-90. New York: Plenum Press.

Burand, J. P. & H. Lu (1997) Replication of a Gonad-Specific Insect Virus in TN-368 Cells in Culture. *J Invertebr Pathol,* Vol. 70, No. 2, (September 1997), pp. 88-95, ISSN 1096-0805

Burand, J. P. & C. P. Rallis (2004) In vivo dose-response of insects to Hz-2V infection. *Virol J,* Vol. 1, No. (2004), pp. 15, ISSN 1743-422X

Chaivisuthangkura, P., C. Tawilert, T. Tejangkura, S. Rukpratanporn, S. Longyant, W. Sithigorngul & P. Sithigorngul (2008) Molecular isolation and characterization of a novel occlusion body protein gene from Penaeus monodon nucleopolyhedrovirus. *Virology,* Vol. 381, No. 2, (November 2008), pp. 261-267, ISSN 1096-0341

Chao, Y. C., S. T. Lee, M. C. Chang, H. H. Chen, S. S. Chen, T. Y. Wu, F. H. Liu, E. L. Hsu & R. F. Hou (1998) A 2.9-kilobase noncoding nuclear RNA functions in the establishment of persistent Hz-1 viral infection. *J Virol,* Vol. 72, No. 3, (March 1998), pp. 2233-2245, ISSN 0022-538X

Chao, Y. C., H. A. Wood, C. Y. Chang, H. J. Lee, W. C. Shen & H. T. Lee (1992) Differential expression of Hz-1 baculovirus genes during productive and persistent viral infections. *J Virol,* Vol. 66, No. 3, (March 1992), pp. 1442-1448, ISSN 0022-538X

Cheng, C. H., S. M. Liu, T. Y. Chow, Y. Y. Hsiao, D. P. Wang, J. J. Huang & H. H. Chen (2002) Analysis of the complete genome sequence of the Hz-1 virus suggests that it is related to members of the *Baculoviridae. J Virol,* Vol. 76, No. 18, (September 2002), pp. 9024-9034, ISSN 0022-538X

Chenna, R., H. Sugawara, T. Koike, R. Lopez, T. J. Gibson, D. G. Higgins & J. D. Thompson (2003) Multiple sequence alignment with the Clustal series of programs. *Nucleic Acids Res,* Vol. 31, No. 13, (July 2003), pp. 3497-3500, ISSN 1362-4962

Crawford, A. M. & C. Sheehan (1985) Replication of *Oryctes* baculovirus in cell culture: viral morphogenesis, infectivity and protein synthesis. *J Gen Virol.,* Vol. 66, No. (1985), pp. 529-539.

de Queiroz, A. & J. Gatesy (2007) The supermatrix approach to systematics. *Trends Ecol Evol,* Vol. 22, No. 1, (January 2007), pp. 34-41, ISSN 0169-5347

Edgar, R. C. (2004) MUSCLE: a multiple sequence alignment method with reduced time and space complexity. *BMC Bioinformatics,* Vol. 5, (August 2004), pp. 113, ISSN 1471-2105

Felsenstein, J. (1985) Confidence limits on phylogenies: an approach using the bootstrap. *Evolution,* Vol. 39, (1985), pp. 783-791.

Garcia-Maruniak, A., A. M. Abd-Alla, T. Z. Salem, A. G. Parker, V. U. Lietze, M. M. van Oers, J. E. Maruniak, W. Kim, J. P. Burand, F. Cousserans, A. S. Robinson, J. M. Vlak, M. Bergoin & D. G. Boucias (2009) Two viruses that cause salivary gland hypertrophy in Glossina pallidipes and Musca domestica are related and form a distinct phylogenetic clade. *J Gen Virol*, Vol. 90, Pt 2, (February 2009), pp. 334-346, ISSN 0022-1317

Garcia-Maruniak, A., J. E. Maruniak, W. Farmerie & D. G. Boucias (2008) Sequence analysis of a non-classified, non-occluded DNA virus that causes salivary gland hypertrophy of Musca domestica, MdSGHV. *Virology*, Vol. 377, No. 1, (July 2008), pp. 184-196, ISSN 0042-6822

Granados, R. R., T. Nguyen & B. Cato (1978) An insect cell line persistently infected with a baculovirus-like particle. *Intervirology*, Vol. 10, No. 5, (1978), pp. 309-317, ISSN 0300-5526

Guttieri, M. C. & J. P. Burand (2001) Location, nucleotide sequence, and regulation of the p51 late gene of the hz-1 insect virus: identification of a putative late regulatory element. *Virus Genes*, Vol. 23, No. 1, (2001), pp. 17-25, ISSN 0920-8569

Guttieri, M. C. & J. P. Burand (1996) Nucleotide sequence, temporal expression, and transcriptional mapping of the p34 late gene of the Hz-1 insect virus. *Virology*, Vol. 223, No. 2, (September 1996), pp. 370-375, ISSN 0042-6822

Herniou, E. A. & J. A. Jehle (2007) Baculovirus phylogeny and evolution. *Curr Drug Targets*, Vol. 8, No. 10, (October 2007), pp. 1043-1050, ISSN 1873-5592

Herniou, E. A., J. A. Olszewski, J. S. Cory & D. R. O'Reilly (2003) The genome sequence and evolution of baculoviruses. *Annu Rev Entomol*, Vol. 48, (2003), pp. 211-234, ISSN 0066-4170

Huger, A. M. (1966) A virus disease of the Indian rhinoceros beetle, Oryctes rhinoceros(Linnaeus), caused by a new type of insect virus, Rhabdionvirus oryctes gen. n., sp. n. *J Invertebr Pathol*, Vol. 8, No. 1, (March 1966), pp. 38-51, ISSN 0022-2011

Huger, A. M. (1985) A new virus disease of crickets (Orthoptera: Gryllidae) causing macronucleosis of fatbody. *J. Invertebr. Pathol.*, Vol. 45, No. 1, (1985), pp. 108-111.

Huger, A. M. & A. Krieg. 1991. Baculoviridae. Nonoccluded Baculoviruses. In *Atlas of Invertebrate Viruses*, eds. J. R. Adams & J. R. Bonami, 287-319. Boca Raton: CRC Press, Inc.

Jackson, T. A., A. M. Crawford & T. R. Glare (2005) *Oryctes* virus – time for a new look at a useful biocontrol agent. *J Invertebr Pathol.*, Vol. 89, No. 1, (2005), pp. 91-94.

Jehle, J. A. & H. Backhaus (1994) Genome organization of the DNA-binding protein gene region of Cryptophlebia leucotreta granulosis virus is closely related to that of nuclear polyhedrosis viruses. *J Gen Virol*, Vol. 75 (Pt 7), (July 1994), pp. 1815-1820. ISSN 0022-1317

Jehle, J. A., G. W. Blissard, B. C. Bonning, J. S. Cory, E. A. Herniou, G. F. Rohrmann, D. A. Theilmann, S. M. Thiem & J. M. Vlak (2006a) On the classification and nomenclature of baculoviruses: a proposal for revision. *Arch Virol*, Vol. 151, No. 7, (July 2006), pp. 1257-1266, ISSN 0304-8608

Jehle, J. A., M. Lange, H. Wang, Z. Hu, Y. Wang & R. Hauschild (2006b) Molecular identification and phylogenetic analysis of baculoviruses from Lepidoptera. *Virology*, Vol. 346, No. 1, (March 2006), pp. 180-193, ISSN 0042-6822

Katoh, K., K. Misawa, K. Kuma & T. Miyata (2002) MAFFT: a novel method for rapid multiple sequence alignment based on fast Fourier transform. *Nucleic Acids Res*, Vol. 30, No. 14, (July 2002), pp. 3059-3066, ISSN 1362-4962

Kelly, D. C., T. Lescott, M. D. Ayres, D. Carey, A. Coutts & K. A. Harrap (1981) Induction of a nonoccluded baculovirus persistently infecting Heliothis zea cells by Heliothis armigera and Trichoplusia ni nuclear polyhedrosis viruses. *Virology*, Vol. 112, No. 1, (July 1981), pp. 174-189, ISSN 0042-6822

Lassmann, T. & E. L. Sonnhammer (2005) Automatic assessment of alignment quality. *Nucleic Acids Res*, Vol. 33, No. 22, (2005), pp. 7120-7128, 1362-4962

Lin, C. L., J. C. Lee, S. S. Chen, H. A. Wood, M. L. Li, C. F. Li & Y. C. Chao (1999) Persistent Hz-1 virus infection in insect cells: evidence for insertion of viral DNA into host chromosomes and viral infection in a latent status. *J Virol*, Vol. 73, No. 1, (January 1999), pp. 128-139, ISSN 0022-538X

Lu, H. & J. P. Burand (2001) Replication of the gonad-specific virus Hz-2V in Ld652Y cells mimics replication in vivo. *J Invertebr Pathol*, Vol. 77, No. 1, (January 2001), pp. 44-50, ISSN 0022-2011

Mayo, M. A. 1995. Unassigned Viruses. In *Virus Taxonomy: The Sixth Report of the International Committee on Taxonomy of Viruses*, eds. F.A. Murphy, C.M. Fauquet, D.H.L. Bishop, S.A. Ghabrial, A.W. Jarvis, G.P. Martelli, M.A. Mayo & M. D. Summers, 504-507. Wien: Springer-Verlag.

Notredame, C., D. G. Higgins & J. Heringa (2000) T-Coffee: A novel method for fast and accurate multiple sequence alignment. *J Mol Biol*, Vol. 302, No. 1, (September 2000), pp. 205-217, ISSN 0022-2836

Payne, C. C. (1974) The isolation and characterization of a virus from Oryctes rhinoceros. *J Gen Virol*, Vol. 25, No. 1, (October 1974), pp. 105-116, ISSN 0022-1317

Payne, C. C., D. Compson & S. M. de looze (1977) Properties of the nucleocapsids of a virus isolated from *Oryctes rhinoceros*. *Virology.*, Vol. 77, No. 1, (1977), pp. 269-280.

Price, S. A., O. R. Bininda-Emonds & J. L. Gittleman (2005) A complete phylogeny of the whales, dolphins and even-toed hoofed mammals (Cetartiodactyla). *Biol Rev Camb Philos Soc*, Vol. 80, No. 3, (August 2005), pp. 445-473, ISSN 1464-7931

Ragan, M. A. (1992) Phylogenetic inference based on matrix representation of trees. *Mol Phylogenet Evol*, Vol. 1, No. 1, (March 1992), pp. 53-58, ISSN 1055-7903

Raina, A. K., J. R. Adams, B. Lupiani, D. E. Lynn, W. Kim, J. P. Burand & E. M. Dougherty (2000) Further characterization of the gonad-specific virus of corn earworm, *Helicoverpa zea*. *J Invertebr Pathol*, Vol. 76, No. 1, (July 2000), pp. 6-12, ISSN 0022-2011

Raina, A. K. & B. Lupiani (2006) Acquisition, persistence, and species susceptibility of the Hz-2V virus☆. *Journal of Invertebrate Pathology*, Vol. 93, No. 2, (2006), pp. 71-74, ISSN 0022-2011

Shackelton, L. A. & E. C. Holmes (2004) The evolution of large DNA viruses: combining genomic information of viruses and their hosts. *Trends Microbiol.*, Vol. 12, No. 10, (2004), pp. 458-465

Slack, J. & B. M. Arif (2007) The baculoviruses occlusion-derived virus: virion structure and function. *Adv Virus Res,* Vol. 69, (2007), pp. 99-165, ISSN 0065-3527

Stamatakis, A. 2006. Phylogenetic Models of Rate Heterogeneity: A High Performance Computing Perspective. In *Proceedings of 20th IEEE/ACM International Parallel and Distributed Processing Symposium (IPDPS2006), High Performance Computational Biology Workshop.* Rhodos, Greece.: Proceedings on CD.

Stamatakis, A., P. Hoover & J. Rougemont (2008) A rapid bootstrap algorithm for the RAxML Web servers. *Syst Biol,* Vol. 57, No. 5, (October 2008), pp. 758-771, ISSN 1076-836X

Swofford, D. L. 2002. PAUP*. Phylogenetic analysis using parsimony (*and other methods). 4, Sinauer Associates, Sunderland, Massachusetts

van Oers, M. M. & J. M. Vlak (2007) Baculovirus genomics. *Curr Drug Targets,* Vol. 8, No. 10, (October 2007), pp. 1051-1068, ISSN 1873-5592

Wang, Y., O. R. Bininda-Emonds, M. M. van Oers, J. M. Vlak & J. A. Jehle (2011) The genome of Oryctes rhinoceros nudivirus provides novel insight into the evolution of nuclear arthropod-specific large circular double-stranded DNA viruses. *Virus Genes,* Vol. 42, No. 3, (June 2011), pp. 444-456, ISSN 1572-994X

Wang, Y., J. P. Burand & J. A. Jehle (2007a) Nudivirus genomics: diversity and classification. *Virologica Sinica,* Vol. 22, No.1, (2007), pp. 128-136

Wang, Y. & J. A. Jehle (2009) Nudiviruses and other large, double-stranded circular DNA viruses of invertebrates: new insights on an old topic. *J Invertebr Pathol,* Vol. 101, No. 3, (July 2009), pp. 187-193, ISSN 1096-0805

Wang, Y., R. G. Kleespies, A. M. Huger & J. A. Jehle (2007b) The genome of Gryllus bimaculatus nudivirus indicates an ancient diversification of baculovirus-related nonoccluded nudiviruses of insects. *J Virol,* Vol. 81, No. 10, (May 2007), pp. 5395-5406, ISSN 0022-538X

Wang, Y., R. G. Kleespies, M. B. Ramle & J. A. Jehle (2008) Sequencing of the large dsDNA genome of Oryctes rhinoceros nudivirus using multiple displacement amplification of nanogram amounts of virus DNA. *J Virol Methods,* Vol. 152, No. 1-2, (September 2008), pp. 106-108, ISSN 0166-0934

Wang, Y., M. M. van Oers, A. M. Crawford, J. M. Vlak & J. A. Jehle (2007c) Genomic analysis of Oryctes rhinoceros virus reveals genetic relatedness to Heliothis zea virus 1. *Arch Virol,* Vol. 152, No. 3, (2007), pp. 519-531, ISSN 0304-8608

Wood, H. A. & J. P. Burand (1986) Persistent and productive infections with the Hz-1 baculovirus. *Curr Top Microbiol Immunol,* Vol. 131, (1986), pp. 119-133, ISSN 0070-217X

Wu, G. A., S. R. Jun, G. E. Sims & S. H. Kim (2009) Whole-proteome phylogeny of large dsDNA virus families by an alignment-free method. *Proc Natl Acad Sci U S A,* Vol. 106, No. 31, (August 2009), pp. 12826-12831, ISSN 1091-6490

Wu, Y. L., C. Y. Liu, C. P. Wu, C. H. Wang, S. T. Lee & Y. C. Chao (2008) Cooperation of ie1 and p35 genes in the activation of baculovirus AcMNPV and HzNV-1 promoters. *Virus Res,* Vol. 135, No. 2, (August 2008), pp. 247-254, ISSN 0168-1702

Wu, Y. L., C. P. Wu, S. T. Lee, H. Tang, C. H. Chang, H. H. Chen & Y. C. Chao (2010) The early gene hhi1 reactivates Heliothis zea nudivirus 1 in latently infected cells. *J Virol*, Vol. 84, No. 2, (January 2010), pp. 1057-1065, ISSN 1098-5514

The Baculoviral Genome

M. Leticia Ferrelli[1], Marcelo F. Berretta[2], Mariano N. Belaich[3],
P. Daniel Ghiringhelli[3], Alicia Sciocco-Cap[2] and Víctor Romanowski[1]

[1]*Instituto de Biotecnología y Biología Molecular,
Facultad de Ciencias Exactas, Universidad Nacional de La Plata, CONICET
[2]Laboratorio de Ingeniería Genética y Biología Celular
y Molecular - Area Virosis de Insectos,
Departamento de Ciencia y Tecnología, Universidad Nacional de Quilmes
[3]Instituto de Microbiología y Zoología Agrícola, INTA Castelar
Argentina*

1. Introduction

The molecular biology of Baculoviruses has drawn a great deal of interest due to the variety of applications of these viruses as: 1) agents for biological control of insect pests (Szewczyk et al., 2006); 2) vectors for expression of recombinant proteins in insect cells (Kost et al., 2005); 3) vehicles for gene transduction of mammal cells (Hu, 2006, 2008); and 4) display systems of recombinant epitopes (Makela et al., 2010).

Baculoviridae is a family of insect-specific viruses, with more than 600 reported species, mainly isolated from Lepidoptera (butterflies and moths) and in some cases from Hymenoptera (sawflies) and Diptera (mosquitoes). Baculoviruses have circular, double-stranded DNA genomes ranging in size from approximately 80 to 180 kbp, depending on the species, that are predicted to encode for up to 180 genes. The viral genome associates with proteins forming a nucleocapsid. This structure is surrounded by a membrane envelope to form a rod-shaped virion (hence, the name of the Family: *baculum* is Latin for rod or stick).

During their biological cycle, most baculoviruses produce two different virion phenotypes: the *budded virus* (BV) appears early in infected cells and is responsible for the dissemination of the disease inside the insect body, whereas the *occluded virus* (OV) is produced in the very late stage of the infection and becomes embedded in a protein matrix forming a distinct structure known as *occlusion body* (OB) which is responsible for the horizontal transmission of the virus. OBs are highly stable and protect the virions from damage in the environment.

1.1 Taxonomy

The polyhedral and ovoidal morphology of the different OBs has been used as an initial taxonomic criterion to group baculoviruses in two genera: Nucleopolyhedrovirus (NPVs) and Granulovirus (GVs). The major protein found in polyhedra is polyhedrin, which is expressed very late in infection. Multiple OVs are embedded within a polyhedron. Also, each OV may contain one or more nucleocapsids. This lead to a grouping of the NPVs as SNPVs (*Single* NPV, one enveloped nucleocapsid per virion) and MNPV (*Multiple* NPV, multiple nucleocapsids per virion). The other genus (Granulovirus) has characteristic OBs that appear as ovoidal granules, with granulin as the major protein component. Usually GVs contain a single virion per OB with only one nucleocapsid (Funk et al., 1997).

More recently, a new classification on the *Baculoviridae* based on DNA sequence data has been proposed and accepted by the ICTV (Carstens and Ball, 2009; Jehle et al., 2006a). It preserves correlation with OB morphology but also reflects host taxonomic classification. Four genera are recognized: *Alphabaculovirus* (NPVs isolated from Lepidoptera); *Betabaculovirus* (GVs isolated from Lepidoptera); *Gammabaculovirus* (NPVs isolated from Hymenoptera) and *Deltabaculovirus* (NPVs isolated from Diptera).

The type baculovirus is *Autographa californica* nucleopolyhedrovirus, AcMNPV, a member of the *Alphabaculovirus* genus (Table 1). The present knowledge about the baculovirus molecular biology is based largely on studies performed with this virus. Consequently, the most of the information presented here is based on AcMNPV.

1.2 Two types of enveloped virions

OBs ingested by susceptible insect larvae are dissolved in the midgut releasing the *occlussion derived virus* (ODV) that initiate the infection of midgut epithelial cells.

Structural differences between BVs and ODVs are due mainly to the origin and composition of the lipoproteic membrane envelope. BVs acquire their envelope from the infected cell membrane (modified by the insertion of viral proteins) during the budding process. On the other hand, ODVs envelope is built at the nuclear stage using the nuclear membrane components and it is much more complex than the envelope of the BVs regarding their protein content (Rohrmann, 2011e).

A distinctive characteristic of the BV phenotype is the presence of a membrane protein that mediates viral entry via an endocytic, pH-dependent mechanism (Blissard and Wenz, 1992; Pearson et al., 2000). There are two types of envelope fusogenic proteins in baculoviruses, GP64 and F. All baculoviruses contain F (the exception being NPVs isolated from hymenoptera) but it does not play a functional role in the early stages of infection in all cases; those expressing GP64 use it as the major player in the early stage of virus entry and infection. Coincident with this difference, NPVs isolated from Lepidoptera have been found to cluster in two phylogenetic subgroups (I and II) based on their polyhedrin sequences as well as the presence or absence of a *gp64* gene, respectively. Those containing GP64 belong to group I (Hefferon et al., 1999; IJkel et al., 2000; Monsma et al., 1996), while those that lack GP64 but have a functional F protein belong to group II (IJkel et al., 2000; Pearson et al., 2000).

1.3 Infectious cycle

The natural infection cycle begins when the insect ingests the OBs contaminating its food (figure 1). Once in the midgut of the insect larvae, the highly alkaline environment contributes to the dissolution of the OBs releasing ODVs. The ODVs must traverse the peritrofic membrane lining the midgut lumen and fuse with epithelial cell membrane, allowing the entry of the nucleocapsids. These make their way to the nucleus, where the transcription starts in a very finely regulated manner initiating a gene transcription cascade (Friesen, 1997; O' Reilly et al., 1992; Romanowski and Ghiringhelli, 2001).

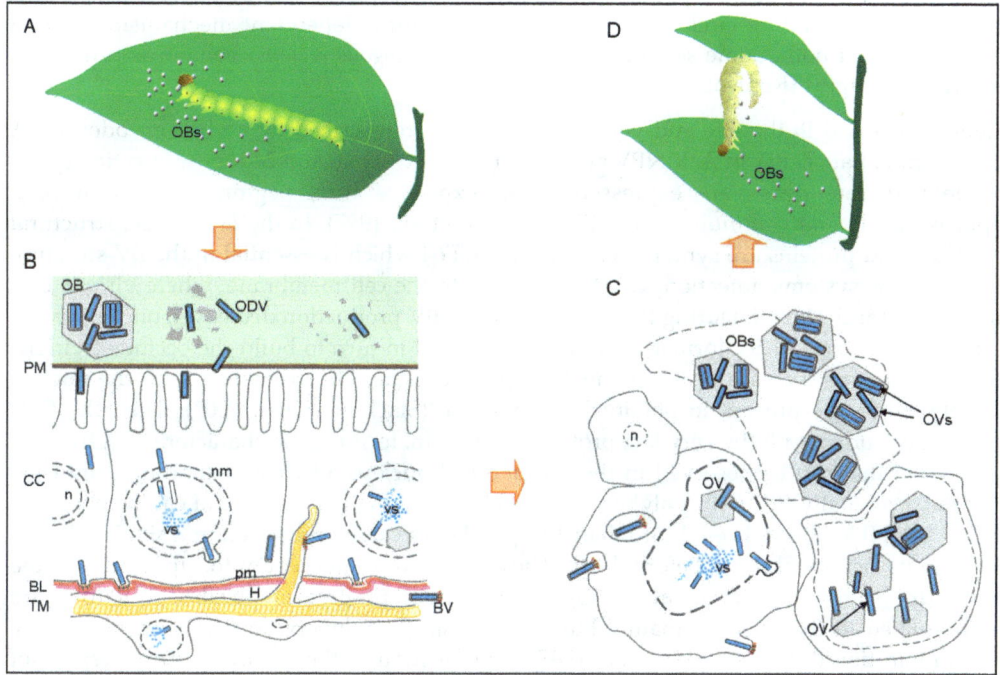

Fig. 1. Baculovirus (Alphabaculovirus) infection cycle. **A.** Larva ingests food contaminated with OBs. **B.** OBs are dissolved in the alkaline midgut releasing ODVs which upon overcoming the PM infect midgut epithelial cells. Newly formed nucleocapsids bud from the plasma membrane and disseminate inside the larval body, via the tracheal cells or directly through the hemolymph. **C.** In the late stage of infection nucleocapsids acquire their envelope from the nuclear membrane forming OVs (arrows), which may contain one or several nucleocapsids, and are occluded within a polyhedrin matrix forming the OBs. **D.** Dead larva full of OBs typically appears hanging in a the upper part of the plant. OB (Occlusion Body); ODV (Occlusion Derived Virion); PM (Peritrophic Membrane); CC (Columnar Cell); n (nucleus); nm (nuclear membrane); vs (virogenic stroma); BL (Basal Lamina); TM (Tracheal Matrix); H (Hemolymph); OV (Occluded Virion). Figure modified from Federici (1997).

Transcription of viral genes occurs in four stages: immediate early, delayed early, late and very late. Genes of the early stages are transcribed by the cell RNA polymerase II. Immediate early genes are transactivated by host cell transcription factors with no participation of virus-encoded proteins, reflecting the empirical observation that naked baculovirus genomic DNA is infective (Burand et al., 1980; Carstens et al., 1980; Hajos et al., 1998). Transcription of delayed early genes requires the activation by viral gene products expressed at the previous stage. Among delayed early gene products, those called LEFs (late expression factors) are required for DNA replication and late transcription (Hefferon and Miller, 2002). After the delayed-early stage viral DNA synthesis occurs within the nucleus of the infected cell, in what is called the virogenic stroma. Baculoviral DNA replication is not totally understood but evidence exists that it may occur by a rolling circle mechanism, recombination-dependent mechanism, or by a combination of both. Some sequences called *hrs* (*homologous regions*) behave as functional replication origins (Rohrmann, 2011c).

Genes expressed in the late and very late phases are transcribed by a virus-encoded RNA polymerase. Late genes in AcMNPV are transcribed between 6 and 24 h post infection (p. i.), while very late genes are expressed in an explosive way, beginning at 18 h p. i. approximately and continuing up to 72 h p.i. (Lu et al., 1997). In the late phase, structural nucleocapsid proteins are synthesized, and also GP64 which is essential in the BV structure for the virus systemic infection. GP64 is targeted to the cell membrane, where virions bud between 10 and 24 h p.i. During the very late phase, BV production decreases; nucleocapsids are no longer used in BV formation and they are used in turn to build the occluded virions (OV). In the specific case of NPVs, nucleocapsids are thought to interact with the nuclear membrane in the process to obtain their envelope (Slack and Arif, 2007). Then the OVs become occluded with the very late protein polyhedrin, forming the characteristic refringent polyhedra that can be observed in the infected cell nucleus. Occlusion continues until the nucleus eventually fills with polyhedra. Typically more than 30 polyhedra can be observed in an AcMNPV infected cell. More than 10^{10} polyhedra can be produced in a single infected larva in its last larval stage, before death. These polyhedra can account for up to 30 % of the larva dry weight (Miller et al., 1983). As occlusion progresses fibrillar structures are accumulated in the nucleus, mainly built from a single polypeptide (P10) expressed very late in infection (Van Der Wilk et al., 1987). The function of these fibrillar structures is not absolutely clear but seems to play a role in the controlled disintegration of larvae (Dong et al., 2007; Van Oers et al., 1994; Williams et al., 1989). In the final stage of the infection, virus encoded enzymes, cathepsin and chitinase, aid in the cuticle rupture and liquefaction of the dead larva, leading to the release polyhedra in the environment and making them avaliable for ingestion by a new insect (Hawtin et al., 1997).

2. Baculovirus genomes

Since the first complete sequence of a baculoviral genome was reported (AcMNPV; Ayres et al., 1994), many baculovirus genomes were sequenced to further improve the understanding of the molecular biology of these viruses. To date, there are 58 fully sequenced baculoviral genomes available in GenBank. Forty one belong to the *Alphabaculovirus* genus, thirteen to the *Betabaculovirus*, three to the *Gammabaculovirus* and one to the *Deltabaculovirus* (Table 1). Baculovirus encode 89 (NeleNPV) to 183 (PsunGV) predicted ORFs, in both strands, apparently with no preferred orientation. Typically, the ORF designated as number 1 is that encoding the major occlusion protein (polyhedrin/granulin) and the following ORFs are numbered sequentially in a clockwise direction. In general, baculovirus genomes have low GC content (<50%). The virus with the lowest GC% is NeleNPV (33.3%).

The generally adopted criterion to predict ORFs is to only consider those that code for a polypeptide at least 50 amino acid long (aa) and minimal overlap with other ORFs. Baculovirus genes are not clustered in the genome by function or the time of transcription. Noteworthy, only one expression unit has been detected to contain an intron (*ie0*) (Chisholm and Henner, 1988), which makes it easier to predict ORFs at the DNA sequence level.

The sequencing of complete genomes allowed estimating the whole baculovirus gene content in about 900 genes. All baculovirus genomes sequenced so far encode for a group of 31 genes, known as the core genes. These genes represent a hallmark of the virus family and may play a role in essential biological functions (Miele et al., 2011). According to their function, the core genes (Table 2) can be classified as belonging to the following categories: replication, transcription, packaging and assembly, cell cycle arrest/interaction with host proteins and oral infectivity.

As most of available genomes belong to baculoviruses specific for lepidopteran insects (*Alpha-* and *Betabaculovirus*), there is a good deal of information to characterize a set of genes associated with specificity for Lepidoptera. Likewise, there are some *Betabaculovirus*-specific genes, not found in NPVs, which may be implicated in the differential pathogenesis displayed by these viruses. It is worth noting that GVs (*Beta-*) are not as well studied as *Alpha-* NPVs at the molecular level because of the lack of proper susceptible insect cell lines. So far, the only GV-specific gene characterized at the functional level is a metalloproteinase of XcGV which has orthologs in all GVs (Ko et al., 2000).

Regarding the gene promoters, there are many baculovirus early genes that are preceded by either a TATA-box or a CAGT initiator motif, or both. These motifs are found also in promoters of the host genome and are characteristic of genes transcribed by the RNA polymerase II of the insect cell. Late and very late genes are expressed by the viral RNA polymerase from promoters containing the DTAAG motif. The occurrence of this motif is less frequent than predicted by stochastic distribution, according to its functional role as initiator of late and very late transcription. Some genes contain both early and late promoter motifs and are expected to be expressed throughout the infection. However, not every predicted ORF is preceded by a known motif, which does not imply that it is not expressed. Other elements have been characterized to play a role in baculovirus transcription such as GATA motifs and distal CGT motif (van Oers and Vlak, 2007).

Traditionally, baculovirus gene functions were studied by constructing deletion mutants upon cotransfection of wild type viral DNA and a transfer vector containing an insertion cassette flanked by fragments of homology to the target region in the genome [*e. g.* (Lee et al., 1998)]. As baculoviruses have been widely used as expression vectors, much effort was made to improve the production of recombinant virus. This led to the construction of the first *bacmid* which is a recombinant AcMNPV genome containing the mini-F origin of replication that allows the maintenance and recombination of the virus in *Escherichia coli* (Luckow et al., 1993). This fact led to a new way of studying baculoviral genes: a specific deletion can be made by recombination in *E. coli* and, upon recovering viral DNA from a bacterial culture, it can be transfected in insect cells to study the effect of the modified genome in the viral cycle (Zhao et al., 2003).

Genus	Virus Name	Acronym	Genome Size (bp)	number of ORFs or CDS	Reference	GenBank Accesion Number
Alphabaculovirus NPVs Group I	*Antheraea pernyi* NPV-Z	AnpeNPV	126.629	147	Nie et al., 2007	DQ486030
	Antheraea pernyi NPV-L2	AnpeNPV	126.246	145	Fan et al., 2007	EF207986
	Anticarsia gemmatalis NPV D2	AgMNPV	132.239	152	Oliveira et al., 2006	DQ813662
	Autographa californica NPV C6	AcMNPV	133.894	156	Ayres et al., 1994	L22858
	Bombyx mandarina NPV	BomanNPV	126.770	141	Xu et al., 2009, unpublished	NC012672
	Bombyx mori NPV T3	BmNPV	128.413	143	Gomi et al., 1999	L33180
	Choristoneura fumiferana DEF NPV	CfDefNPV	131.160	149	Lauzon et al., 2005	AY327402
	Choristoneura fumiferana MNPV	CfMNPV	129.593	146	de Jong et al., 2005	AF512031
	Epiphyas postvittana NPV	EppoNPV	118.584	136	Hyink et al., 2002	AY043265
	Hyphantria cunea NPV	HycuNPV	132.959	148	Ikeda et al., 2006	AP009046
	Maruca vitrata MNPV	MaviMNPV	111.953	126	Chen et al., 2008	EF125867
	Orgyia pseudotsugata MNPV	OpMNPV	131.995	152	Ahrens et al., 1997	U75930
	Plutella xylostella MNPV CL3	PlxyMNPV	134.417	152	Harrison and Lynn, 2007	DQ457003
	Rachiplusia ou MNPV	RoMNPV	131.526	149	Harrison and Bonning, 2003	AY145471
Alphabaculovirus NPVs Group II	*Adoxophyes honmai* NPV ADN001	AdhoNPV	113.220	125	Nakai et al., 2003	AP006270
	Adoxophyes orana NPV	AdorNPV	111.724	121	Hilton and Winstanley, 2008a	EU591746
	Agrotis ipsilon MNPV	AgipMNPV	155.122	163	Harrison, 2009	EU839994
	Agrotis segetum NPV	AgseNPV	147.544	153	Jakubowska et al., 2006	DQ123841
	Apocheima cinerarium NPV	ApciNPV	123876	118	Zhang et. al, unpublished	FJ914221
	Chrysodeixis chalcites NPV	ChchNPV	149.622	151	van Oers et al., 2005	AY864330
	Clanis bilineata NPV DZ1	ClbiNPV	135.454	139	Zhu et al., 2009	DQ504428
	Ecotropis obliqua NPV A1	EcobNPV	131.204	126	Ma et al., 2007	DQ837165
	Euproctis pseudoconspersa NPV	EupsNPV	141.291	139	Tang et al., 2009	NC_012639
	Helicoverpa armigera MNPV	HearMNPV	154.196	162	Tang et al., 2008, unpublished	EU730893
	Helicoverpa armigera NPV C1	HearSNPV	130.759	137	Zhang et al., 2005	AF303045
	Helicoverpa armigera NPV G4	HearSNPV	131.405	135	Chen et al., 2001	AF271059
	Helicoverpa armigera SNPV NNg1	HearSNPV	132.425	143	Ogembo et al., 2009	AP010907
	Helicoverpa zea SNPV	HzSNPV	130.869	139	Chen et al., 2002	AF334030
	Leucania separata NPV AH1	LeseNPV	168.041	169	Xiao and Qi, 2007	AY394490
	Lymantria dispar NPV	LdMNPV	161.046	164	Kuzio et al., 1999	AF081810
	Lymantria xylina MNPV	LyxyMNPV	156.344	157	Nai et al., 2010	GQ202541
	Mamestra configurata NPV A	MacoNPV A-90-2	155.060	169	Li et al., 2002b	U59461

Genus	Virus Name	Acronym	Genome Size (bp)	number of ORFs or CDS	Reference	GenBank Accesion Number
Alphabaculovirus NPVs Group II	Mamestra configurata NPV A	MacoNPV A-90-4	153656	168	Li et al., 2005	AF539999
	Mamestra configurata NPV B	MacoNPV B	158.482	168	Li et al., 2002a	AY126275
	Orgyia leucostigma NPV CSF-77	OrleNPV	156.179	135	Eveleigh et al., 2008, unpublished	EU309041
	Spodoptera exigua NPV	SeMNPV	135.611	139	Ijkel et al., 1999	AF169823
	Spodoptera frugiperda MNPV	SfMNPV 19	132.565	141	Wolff et al., 2008	EU258200
	Spodoptera frugiperda MNPV 3AP2	SfMNPV 3AP2	131.330	142	Harrison et al., 2008	EF035042
	Spodoptera litura NPV G2	SpltMNPV	139.342	141	Pang et al., 2001	AF325155
	Spodoptera litura NPV II	SpltNPV II	148.634	147	Li et al., 2008, unpublished	EU780426
	Trichoplusia ni SNPV	TnSNPV	134.394	145	Willis et al., 2005	DQ017380
Betabaculovirus	Adoxophyes orana GV	AdorGV	99.657	119	Wormleaton et al., 2003	AF547984
	Agrotis segetum GV	AgseGV	131.680	132	Xiulian et al., 2004, unpublished	AY522332
	Choristoneura occidentalis GV	ChocGV	104.710	116	Escasa et al., 2006	DQ333351
	Clostera anachoreta GV	ClanGV	101487	123	Liang et al., 2011	HQ116624
	Cryptophlebia leucotreta GV CV3	CrleGV	110.907	128	Lange and Jehle, 2003	AY229987
	Cydia pomonella GV	CpGV	123.500	143	Luque et al., 2001	U53466
	Helicoverpa armigera GV	HearGV	169.794	179	Harrison and Popham, 2008	EU255577
	Phthorimaea operculella GV	PhopGV	119.217	130	Croizier et al., 2002, unpublished	AF499596
	Pieris rapae GV	PrGV	108.592	120	Zhang et al., 2010, unpublished	NC_013797
	Plutella xylostella GV K1	PlxyGV K1	100.999	120	Hashimoto et al., 2000	AF270937
	Pseudelatia unipuncta GV	PsunGV	176.677	183	Li et al., 2008, unpublished	EU678671
	Spodoptera litura GV K1	SpltGV K1	124.121	136	Wang et al., 2007, unpublished	DQ288858
	Xestia c-nigrum GV	XecnGV	178.733	181	Hayakawa et al., 1999	AF162221
γ	Neodiprion abietis NPV	NeabNPV	84.264	93	Duffy et al., 2006	DQ317692
	Neodiprion sertifer NPV	NeseNPV	86.462	90	Garcia-Maruniak et al., 2004	AY430810
	Neodiprion lecontei NPV	NeleNPV	81.755	89	Lauzon et al., 2004	AY349019
δ	Culex nigripalpus NPV	CuniNPV	108.252	109	Afonso et al., 2001	AF403738

Table 1. **Fully sequenced baculovirus genomes.** ORF (Open Reading Frame) or CDS (protein CoDing Sequence): defined by a start codon ATG followed by at least 50 codons before a stop codon in frame. Gamma and Deltabaculovirus genera are indicated by greek characters.

3. Replication genes

Replication of baculovirus genome is poorly understood. As mentioned above, a rolling circle mechanism has been proposed but there are evidences of recombination being involved as well. Baculovirus genomes contain multiple origins of replication. Sequences that act as origins are called *hrs* (for homologous regions) and are dispersed throughout the genome (explained in section 8). In addition, *non-hr* origins were also found, present only once per genome.

Several viral factors have been demonstrated to be essential for viral replication and others to be stimulatory. IE-1, a known activator of early transcription (see below), was found to be necessary for plasmid replication in transient assays. IE-1 binds to *hr* sequences but it is not clear if this is a requirement for initiation of DNA replication. The other proteins essential for DNA synthesis in AcMNPV are DNA polymerase, DNA helicase, LEF-1, LEF-2 and LEF-3. In addition to polymerization activity by DNA polymerase and DNA unwinding by DNA helicase, a primase activity is associated with LEF-1, and LEF-2 as a primase accessory factor. LEF-3 is a single-stranded DNA binding protein (Mikhailov, 2003; Rohrmann, 2011c).

These proteins were found in all baculoviruses sequenced to date but some other proteins have been identified to have an influence on DNA replication. These are P35, IE-2; PE38; LEF-7; VLF-1, Alcaline Exonuclease (AN); DBP, LEF-11, ME53 and PCNA (Mikhailov, 2003).

Some baculovirus genomes code also for other proteins that may be involved in DNA replication like DNA ligase, and a second helicase. In addition, genes encoding enzymes related to DNA repair: photolyase (present in some group II nucleopolyhedrovirus; (Xu et al., 2008), Ac79 (homolog to UvrC endonuclease superfamily), V-trex exonuclease (present in AgMNPV and CfMNPV), polyADP-ribose polymerase (PARP, found in AgMNPV) and polyADP-ribose glycohydrolase (PARG, present in all sequenced group II NPVs) (Rohrmann, 2011c). Nucleotide biosynthesis seems to be another aspect of DNA replication that some baculoviruses may influence since they have genes for ribonucleotide reductase subunits and dUTPase, both related to dTTP biosynthesis (Herniou et al., 2003).

4. Transcription genes

Transcription of baculovirus genes occurs in several temporal stages. As mentioned above, early genes are transcribed by the host RNA pol II and after DNA replication, late gene transcription proceeds through the action of a viral RNA polymerase. One of the first proteins to be transcribed is IE-1, which functions as a transcriptional activator of itself and delayed early genes. It is known that IE-1 binds to *hr* sequences as a dimer and it is thought that this complex interacts with the host transcription machinery to enhance expression of early genes. Although there are no other recognizable IE-1 binding sites, an *hr*-independent mechanism of transactivation is likely to occur also (Friesen, 1997). Part of the IE-1 population is called IE-0 which is the translation product of the only spliced mRNA described in AcMNPV. IE-1 orthologs appear in all *Alpha* and *Betabaculovirus* genomes. Other transcription factors encoded by AcMNPV and other baculoviruses that were found to transactivate early genes are IE-2 and PE-38 (Cohen et al., 2009; Rohrmann, 2011d).

In the late stage of the infection additional genes are implicated in transcription. Viral RNA polymerase is made of four subunits coded by four core genes: *lef-4*, *lef-8*, *lef-9* and *p47* (Guarino et al., 1998). LEF-8 and LEF-9 have motifs common to the largest subunits of bacterial and eukaryotic RNA polymerases. LEF-8 contains the essential C-terminal region conserved in RNA polymerases, while the rest of the polypeptide shows no sequence homology to other known RNA polymerases. LEF-9 contains the Mg^{2+} binding site of the catalytic centre found in other RNA polymerases. LEF-4 is an RNA capping enzyme and P47 does not show homology with other RNA polymerase subunits (van Oers and Vlak, 2007). Two other core genes are implicated in late transcription: *lef-5* and *vlf-1*. LEF-5 appears to be an initiation factor in AcMNPV (Guarino et al., 2002). VLF-1 (Very Late Expression Factor-1) is involved in the expression of the very late genes *polyhedrin* (*polh*) and *p10*. VLF-1 was found to interact with the so-called "burst sequence" present downstream of the very late genes triggering their hyperexpression (Yang and Miller, 1999). Other proteins required for late transcription as revealed by transient expression assays are LEF-6 (a putative mRNA export factor), LEF-10, LEF-12 and PP31 (Rohrmann, 2011b). Additional proteins may be involved: a methyltrasferase (Ac69), probably implicated in mRNA capping (Wu and Guarino, 2003), an ADP-ribose pyrophosphatase of the nudix superfamily (Ge et al., 2007) which is a putative decapping enzyme, LEF-2 which apart from being an essential replication factor (see above) is also implicated in the very late transcription (Merrington et al., 1996); and PK1 (Mishra et al., 2008)

5. Structural genes

While nucleocapsids are essentially the same in both baculovirus phenotypes, BVs and ODVs differ in the origin and protein composition of their envelope. Moreover, ODVs are occluded in a proteinaceus matrix forming the OBs, which are essential structures for maintenance of orally infective virus. No structural protein is needed for the initiation of transcription once viral DNA enters the nucleus of the cell; therefore the structural proteins found in virions are supposed to focus on overcoming the barriers for cell entry. In a simmilar fashion some OB proteins are involved in facilitating horizontal transmission and invasion of the midgut altogether. As expected, baculovirus genome encodes many genes for proteins that are included in the the virion and OB structures, as well as genes whose products may not be present in the final structure but are important for its assembly (Funk et al., 1997).

5.1 Occlusion body

Baculovirus OB is formed by the major occlusion body protein polyhedrin, for NPVs, or granulin, for GVs. Polyhedrin and granulin are closely related. The occlusion body protein for the only dipteran baculovirus completely sequenced, CuniNPV, does not show sequence homology to its lepidopteran counterparts and is a much larger protein. Hymenopteran baculoviruses occlusion protein is homologous to Alphabaculoviruses polyhedrin (Garcia-Maruniak et al., 2004). *Polyhedrin* is a very late gene that is expressed at very high levels. This characteristic has been exploited for the expression of recombinant proteins in insect cells. In the natural cycle of a NPV, polyhedrin forms a crystalline cubic lattice that surrounds the ODVs. The structure of the polyhedron was recently determined (Ji et al., 2010).

Other proteins apart from the major occlusion protein are present in the structure of the OB or play a role in its morphogenesis. The polyhedron is surrounded by a protein layer which provides the OB with a smooth, sealed surface that enhances its stability (Gross et al., 1994). The viral protein responsible of this envelope is the calyx/PE. During OBs formation calyx/PE is found associated with fibrillar structures formed by P10, the other protein that is highly expressed at the very late phase. Although it is not part of the OB, P10 plays a role in its correct morphogenesis (Williams et al., 1989).

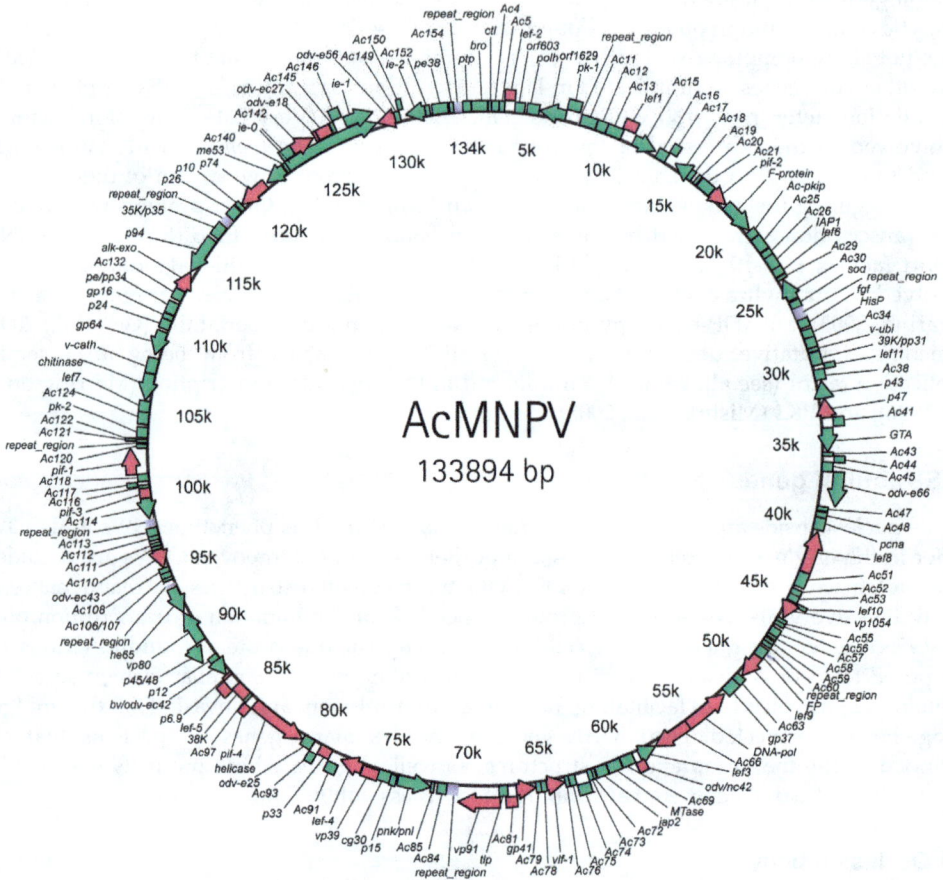

Fig. 2. Baculovirus genome (AcMNPV). Core genes are shown in pink.

Depending on the species there may be other proteins associated with the polyhedron: enhancin/viral enhancing factor (Vef) and proteinases. Enhancins are metalloproteinases that help disrupt the peritrophic membrane (PM) of the insect midgut. PM is the first barrier baculoviruses must overcome when ingested in order to get midgut epithelial cells. PM is made of mucin proteins and chitin. Enhancin degrades mucin helping this way ODVs pass through the PM (Wang and Granados, 1997). Not all the baculoviruses encode for enhancins but, for example, XcGV has four copies (Hayakawa et al., 1999). Alkaline proteinases were found associated to the OBs that may aid in the dissolution of OBs and subsequent ODVs release. However since there is not such a gene identified in baculovirus genomes, those could be bacterial or insect contaminants present in the OB preparation (Rohrmann, 2011e).

5.2 BV and ODV

As mentioned above, although both BVs and ODVs carry the same genetic information there are several differences between them in function and structure. BVs are the first virion phenotype produced in an infected cell and consist of a nucleocapsid which acquire their envelope as they bud from the cell membrane previously modified with the GP64 (group I *Alphabaculovirus*) or F protein (rest of the *Baculoviridae*). On the other hand OVs obtain their envelope from the nuclear membrane, may include several nucleocapsids per virion -in the case of MNPVs- and their protein content seems to be more complex than that of BVs.

Genomic DNA associates with proteins to form nucleocapsids. A small basic protein, P6.9, directly interacts with DNA and is involved in the assembly of highly condensed DNA (Kelly et al., 1983). VP39 is the major nucleocapsid protein and, along with P6.9, is a core gene. Both proteins are two of the three most abundant proteins in AcMNPV BV, being GP64 the third one (Wang et al., 2010b). VLF-1 is also a core gene and it was first described as the factor necessary for the expression of very late genes. But later it was shown that VLF-1 is present in both, BV and ODV, localizing at one end of the nucleocapsid. This protein belongs to the lambda integrase family and is involved in the production of nucleocapsids (Vanarsdall et al., 2006).

Other core gene products are GP41 (tegument protein), 38K, P49 and ODV-EC27 (Table 2). They seem to be associated with the nucleocapsid, and consequently found in both BVs and ODVs.

	AcMNPV ORF number	References
Replication		
lef-1	14	(Evans et al., 1997)
lef-2	6	(Evans et al., 1997)
DNApol	65	(McDougal and Guarino, 1999)
Helicase	95	(McDougal and Guarino, 2000)
Transcription		
lef-4	90	(Jin et al., 1998)
lef-8	50	(Titterington et al., 2003)
lef-9	62	(Iorio et al., 1998)
p47	40	(Guarino et al., 1998)
lef-5	99	(Guarino et al., 2002)
Oral infectivity		
p74	138	(Faulkner et al., 1997)
pif-1	119	(Kikhno et al., 2002)
pif-2	22	(Pijlman et al., 2003)
pif-3	115	(Ohkawa et al., 2005)
pif-4	96	(Fang et al., 2009)
pif-5/odv-e56	148	(Sparks et al., 2011)

	AcMNPV ORF number	References
Cell cycle arrest and/or interaction with host		
odv-e27	144	(Belyavskyi et al., 1998)
ac81	81	(Chen et al., 2007)
Packaging, assembly, and release		
p6.9	100	(Wang et al., 2010a)
vp39	89	(Thiem and Miller, 1989)
vlf-1	77	(Vanarsdall et al., 2006)
alk-exo	133	(Mikhailov et al., 2003)
vp1054	54	(Olszewski and Miller, 1997a)
vp91	83	(Russell and Rohrmann, 1997)
gp41	80	(Olszewski and Miller, 1997b)
38k	98	(Wu et al., 2006)
p33	92	(Wu and Passarelli, 2010)
odv-ec43	109	(Fang et al., 2003)
p49	142	(McCarthy et al., 2008)
odv-nc42	68	(Li et al., 2008)
odv-e18	143	(McCarthy and Theilmann, 2008)
desmoplakin	66	(Ke et al., 2008)

Table 2. Baculovirus core genes

Proteins included in ODV and BV structures of some baculoviruses have been identified by high throughput techniques based on mass spectrometry. Those are the cases of the ODVs of AcMNPV (*Alphabaculovirus*) (Braunagel et al., 2003), CuniNPV (*Deltabaculovirus*) (Perera et al., 2007), HearNPV (*Alphabaculovirus*) (Deng et al., 2007) and PrGV (*Betabaculovirus*) (Wang et al., 2011), and the BVs of AcMNPV (Wang et al., 2010b). These studies demonstrated that baculovirus virions are complex: in addition to *ca.*, 40 virally encoded proteins host proteins may be present as well.

5.3 *Per os* infectivity factors

Per os infectivity factors (PIFs) are baculovirus proteins essential for oral infection of insect hosts but not relevant in cell culture propagation. Six proteins have been described to play this role and are encoded by 6 core genes *p74, pif-1, pif-2, pif-3, pif-4* and *pif-5* (*odv-e56*). PIF-1, PIF-2 and PIF-3 form a stable complex on the surface of AcMNPV ODV in association with P74. It was proposed that these four proteins form an evolutionarily conserved complex on ODV surface that may play an essential role in the initial stage of infection (Peng et al., 2010). PIF-4 was found to be essential for oral infection of AcMNPV in *Trichoplusia ni* larvae (Fang et al., 2009). In recent studies ODV-E56 was demonstrated to be a *PIF* (PIF-5) in AcMNPV (Sparks et al., 2011) and BmNPV (Xiang et al., 2011).

6. Auxiliary genes

Baculovirus whole gene content is wide and diverse. As already noted, there is a group of 31 core genes that are present in all the baculoviruses sequenced to date. However, each particular baculovirus species codes for many more than 31 genes. A recent study determined the whole gene content based on the information of 57 baculovirus genomes and came up to a sum of 895 different ORFs (Miele et al., 2011). This means that there may be genes that are not essential but capable of modulating the infection of viruses with a particular gene subset. Moreover, some genes might have evolved a particular function and play a role only in the context of species-specific virus-host interactions. Those genes are commonly categorized as auxiliary genes. Other genes that could be included in this group, may participate in processes other than replication and transcription or may code for structural genes essential for a particular virus to succeed in the infection of a specific host.

This section focuses on some of the auxiliary genes that are widely distributed in the family and/or their function has been described.

6.1 Genes affecting cellular metabolism

To succeed in infection a virus needs to circumvent host cell apoptosis. It is well known that apoptosis is one of the mechanisms an organism uses to clear an infection: a cell detected as being infected is set to die. All baculoviruses encode anti-apoptotic genes to counteract this cell response in order to complete their replicative cycle. There are two types of antiapoptotic genes in baculoviruses: P35/P49 homologs and IAPs. P35/P49 function directly inhibiting the effector action of caspases and they have been found in some NPVs and one GV (Escasa et al., 2006). IAPs are metalloproteinases that act upstream P35 in the apoptotic pathway (van Oers and Vlak, 2007).

A gene coding for a superoxide dismutase (*sod*) is widely distributed among baculoviruses. Its function seems to be the removal of free radicals in infected hemocytes, which are superoxide producers (Rohrmann, 2011a).

Most lepidopteran baculoviruses encode a viral ubiquitin. It was suggested that baculoviruses carry this gene in order to inhibit steps in the host degradative pathways in a strategy to stabilize viral proteins that otherwise would be short-lived (Haas et al., 1996).

Most of the baculovirus genes are present in a single copy in the genome. But there is the special case of *bro* (*Baculovirus repeated orf*) genes that are a multigene family present in several baculoviruses. They appear in different number of copies: from 0 to 16, in the LdMNPV genome. Most *bro* genes share a core sequence but show different degrees of similarity in other regions (Kuzio et al., 1999). Although *bro* genes are similar among them, they have no homology with other known proteins, making it difficult to predict their function. A study of BmNPV *bro* genes showed that these proteins have DNA binding activity, preferentially to single stranded DNA, and two of them were speculated to function as DNA binding proteins that influence host DNA replication and/or transcription (Zemskov et al., 2000).

6.2 Genes affecting the insect host as an organism

Baculoviruses that infect lepidoptera are characterized by the systemic infection of the host rather than being restricted to the midgut epithelial cells. It was proposed that, in order to spread from this primary site of infection, they use the insect tracheal system. Fibroblast growth factor (FGF) involvement in the attraction and motility of tracheal cells has been well studied in *Drosophila melanoganster* (Sutherland et al., 1996). Alpha and Betabaculoviruses carry viral *fgf* homologs (*v-fgf*) in their genomes. Conversely, this gene is absent in Gamma and Deltabaculoviruses which cause midgut-restricted infections. Interestingly, it was found that the presence of *v-fgf* accelerates larval death as knockouts of these genes in AcMNPV and BmNPV caused a retardation in host death compared to infection with wild type viruses (Passarelli, 2011).

Several lepidopteran baculoviruses code for a protein designated GP37, which is homologous to fusolin, encoded by entomopoxviruses. Fusolin, as well as GP37, is a glycoprotein that contains chitin binding domains. Fusolin was demonstrated to form spindle-like bodies that enhance the entomopoxvirus oral infection in host larvae. The mechanism of action for these spindles appears to be associated with the disruption of the peritrophic membrane (PM) allowing the virions to reach the midgut epithelial cells (Mitsuhashi et al., 2007). Except for the case of CfDEFMNPV (Li et al., 2000), in baculoviruses no spindle bodies have been observed, although GP37 was found to be associated with OBs in AcMNPV (Vialard et al., 1990). On the other hand, in OpMNPV and MbMNPV this protein was found in cytoplasmic inclusion bodies that accumulate late in infection (Gross et al., 1993; Phanis et al., 1999). In SpliMNPV infected cells GP37 was found to localize in the cytoplasm and the nucleus as well as in the envelopes of BVs and ODVs. Its chitin binding capacity was demonstrated suggesting that it may bind to the chitin component of the PM (Li et al., 2003).

Another strategy for the baculoviruses success is the delay of larval molting. Ecdysteroid-UDP-glycosytransferase (EGT) mediates the inactivation of molting hormones (ecdysone) in insects. The *egt* gene is present in most baculovirus genomes. The virus benefits from the presence of this gene product that prevents the infected larva from molting as it keeps feeding, thus allowing higher virus progeny yields (O'Reilly and Miller, 1989).

In the final stage of infection, after insect death, the larva liquefies releasing baculovirus OBs to the environment. This liquefaction is mediated by the two viral-encoded enzymes: cathepsin and chitinase. Cathepsin is a protease that acts together with chitinase disrupting the insect exoskeleton and promoting the release and spread of progeny virus (Hawtin et al., 1997).

7. Host range

One of the characteristic features of baculoviruses is their narrow host range. Due to their exquisite specificity, most baculoviruses can be regarded as "magic bullets" targeting a single host organism and, therefore, are excellent candidates for biological pest control. From the environmental point of view baculoviruses are safe alternatives for pest control as their host range is generally restricted to one insect species, not affecting other organisms. But from the economical point of view the narrow host range represents a disadvantage when more than one pest is to be controlled simultaneously in the same field.

On the other hand, as a consequence of their narrow host range, baculoviruses are innocuous to vertebrates. Moreover, as they are able to enter mammalian cells, they have been widely studied as viral vectors for gene therapy (Hu, 2006). For these reasons, baculoviral genes affecting host range and the interaction with the host are an important object of study.

One of the first studies in this field was conducted on two closely related viruses, *i.e.* AcMNPV and BmNPV. Despite the high similarity of these viruses their host specificities do not overlap. AcMNPV infects Sf-9 cells (derived from *Spodoptera frugiperda*) but not BmN cells (derived from *Bombyx mori*). Conversely, BmNPV does infect BmN cells but not Sf-9 cells. In coinfection assays a recombinant BmNPV was obtained that could replicate in both cell lines. The characterization of this virus revealed that its altered host range was due to a recombinant sequence in the helicase gene (Maeda et al., 1993), being a single amino acid change enough for this phenotypic change (Kamita and Maeda, 1997).

Another example of host range expansion due to a single gene was the case of AcMNPV modified by the insertion of a LdMNPV gene, the *host range factor 1* (*hrf-1*). This modification allowed AcMNPV to replicate in *Lymantria dispar* cells and larvae (Thiem et al., 1996). HRF-1 is present in the genome of LdMNPV and OpMNPV, both of which are able to replicate in Ld652Y cells, derived from *L. dispar*. Moreover, other NPVs modified by the incorporation of this factor were found to replicate in these cells, that are non-permissive for the corresponding wild type viruses. It was suggested that this factor is important in the progression of the infection after DNA replication and that the global protein synthesis shutoff is the major factor that restricts NPV replication in Ld652Y cells, being HRF-1 a crucial viral factor that counteracts this antiviral mechanism active in NPV-infected Ld652Y cells (Ishikawa et al., 2004).

Other genes that play a role in baculovirus host range have been detected and studied: *host cell factor 1* (*hcf-1*), *p35*, *iap* and *lef-7* (Miller and Lu, 1997; Thiem, 1997).

8. Homologous regions and replication origins

Homologous regions (*hrs*) are repeated sequences present in baculovirus genomes that vary widely in terms of length, sequence and copy number between species (Berretta and Romanowski, 2008). They occur also in other viruses of invertebrates that appear to be phylogenetically related to baculoviruses (van Oers and Vlak, 2007). In general, each repeat consists of an imperfect palindrome and a number of repeats with similar sequences are distributed in the genome as singletons or arranged in tandem with variable number of copies. *Hrs* have been found in genomes of the four genera of the current baculovirus classification, including all non-lepidopteran species with fully sequenced genomes. However, *hrs* could not be found in the genomes of some species such as TnSNPV (Willis et al., 2005), ChchNPV (van Oers et al., 2005), and AgseGV (Hilton and Winstanley, 2008b). *Hrs* are A-T rich compared to the overall genome nucleotide composition. They represent part of the non-coding regions that account for less than 10% of baculovirus genomes, although in some GVs, they overlap predicted genes likely to be transcribed (Hilton and Winstanley, 2008). AcMNPV has nine *hrs* that contain a total of 38 repeats with a copy number ranging from one to eight (Ayres et al., 1994). Each repeat consists of a 28 bp-long imperfect palindrome that diverges slightly from a consensus sequence. Similar to *hrs* found in other NPVs, AcMNPV *hrs* displaying several palindromic repeats, have a modular organization in which each palindrome is embedded within a direct or inverted repeat in tandem. Frequently, *hr* palindromes are bisected by a restriction enzyme site (eg EcoRI in AcMNPV *hrs*). In GVs the majority of *hrs* are less structured, although imperfect palindromes may be as long as *ca.* 300 bp. Repeats are poorly conserved except for 13 bp at their ends (Hilton and Winstanley, 2008b). Regarding to their function, *hrs* act as enhancers of transcription of early genes in those NPVs in which they were studied and there are indirect evidences that they serve as origins of replication in NPVs and GVs. Non-homologous sequences within many NPV *hrs* have motifs known to bind cellular transcription factors of the bZIP family (Landais et al., 2006) but the enhancing activity of *hrs* depends primarily on viral factor IE-1 binding to palindromic repeats. In AcMNPV, IE-1 binds to the 28-mer element as a dimer and this interaction stimulates transcription of *cis*-linked promoters that are responsive to the RNA pol II activity in transient assays (Rodems and Friesen, 1995).

The first evidence that *hrs* are putative origins of DNA replication was the accumulation of *hrs* in defective genomes obtained by serial passages of AcMNPV in cultured cells (Kool et al., 1993). These viral particles have genomes smaller than the wild type virus, which means a replicative advantage for those retaining *ori* sequences in the molecule. In infection-dependent replication assays, performed in different virus/permissive cell line systems, *hrs* were found to confer plasmids the ability to replicate (Broer et al., 1998; Hilton and Winstanley, 2008b; Pearson et al., 1992). It was observed that viruses promote replication of *hr*-containing plasmids only when the *hr* comes from the same or a closely related viral species. The stringency of this specificity is higher than that observed associated with the function of *hrs* as enhancers (Berretta and Passarelli, 2006). This may come as a result of more viral factors involved in the replication process and possible interactions thereof. IE-1 binding to *hr* sequences is also thought to play a role in replication possibly by recruiting the components of the replication machinery (Nagamine et al., 2006). Since the replication mechanism of baculoviruses is not well understood, the function of *hrs* as origins during the infective cycle remains to be confirmed. Deletion of up to two *hrs* from AcMNPV did not impair replication of the virus (Carstens and Wu, 2007). Moreover, there are other *non-hr* sequences that function as *oris* in transient assays, including promoters of early genes (Kool et al., 1993; Wu and Carstens, 1996). *Hrs* may produce cruciform structures in the DNA, although *in vitro* studies were unable to detect such forms in AcMNPV imperfect palindromes. This kind of branched structures are likely to form if baculovirus replication involves recombination events. VLF-1 protein was found to bind cruciform DNA as well as certain *hr* sequences; this capacity is consistent with its requirement during the DNA packaging process (Rohrmann, 2011c). Whether *hrs* participate in the final stages of genome processing is not known. Consistent with this possibility it has been suggested that hrs constitute factors of genome plasticity as mediators of intra- and inter-molecular recombination events during baculovirus evolution (van Oers and Vlak, 2007).

9. Baculoviral microRNAs

MicroRNAs (miRNAs) are small non coding RNAs that play a role in the regulation of the expression of genes in a wide variety of cellular processes. Typically they are molecules of about 22 nucleotides obtained by the processing of a longer primary RNA (pri-miRNA). In most cases this pri-miRNA is transcribed by the RNA polymerase II, and contains a ~80 nt hairpin that is recognized by the RNaseIII-like enzyme Drosha that removes it from the pri-miRNA to give the pre-miRNA. Pre-miRNA is exported to the cytosol. Once there, it is processed by Dicer which cleaves the terminal loop of the hairpin. One strand of the remaining dsRNA is incorporated by the RISC complex in order to target a specific mRNA and inhibit its translation. Recently it was found that viruses also encode miRNAs. These are from DNA virus families and were first discovered in herpesvirus (Grundhoff and Sullivan, 2011).

More recently, microRNAs were discovered in baculovirus; Singh et al., (2010) demonstrated that BmNPV encodes four miRNAs by sequencing small RNAs followed by *in silico* analysis and validation using other techniques. Micro RNAs were searched in two different tissues of infected larvae. As the genome of *Bombyx mori* is available it was possible to discriminate the miRNAs encoded by the virus from those encoded by the host. Other related baculoviruses were searched to see if these miRNAs were conserved. All four miRNAs were found to be present with 100% identity in AcMNPV, BomaNPV and PlxyMNPV. Three miRNAs were conserved in RoMNPV and one in MaviNPV. This conservation is strongly suggesting that the miRNAs play some kind of crucial role in the viral cycle. Regarding their targeting, the *in silico* analysis revealed that these miRNAs have more than one target that could be either viral or host-cell in origin. Primarily, miRNAs bind to 3′UTR of target mRNA, but there are recent reports of miRNAs binding to 5′UTR or the coding sequence triggering the translation repression, as well. Two of the predicted viral targets of BmNPV miRNAs are *dna binding protein* and *chitinase* mRNAs targeted by two different miRNAs (*bmnpv-miR-3* and *bmnpv-miR-2*). Interestingly, they were found to bind to the complementary region from which they were transcribed. Other viral targets are *bro-I*, *bro-III*, *lef-8*, *fusolin*, *DNA polymerase*, *p25* and *ORF 3* of BmNPV. Another interesting finding was that the computationally identified cellular targets such as prophenoloxidase and hemolin are related to different antiviral host defense mechanisms. Other important cellular targets were GTP binding nuclear protein Ran, DEAD box polypeptides and eukaryotic translation initiation factors that play an important role in small RNA-mediated gene regulation. It was proposed that these viral miRNAs are important for regulating cellular activities in order to easily establish infection in the host (Singh et al., 2010).

10. Baculovirus phylogeny

Before the advent of rapid automatic sequencing methods, when only a restricted number of complete genomes sequences was available, baculovirus phylogeny studies were performed using single homologous genes. Initially, the preferred gene product was polyhedrin/granulin, the major occlusion body protein. It is highly expressed; therefore, easily purified and its N-terminal region could be sequenced. Also, as it is a conserved protein it was easy to identify in new baculovirus isolations (Herniou and Jehle, 2007). These first phylogenetic studies revealed that baculoviruses were divided in 4 different groups (Rohrmann, 1986): (i) dipteran-specific baculovirus with OB protein unrelated to Polh/Gran; (ii) hymenopteran-specific baculovirus with OB protein being poorly related to Polh/Gran; (iii) lepidopteran nucleopolyhedroviruses and (iv) granuloviruses. The analyses of Polh/Gran also revealed a subgrouping of lepidopteran NPVs in groups I and II (Zanotto et al., 1993). Interestingly, this separation was correlated with the different utilization of fusogenic protein of the BV, GP64 or F, respectively (Lung et al., 2002).

The use of single genes to infer phylogeny must follow, at least, two criteria: the gene must be present in all members of the virus family and its level of conservation must reflect evolutionary distance (Herniou and Jehle, 2007). The studies using genes such as *lef-8* and *pif-2* supported the grouping mentioned above (Herniou et al., 2004). When several complete genomes became available better phylogenetic analyses could be undertaken based on the sequence of all genes that were present all the genomes (Herniou et al., 2001). One approach consists in concatenating the amino acid sequences of all these gene products to perform the analysis. This approach is convenient because each gene contributes to the overall phylogenetic signal and a synergistic effect is produced by the combination of all the signals (Herniou et al., 2003). The first report comprising whole-genome data was based on nine complete genomes available at that moment, which only represented lepidopteran baculoviruses and 63 common genes were detected and employed in the analysis (Herniou et al., 2001). When more genomes became available, especially those from non-lepidopteran baculoviruses, a group of about 30 genes were found to be present in all the baculoviruses, allowing to perform more significant phylogenetic studies. One important consequence of this increasing amount of sequence data was the proposal of a new classification of the *Baculoviridae*, based on 29 core genes among 29 baculovirus genomes, including the dipteran and hymenopteran ones (Jehle et al., 2006a). A recent report utilized 57 baculovirus genomes of which a group of 31 core genes was determined and used to perform an up-to-date phylogeny (Miele et al., 2011). In this report the cladogram obtained reproduced the current baculovirus classification. Also it was consistently reproduced the separation of Alphabaculoviruses in groups I and II as well as the subdivision of group I in clades Ia and Ib previously reported (Herniou and Jehle, 2007; Jehle et al., 2006b). The Betabaculovirus genus clade also reveals a subdivision in two groups (Miele et al., 2011).

11. Transposable elements

Transposons have been found in almost all eukaryotic organisms, being a central component in many genomes (Wicker et al., 2007). These sequences, also known as transposable elements (TEs), are characterized by the ability to move and replicate through various mechanisms according to their genetic nature. In view of this, TEs are not innocuous for genomes, because their activity may affect the genetic endowment of a species. Their prominent role in biological evolution has been thoroughly reviewed and the conclusion is that TEs provide plasticity to the genomes and are an important source of variability.

In Eukarya, TEs show a great diversity in gene content, size and mechanism of transposition. According to shared characteristics, these sequences are classified into two main groups: Class I (retrotransposons) and Class II (DNA transposons). A crucial difference between them resides in the existence of an RNA intermediate in the Class I TEs. Other properties are also used to subdivide into subclasses, including the size of the target site duplication, the occurrence and gene content (Wicker et al., 2007). Retrotransposons can be grouped into two subclasses: the LTR retrotransposons and the non-LTR retrotransposons or retroposons (Capy, 2005). This is mainly based on the presence/absence of LTRs (Long Terminal Repeats), but other features are also considered. In all cases, reverse transcription processes are involved. On the other hand, Class II transposons or DNA TEs are mobilized in the genomes using a single or double-stranded DNA intermediate. These sequences can be divided into three major subclasses: those that excise as dsDNA and reinsert elsewhere in the genome ("cut-and-paste" transposons); (ii) those that utilize a mechanism related to rolling-circle replication (helitrons), and Mavericks, whose mechanism of transposition is not yet well understood, but that likely replicate using a self-encoded DNA polymerase (Feschotte and Pritham, 2007).

Because of their biological activity, TEs behave like selfish sequences that impact on genomic architecture. However, it has also been reported that some TEs participate in other biological functions such as transcription, translation and DNA replication, localization and movement (Ponicsan et al., 2010; von Sternberg and Shapirob, 2005). In any case, TEs can be mobilized within a genome or between genomes. It is at this latter point where viruses take a leading role, because they can be recipients of TEs and transport them to other individuals in subsequent infections. In particular, the genomes of baculoviruses can be the targets for the isertion of different insect TEs when they replicate in the host cells. Taking into account that one of the main sources of genome variability in viruses with large dsDNA genomes are structural mutations, the possible sequence rearrangements produced by transposition processes (gene interruption, deletions, inversions, translocations, etc.) could actively participate in their evolution (Herniou et al., 2001).

TEs have also been exploided for genetic modification in the laboratory. One of the transposons most widely used in biotechnology is probably piggyBac, an insect DNA TE. This sequence was identified in AcMNPV propagated in TN-368 cell line (Fraser et al., 1985). The trans-mobilization between host chromosome and virus genome was discovered because the transposition occurred into 25k gene, producing a distinct "few polyhedra" phenotype. Later, other reports acknowledged the presence of TEs in baculoviral genomes, including the description of TED -a retrotransposon in AcMNPV- and TC14.7 -a DNA TE in CpGV (Friesen and Nissen, 1990; Jehle et al., 1995). Two additional DNA TEs have been described, one from CpGV designated TCp3.2 (Jehle et al, 1997), and the other, a new piggyBac-related transposon isolated from AgMNPV and designated IDT for *iap* disruptor transposon (Carpes et al., 2009). Considering these evidences, gene transfer processes could be more common than initially realized.

TEs may play an important role in baculovirus biology and evolution. They can provide mechanisms for horizontal transfer of genes between virus species replicating in the same host cell, between the host genome and the viral genome, or between this and the genome of other entities such as pathogenic bacteria. The high similarity between baculovirus and insect sequences (*egt* and *sod* genes), or between baculovirus and other pathogens (v-*chitinase* gene) could be the consequence of transposition events that were selected during evolution of baculoviruses.

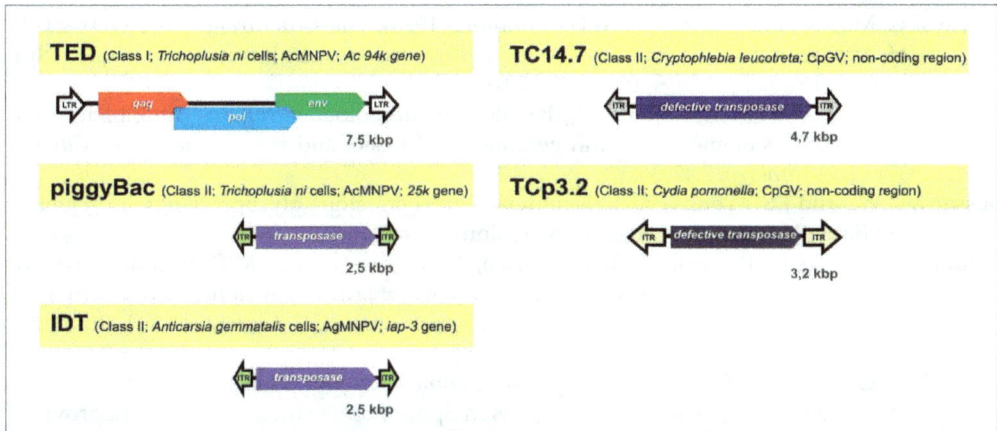

Fig. 3. **Types of transposable elements** (TE) found and described in baculovirus genomes. ORFs contained in each transposon are shown as coloured block arrows indicating their predicted function. The size of the TE is indicated in kilobasepairs; class of transposon, the species of donor insect genome, baculovirus species, viral gene sequence interrupted by the TE are indicated in brackets. Signature sequences for different TEs are indicated by arrows: LTR (Long Terminal Repeats) and ITR (Inverted Terminal Repeats).

12. Concluding remarks

Baculoviruses are a family of insect specific viruses with quite diverse and interesting applications. Therefore the knowledge of their gene content and molecular biology is a matter of growing interest. For instance, discovery and characterization of genes implicated in host range are subject of investigations for improvement of their application as designer biopesticides. Another focus of studies is interaction of baculoviruses with non target cells (*e.g.* mammalian cells) to assess the biosafety of using them for efficient gene transduction in therapeutic applications. As more baculovirus full genome sequences become available (especilly dipteran and hymenopteran-specific viruses), the bioinformatic analysis and experimental validation will help to establish a better defined set of genes characteristic of the family and those that are involved with the host specificity. In addition a more robust and detailed evolutionary tree will be probably assembled.

13. References

Afonso, C.L., E.R. Tulman, Z. Lu, C.A. Balinsky, B.A. Moser, J.J. Becnel, D.L. Rock, and G.F. Kutish (2001). Genome sequence of a baculovirus pathogenic for Culex nigripalpus. *J Virol.* 75:11157-11165.

Ahrens, C.H., R.L. Russell, C.J. Funk, J.T. Evans, S.H. Harwood, and G.F. Rohrmann (1997). The sequence of the Orgyia pseudotsugata multinucleocapsid nuclear polyhedrosis virus genome. *Virology.* 229:381-399.

Ayres, M.D., S.C. Howard, J. Kuzio, M. Lopez-Ferber, and R.D. Possee (1994). The complete DNA sequence of Autographa californica nuclear polyhedrosis virus. *Virology.* 202:586-605.

Belyavskyi, M., S.C. Braunagel, and M.D. Summers (1998). The structural protein ODV-EC27 of Autographa californica nucleopolyhedrovirus is a multifunctional viral cyclin. *Proc Natl Acad Sci U S A.* 95:11205-11210.

Berretta, M., and V. Romanowski (2008). Baculovirus homologous regions (hrs): pleiotropic functional cis elements in viral genomes and insect and mammalian cells. *Current Topics in Virology.* 7:47-56.

Blissard, G.W., and J.R. Wenz (1992). Baculovirus gp64 envelope glycoprotein is sufficient to mediate pH-dependent membrane fusion. *J Virol.* 66:6829-6835.

Braunagel, S.C., W.K. Russell, G. Rosas-Acosta, D.H. Russell, and M.D. Summers (2003). Determination of the protein composition of the occlusion-derived virus of Autographa californica nucleopolyhedrovirus. *Proc Natl Acad Sci U S A.* 100:9797-9802.

Broer, R., J.G. Heldens, E.A. van Strien, D. Zuidema, and J.M. Vlak (1998). Specificity of multiple homologous genomic regions in Spodoptera exigua nucleopolyhedrovirus DNA replication. *J Gen Virol.* 79 (Pt 6):1563-1572.

Burand, J.P., M.D. Summers, and G.E. Smith (1980). Transfection with baculovirus DNA. *Virology.* 101:286-290.

Capy, P. (2005). Classification and nomenclature of retrotransposable elements. *Cytogenet Genome Res.* 110:457-461.

Carpes, M.P., J.F. Nunes, T.L. Sampaio, M.E.B. Castro, P.M.A. Zanotto, and B.M. Ribeiro (2009). Molecular analysis of a mutant Anticarsia gemmatalis multiple nucleopolyhedrovirus (AgMNPV) shows an interruption of an inhibitor of apoptosis gene (iap-3) by a new class-II piggyBac-related insect transposon. *Insect Molecular Biology.* 18:747-757.

Carstens, E.B., and L.A. Ball (2009). Ratification vote on taxonomic proposals to the International Committee on Taxonomy of Viruses (2008). *Arch Virol.* 154:1181-1188.

Carstens, E.B., S.T. Tjia, and W. Doerfler (1980). Infectious DNA from Autographa californica nuclear polyhedrosis virus. *Virology.* 101:311-314.

Carstens, E.B., and Y. Wu (2007). No single homologous repeat region is essential for DNA replication of the baculovirus Autographa californica multiple nucleopolyhedrovirus. *J Gen Virol.* 88:114-122.

Cohen, D., M. Marek, B. Davies, J. Vlak, and M. van Oers (2009). Encyclopedia of Autographa californica nucleopolyhedrovirus genes. *Virologica Sinica.* 24:359-414.

Chen, H.Q., K.P. Chen, Q. Yao, Z.J. Guo, and L.L. Wang (2007). Characterization of a late gene, ORF67 from Bombyx mori nucleopolyhedrovirus. *FEBS Lett.* 581:5836-5842.

Chen, X., I.J. WF, R. Tarchini, X. Sun, H. Sandbrink, H. Wang, S. Peters, D. Zuidema, R.K. Lankhorst, J.M. Vlak, and Z. Hu (2001). The sequence of the Helicoverpa armigera single nucleocapsid nucleopolyhedrovirus genome. *J Gen Virol.* 82:241-257.

Chen, X., W.J. Zhang, J. Wong, G. Chun, A. Lu, B.F. McCutchen, J.K. Presnail, R. Herrmann, M. Dolan, S. Tingey, Z.H. Hu, and J.M. Vlak (2002). Comparative analysis of the complete genome sequences of Helicoverpa zea and Helicoverpa armigera single-nucleocapsid nucleopolyhedroviruses. *J Gen Virol*. 83:673-684.

Chen, Y.R., C.Y. Wu, S.T. Lee, Y.J. Wu, C.F. Lo, M.F. Tsai, and C.H. Wang (2008). Genomic and host range studies of Maruca vitrata nucleopolyhedrovirus. *J Gen Virol*. 89:2315-2330.

Chisholm, G.E., and D.J. Henner (1988). Multiple early transcripts and splicing of the Autographa californica nuclear polyhedrosis virus IE-1 gene. *J Virol*. 62:3193-3200.

de Jong, J.G., H.A. Lauzon, C. Dominy, A. Poloumienko, E.B. Carstens, B.M. Arif, and P.J. Krell (2005). Analysis of the Choristoneura fumiferana nucleopolyhedrovirus genome. *J Gen Virol*. 86:929-943.

Deng, F., R. Wang, M. Fang, Y. Jiang, X. Xu, H. Wang, X. Chen, B.M. Arif, L. Guo, and Z. Hu (2007). Proteomics analysis of Helicoverpa armigera single nucleocapsid nucleopolyhedrovirus identified two new occlusion-derived virus-associated proteins, HA44 and HA100. *J Virol*. 81:9377-9385.

Dong, C., F. Deng, D. Li, H. Wang, and Z. Hu (2007). The heptad repeats region is essential for AcMNPV P10 filament formation and not the proline-rich or the C-terminus basic regions. *Virology*. 365:390-397.

Duffy, S.P., A.M. Young, B. Morin, C.J. Lucarotti, B.F. Koop, and D.B. Levin (2006). Sequence analysis and organization of the Neodiprion abietis nucleopolyhedrovirus genome. *J Virol*. 80:6952-6963.

Escasa, S.R., H.A. Lauzon, A.C. Mathur, P.J. Krell, and B.M. Arif (2006). Sequence analysis of the Choristoneura occidentalis granulovirus genome. *J Gen Virol*. 87:1917-1933.

Evans, J.T., D.J. Leisy, and G.F. Rohrmann (1997). Characterization of the interaction between the baculovirus replication factors LEF-1 and LEF-2. *J Virol*. 71:3114-3119.

Fan, Q., S. Li, L. Wang, B. Zhang, B. Ye, Z. Zhao, and L. Cui (2007). The genome sequence of the multinucleocapsid nucleopolyhedrovirus of the Chinese oak silkworm Antheraea pernyi. *Virology*. 366:304-315.

Fang, M., Y. Nie, S. Harris, M.A. Erlandson, and D.A. Theilmann (2009). Autographa californica multiple nucleopolyhedrovirus core gene ac96 encodes a per os infectivity factor (PIF-4). *J Virol*. 83:12569-12578.

Fang, M., H. Wang, H. Wang, L. Yuan, X. Chen, J.M. Vlak, and Z. Hu (2003). Open reading frame 94 of Helicoverpa armigera single nucleocapsid nucleopolyhedrovirus encodes a novel conserved occlusion-derived virion protein, ODV-EC43. *J Gen Virol*. 84:3021-3027.

Faulkner, P., J. Kuzio, G.V. Williams, and J.A. Wilson (1997). Analysis of p74, a PDV envelope protein of Autographa californica nucleopolyhedrovirus required for occlusion body infectivity in vivo. *J Gen Virol*. 78 (Pt 12):3091-3100.

Federici, B.A. (1997). Baculovirus Pathogenesis. *In* The Baculoviruses. L.K. Miller, editor. Plenum Press, New York and London. 33-59.

Feschotte, C., and E.J. Pritham (2007). DNA transposons and the evolution of eukaryotic genomes. *Annu Rev Genet*. 41:331-368.

Fraser, M.J., J.S. Brusca, G.E. Smith, and M.D. Summers (1985). Transposon-mediated mutagenesis of a baculovirus. *Virology*. 145:356-361.

Friesen, P.D. (1997). Regulation of baculovirus early gene expression. *In* The Baculoviruses. L.K. Miller, editor. Plenum Press, New York and London. 141-170.

Friesen, P.D., and M.S. Nissen (1990). Gene organization and transcription of TED, a lepidopteran retrotransposon integrated within the baculovirus genome. *Mol Cell Biol.* 10:3067-3077.

Funk, C.J., S.C. Braunagel, and G.F. Rohrmann (1997). Baculovirus structure. *In* The Baculoviruses. L.K. Miller, editor. Plenum Press, New York and London. 7-32.

Garcia-Maruniak, A., J.E. Maruniak, P.M. Zanotto, A.E. Doumbouya, J.C. Liu, T.M. Merritt, and J.S. Lanoie (2004). Sequence analysis of the genome of the Neodiprion sertifer nucleopolyhedrovirus. *J Virol.* 78:7036-7051.

Ge, J., Z. Wei, Y. Huang, J. Yin, Z. Zhou, and J. Zhong (2007). AcMNPV ORF38 protein has the activity of ADP-ribose pyrophosphatase and is important for virus replication. *Virology.* 361:204-211.

Gomi, S., K. Majima, and S. Maeda (1999). Sequence analysis of the genome of Bombyx mori nucleopolyhedrovirus. *J Gen Virol.* 80:1323-1337.

Gross, C.H., R.L.Q. Russell, and G.F. Rohrmann (1994). Orgyia Pseudotsugata Baculovirus p10 and Polyhedron Envelope Protein Genes: Analysis of their Relative Expression Levels and Role in Polyhedron Structure. *Journal of General Virology.* 75:1115-1123.

Gross, C.H., G.M. Wolgamot, R.L. Russell, M.N. Pearson, and G.F. Rohrmann (1993). A 37-kilodalton glycoprotein from a baculovirus of *Orgyia pseudotsugata* is localized to cytoplasmic inclusion bodies. *J Virol.* 67:469-475.

Grundhoff, A., and C.S. Sullivan (2011). Virus-encoded microRNAs. *Virology.* 411:325-343.

Guarino, L.A., W. Dong, and J. Jin (2002). In vitro activity of the baculovirus late expression factor LEF-5. *J Virol.* 76:12663-12675.

Guarino, L.A., B. Xu, J. Jin, and W. Dong (1998). A virus-encoded RNA polymerase purified from baculovirus-infected cells. *J Virol.* 72:7985-7991.

Haas, A.L., D.J. Katzung, P.M. Reback, and L.A. Guarino (1996). Functional characterization of the ubiquitin variant encoded by the baculovirus Autographa californica. *Biochemistry.* 35:5385-5394.

Hajos, J.P., D. Zuidema, P. Kulcsar, J.G. Heldens, P. Zavodszky, and J.M. Vlak (1998). Recombination of baculovirus DNA following lipofection of insect larvae. *Arch Virol.* 143:2045-2050.

Harrison, R.L. (2009). Genomic sequence analysis of the Illinois strain of the Agrotis ipsilon multiple nucleopolyhedrovirus. *Virus Genes.* 38:155-170.

Harrison, R.L., and B.C. Bonning (2003). Comparative analysis of the genomes of Rachiplusia ou and Autographa californica multiple nucleopolyhedroviruses. *J Gen Virol.* 84:1827-1842.

Harrison, R.L., and D.E. Lynn (2007). Genomic sequence analysis of a nucleopolyhedrovirus isolated from the diamondback moth, Plutella xylostella. *Virus Genes.* 35:857-873.

Harrison, R.L., and H.J. Popham (2008). Genomic sequence analysis of a granulovirus isolated from the Old World bollworm, Helicoverpa armigera. *Virus Genes.* 36:565-581.

Harrison, R.L., B. Puttler, and H.J. Popham (2008). Genomic sequence analysis of a fast-killing isolate of Spodoptera frugiperda multiple nucleopolyhedrovirus. *J Gen Virol.* 89:775-790.

Hashimoto, Y., T. Hayakawa, Y. Ueno, T. Fujita, Y. Sano, and T. Matsumoto (2000). Sequence analysis of the Plutella xylostella granulovirus genome. *Virology.* 275:358-372.

Hawtin, R.E., T. Zarkowska, K. Arnold, C.J. Thomas, G.W. Gooday, L.A. King, J.A. Kuzio, and R.D. Possee (1997). Liquefaction of Autographa californica nucleopolyhedrovirus-infected insects is dependent on the integrity of virus-encoded chitinase and cathepsin genes. *Virology.* 238:243-253.

Hayakawa, T., R. Ko, K. Okano, S.I. Seong, C. Goto, and S. Maeda (1999). Sequence analysis of the Xestia c-nigrum granulovirus genome. *Virology.* 262:277-297.

Hefferon, K.L., and L.K. Miller (2002). Reconstructing the replication complex of AcMNPV. *Eur J Biochem.* 269:6233-6240.

Hefferon, K.L., A.G. Oomens, S.A. Monsma, C.M. Finnerty, and G.W. Blissard (1999). Host cell receptor binding by baculovirus GP64 and kinetics of virion entry. *Virology.* 258:455-468.

Herniou, E.A., and J.A. Jehle (2007). Baculovirus phylogeny and evolution. *Curr Drug Targets.* 8:1043-1050.

Herniou, E.A., T. Luque, X. Chen, J.M. Vlak, D. Winstanley, J.S. Cory, and D.R. O'Reilly (2001). Use of whole genome sequence data to infer baculovirus phylogeny. *J Virol.* 75:8117-8126.

Herniou, E.A., J.A. Olszewski, J.S. Cory, and D.R. O'Reilly (2003). The genome sequence and evolution of baculoviruses. *Annu Rev Entomol.* 48:211-234.

Herniou, E.A., J.A. Olszewski, D.R. O'Reilly, and J.S. Cory (2004). Ancient coevolution of baculoviruses and their insect hosts. *J Virol.* 78:3244-3251.

Hilton, S., and D. Winstanley (2008a). Genomic sequence and biological characterization of a nucleopolyhedrovirus isolated from the summer fruit tortrix, Adoxophyes orana. *J Gen Virol.* 89:2898-2908.

Hilton, S., and D. Winstanley (2008b). The origins of replication of granuloviruses. *Arch Virol.* 153:1527-1535.

Hu, Y.C. (2006). Baculovirus vectors for gene therapy. *Adv Virus Res.* 68:287-320.

Hu, Y.C. (2008). Baculoviral vectors for gene delivery: a review. *Curr Gene Ther.* 8:54-65.

Hyink, O., R.A. Dellow, M.J. Olsen, K.M. Caradoc-Davies, K. Drake, E.A. Herniou, J.S. Cory, D.R. O'Reilly, and V.K. Ward (2002). Whole genome analysis of the Epiphyas postvittana nucleopolyhedrovirus. *J Gen Virol.* 83:957-971.

IJkel, W., M. Westenberg, R.W. Goldbach, G.W. Blissard, J.M. Vlak, and D. Zuidema (2000). A novel baculovirus envelope fusion protein with a proprotein convertase cleavage site. *Virology.* 275:30-41.

Ijkel, W.F.J., E.A. van Strien, J.G. Heldens, R. Broer, D. Zuidema, R.W. Goldbach, and J.M. Vlak (1999). Sequence and organization of the Spodoptera exigua multicapsid nucleopolyhedrovirus genome. *J Gen Virol.* 80 (Pt 12):3289-3304.

Ikeda, M., M. Shikata, N. Shirata, S. Chaeychomsri, and M. Kobayashi (2006). Gene organization and complete sequence of the Hyphantria cunea nucleopolyhedrovirus genome. *J Gen Virol.* 87:2549-2562.

Iorio, C., J.E. Vialard, S. McCracken, M. Lagace, and C.D. Richardson (1998). The late expression factors 8 and 9 and possibly the phosphoprotein p78/83 of Autographa californica multicapsid nucleopolyhedrovirus are components of the virus-induced RNA polymerase. *Intervirology.* 41:35-46.

Ishikawa, H., M. Ikeda, C.A. Alves, S.M. Thiem, and M. Kobayashi (2004). Host range factor 1 from Lymantria dispar Nucleopolyhedrovirus (NPV) is an essential viral factor required for productive infection of NPVs in IPLB-Ld652Y cells derived from L. dispar. *J Virol*. 78:12703-12708.

Jakubowska, A.K., S.A. Peters, J. Ziemnicka, J.M. Vlak, and M.M. van Oers (2006). Genome sequence of an enhancin gene-rich nucleopolyhedrovirus (NPV) from Agrotis segetum: collinearity with Spodoptera exigua multiple NPV. *J Gen Virol*. 87:537-551.

Jehle, J.A., G.W. Blissard, B.C. Bonning, J.S. Cory, E.A. Herniou, G.F. Rohrmann, D.A. Theilmann, S.M. Thiem, and J.M. Vlak (2006a). On the classification and nomenclature of baculoviruses: a proposal for revision. *Arch Virol*. 151:1257-1266.

Jehle, J.A., E. Fritsch, A. Nickel, J. Huber, and H. Backhaus (1995). TCl4.7: a novel lepidopteran transposon found in Cydia pomonella granulosis virus. *Virology*. 207:369-379.

Jehle, J.A., M. Lange, H. Wang, Z. Hu, Y. Wang, and R. Hauschild (2006b). Molecular identification and phylogenetic analysis of baculoviruses from Lepidoptera. *Virology*. 346:180-193.

Ji, X., G. Sutton, G. Evans, D. Axford, R. Owen, and D.I. Stuart (2010). How baculovirus polyhedra fit square pegs into round holes to robustly package viruses. *EMBO J*. 29:505-514.

Jin, J., W. Dong, and L.A. Guarino (1998). The LEF-4 subunit of baculovirus RNA polymerase has RNA 5'-triphosphatase and ATPase activities. *J Virol*. 72:10011-10019.

Kamita, S.G., and S. Maeda (1997). Sequencing of the putative DNA helicase-encoding gene of the Bombyx mori nuclear polyhedrosis virus and fine-mapping of a region involved in host range expansion. *Gene*. 190:173-179.

Ke, J., J. Wang, R. Deng, and X. Wang (2008). Autographa californica multiple nucleopolyhedrovirus ac66 is required for the efficient egress of nucleocapsids from the nucleus, general synthesis of preoccluded virions and occlusion body formation. *Virology*.

Kelly, D.C., D.A. Brown, M.D. Ayres, C.J. Allen, and I.O. Walker (1983). Properties of the Major Nucleocapsid Protein of Heliothis zea Singly Enveloped Nuclear Polyhedrosis Virus. *Journal of General Virology*. 64:399-408.

Kikhno, I., S. Gutierrez, L. Croizier, G. Croizier, and M.L. Ferber (2002). Characterization of pif, a gene required for the per os infectivity of Spodoptera littoralis nucleopolyhedrovirus. *J Gen Virol*. 83:3013-3022.

Ko, R., K. Okano, and S. Maeda (2000). Structural and functional analysis of the Xestia c-nigrum granulovirus matrix metalloproteinase. *J Virol*. 74:11240-11246.

Kool, M., J.T. Voeten, R.W. Goldbach, J. Tramper, and J.M. Vlak (1993). Identification of seven putative origins of Autographa californica multiple nucleocapsid nuclear polyhedrosis virus DNA replication. *J Gen Virol*. 74:2661-2668.

Kost, T.A., J.P. Condreay, and D.L. Jarvis (2005). Baculovirus as versatile vectors for protein expression in insect and mammalian cells. *Nat Biotechnol*. 23:567-575.

Kuzio, J., M.N. Pearson, S.H. Harwood, C.J. Funk, J.T. Evans, J.M. Slavicek, and G.F. Rohrmann (1999). Sequence and analysis of the genome of a baculovirus pathogenic for Lymantria dispar. *Virology*. 253:17-34.

Landais, I., R. Vincent, M. Bouton, G. Devauchelle, M. Duonor-Cerutti, and M. Ogliastro (2006). Functional analysis of evolutionary conserved clustering of bZIP binding sites in the baculovirus homologous regions (hrs) suggests a cooperativity between host and viral transcription factors. *Virology.* 344:421-431.

Lange, M., and J.A. Jehle (2003). The genome of the Cryptophlebia leucotreta granulovirus. *Virology.* 317:220-236.

Lauzon, H.A., P.B. Jamieson, P.J. Krell, and B.M. Arif (2005). Gene organization and sequencing of the Choristoneura fumiferana defective nucleopolyhedrovirus genome. *J Gen Virol.* 86:945-961.

Lauzon, H.A., C.J. Lucarotti, P.J. Krell, Q. Feng, A. Retnakaran, and B.M. Arif (2004). Sequence and organization of the Neodiprion lecontei nucleopolyhedrovirus genome. *J Virol.* 78:7023-7035.

Lee, J.-C., H.-H. Chen, and Y.-C. Chao (1998). Persistent Baculovirus Infection Results from Deletion of the Apoptotic Suppressor Gene p35. *J. Virol.* 72:9157-9165.

Li, G., J. Wang, R. Deng, and X. Wang (2008). Characterization of AcMNPV with a deletion of ac68 gene. *Virus Genes.* 37:119-127.

Li, L., C. Donly, Q. Li, L.G. Willis, B.A. Keddie, M.A. Erlandson, and D.A. Theilmann (2002a). Identification and genomic analysis of a second species of nucleopolyhedrovirus isolated from Mamestra configurata. *Virology.* 297:226-244.

Li, L., Q. Li, L.G. Willis, M. Erlandson, D.A. Theilmann, and C. Donly (2005). Complete comparative genomic analysis of two field isolates of Mamestra configurata nucleopolyhedrovirus-A. *J Gen Virol.* 86:91-105.

Li, Q., C. Donly, L. Li, L.G. Willis, D.A. Theilmann, and M. Erlandson (2002b). Sequence and organization of the Mamestra configurata nucleopolyhedrovirus genome. *Virology.* 294:106-121.

Li, X., J. Barrett, A. Pang, R.J. Klose, P.J. Krell, and B.M. Arif (2000). Characterization of an overexpressed spindle protein during a baculovirus infection. *Virology.* 268:56-67.

Li, Z., C. Li, K. Yang, L. Wang, C. Yin, Y. Gong, and Y. Pang (2003). Characterization of a chitin-binding protein GP37 of *Spodoptera litura* multicapsid nucleopolyhedrovirus. *Virus Res.* 96:113-122.

Liang, Z., X. Zhang, X. Yin, S. Cao, and F. Xu (2011). Genomic sequencing and analysis of Clostera anachoreta granulovirus. *Arch Virol.* 156:1185-1198

Lu, A., L.K. Miller, P. Krell, J.M. Vlak, and G. Rohrmann (1997). Baculovirus DNA Replication. *In* The Baculoviruses. L.K. Miller, editor. Plenum Press, New York and London.

Luckow, V.A., S.C. Lee, G.F. Barry, and P.O. Olins (1993). Efficient generation of infectious recombinant baculoviruses by site-specific transposon-mediated insertion of foreign genes into a baculovirus genome propagated in Escherichia coli. *J Virol.* 67:4566-4579.

Lung, O., M. Westenberg, J.M. Vlak, D. Zuidema, and G.W. Blissard (2002). Pseudotyping Autographa californica multicapsid nucleopolyhedrovirus (AcMNPV): F proteins from group II NPVs are functionally analogous to AcMNPV GP64. *J Virol.* 76:5729-5736.

Luque, T., R. Finch, N. Crook, D.R. O'Reilly, and D. Winstanley (2001). The complete sequence of the Cydia pomonella granulovirus genome. *J Gen Virol.* 82:2531-2547.

Ma, X.C., J.Y. Shang, Z.N. Yang, Y.Y. Bao, Q. Xiao, and C.X. Zhang (2007). Genome sequence and organization of a nucleopolyhedrovirus that infects the tea looper caterpillar, Ectropis obliqua. *Virology.* 360:235-246.

Maeda, S., S.G. Kamita, and A. Kondo (1993). Host range expansion of Autographa californica nuclear polyhedrosis virus (NPV) following recombination of a 0.6-kilobase-pair DNA fragment originating from Bombyx mori NPV. *J Virol.* 67:6234-6238.

Mäkelä, A.R., W. Ernst, R. Grabherr, and C. Oker-Blom (2010). Baculovirus-based display and gene delivery systems. *Cold Spring Harb* Protoc. 2010; doi:10.1101/pdb top72.

McCarthy, C.B., X. Dai, C. Donly, and D.A. Theilmann (2008). Autographa californica multiple nucleopolyhedrovirus ac142, a core gene that is essential for BV production and ODV envelopment. *Virology.* 372:325-339.

McCarthy, C.B., and D.A. Theilmann (2008). AcMNPV ac143 (odv-e18) is essential for mediating budded virus production and is the 30th baculovirus core gene. *Virology.*

McDougal, V.V., and L.A. Guarino (1999). Autographa californica nuclear polyhedrosis virus DNA polymerase: measurements of processivity and strand displacement. *J Virol.* 73:4908-4918.

McDougal, V.V., and L.A. Guarino (2000). The Autographa californica nuclear polyhedrosis virus p143 gene encodes a DNA helicase. *J Virol.* 74:5273-5279.

Merrington, C.L., P.A. Kitts, L.A. King, and R.D. Possee (1996). An Autographa californica nucleopolyhedrovirus lef-2 mutant: consequences for DNA replication and very late gene expression. *Virology.* 217:338-348.

Miele, S.A.B., M.J. Garavaglia, M.N. Belaich, and P.D. Ghiringhelli (2011). Baculovirus: Molecular Insights on their Diversity and Conservation. *Int J Evol Biol.* 2011:15.

Mikhailov, V.S. (2003). Replication of the Baculovirus Genome. *Molecular Biology.* 37:250-259.

Mikhailov, V.S., K. Okano, and G.F. Rohrmann (2003). Baculovirus alkaline nuclease possesses a 5'→3' exonuclease activity and associates with the DNA-binding protein LEF-3. *J Virol.* 77:2436-2444.

Miller, L.K., A.J. Lingg, and L.A. Bulla, Jr. (1983). Bacterial, Viral, and Fungal Insecticides. *Science.* 219:715-721.

Miller, L.K., and A. Lu (1997). The Molecular Basis of Baculovirus Host Range. *In* The Baculoviruses. L.K. Miller, editor. Plenum Press, New York and London. 217-235.

Mishra, G., P. Chadha, and R.H. Das (2008). Serine/threonine kinase (pk-1) is a component of Autographa californica multiple nucleopolyhedrovirus (AcMNPV) very late gene transcription complex and it phosphorylates a 102 kDa polypeptide of the complex. *Virus Research.* 137:147-149.

Mitsuhashi, W., H. Kawakita, R. Murakami, Y. Takemoto, T. Saiki, K. Miyamoto, and S. Wada (2007). Spindles of an entomopoxvirus facilitate its infection of the host insect by disrupting the peritrophic membrane. *J Virol.* 81:4235-4243.

Monsma, S.A., A.G. Oomens, and G.W. Blissard (1996). The GP64 envelope fusion protein is an essential baculovirus protein required for cell-to-cell transmission of infection. *J Virol.* 70:4607-4616.

Nagamine, T., Y. Kawasaki, and S. Matsumoto (2006). Induction of a subnuclear structure by the simultaneous expression of baculovirus proteins, IE1, LEF3, and P143 in the presence of hr. *Virology.* 352:400-407.

Nai, Y.S., C.Y. Wu, T.C. Wang, Y.R. Chen, W.H. Lau, C.F. Lo, M.F. Tsai, and C.H. Wang (2010). Genomic sequencing and analyses of Lymantria xylina multiple nucleopolyhedrovirus. *BMC Genomics.* 11:116.

Nakai, M., C. Goto, W. Kang, M. Shikata, T. Luque, and Y. Kunimi (2003). Genome sequence and organization of a nucleopolyhedrovirus isolated from the smaller tea tortrix, Adoxophyes honmai. *Virology*. 316:171-183.

Nie, Z.M., Z.F. Zhang, D. Wang, P.A. He, C.Y. Jiang, L. Song, F. Chen, J. Xu, L. Yang, L.L. Yu, J. Chen, Z.B. Lv, J.J. Lu, X.F. Wu, and Y.Z. Zhang (2007). Complete sequence and organization of Antheraea pernyi nucleopolyhedrovirus, a dr-rich baculovirus. *BMC Genomics*. 8:248.

O' Reilly, D.R., L.K. Miller, and V.A. Luckow (1992). Baculovirus Expression Vectors. A Laboratory Manual. Oxford University Press, New York.

O'Reilly, D.R., and L.K. Miller (1989). A baculovirus blocks insect molting by producing ecdysteroid UDP-glucosyl transferase. *Science*. 245:1110-1112.

Ogembo, J.G., B.L. Caoili, M. Shikata, S. Chaeychomsri, M. Kobayashi, and M. Ikeda (2009). Comparative genomic sequence analysis of novel Helicoverpa armigera nucleopolyhedrovirus (NPV) isolated from Kenya and three other previously sequenced Helicoverpa spp. NPVs. *Virus Genes*. 39:261-272.

Ohkawa, T., J.O. Washburn, R. Sitapara, E. Sid, and L.E. Volkman (2005). Specific binding of Autographa californica M nucleopolyhedrovirus occlusion-derived virus to midgut cells of Heliothis virescens larvae is mediated by products of pif genes Ac119 and Ac022 but not by Ac115. *J Virol*. 79:15258-15264.

Oliveira, J.V., J.L. Wolff, A. Garcia-Maruniak, B.M. Ribeiro, M.E. de Castro, M.L. de Souza, F. Moscardi, J.E. Maruniak, and P.M. Zanotto (2006). Genome of the most widely used viral biopesticide: *Anticarsia gemmatalis* multiple nucleopolyhedrovirus. *J Gen Virol*. 87:3233-3250.

Olszewski, J., and L.K. Miller (1997a). Identification and characterization of a baculovirus structural protein, VP1054, required for nucleocapsid formation. *J Virol*. 71:5040-5050.

Olszewski, J., and L.K. Miller (1997b). A role for baculovirus GP41 in budded virus production. *Virology*. 233:292-301.

Pang, Y., J. Yu, L. Wang, X. Hu, W. Bao, G. Li, C. Chen, H. Han, S. Hu, and H. Yang (2001). Sequence analysis of the Spodoptera litura multicapsid nucleopolyhedrovirus genome. *Virology*. 287:391-404.

Passarelli, A.L. (2011). Barriers to success: How baculoviruses establish efficient systemic infections. *Virology*. 411: 383-392.

Pearson, M., R. Bjornson, G. Pearson, and G. Rohrmann (1992). The Autographa californica baculovirus genome: evidence for multiple replication origins. *Science*. 257:1382-1384.

Pearson, M.N., C. Groten, and G.F. Rohrmann (2000). Identification of the lymantria dispar nucleopolyhedrovirus envelope fusion protein provides evidence for a phylogenetic division of the Baculoviridae. *J Virol*. 74:6126-6131.

Peng, K., M.M. van Oers, Z. Hu, J.W. van Lent, and J.M. Vlak (2010). Baculovirus per os infectivity factors form a complex on the surface of occlusion-derived virus. *J Virol*. 84:9497-9504.

Perera, O., T.B. Green, S.M. Stevens, Jr., S. White, and J.J. Becnel (2007). Proteins associated with Culex nigripalpus nucleopolyhedrovirus occluded virions. *J Virol*. 81:4585-4590.

Phanis, C.G., D.P. Miller, S.C. Cassar, M. Tristem, S.M. Thiem, and D.R. O'Reilly (1999). Identification and expression of two baculovirus *gp37* genes. *J Gen Virol.* 80:1823-1831.

Pijlman, G.P., A.J. Pruijssers, and J.M. Vlak (2003). Identification of pif-2, a third conserved baculovirus gene required for per os infection of insects. *J Gen Virol.* 84:2041-2049.

Ponicsan, S.L., J.F. Kugel, and J.A. Goodrich (2010). Genomic gems: SINE RNAs regulate mRNA production. *Curr Opin Genet Dev.* 20:149-155.

Rodems, S.M., and P.D. Friesen (1995). Transcriptional enhancer activity of hr5 requires dual-palindrome half sites that mediate binding of a dimeric form of the baculovirus transregulator IE1. *J Virol.* 69:5368-5375.

Rohrmann, G. (2011a). The AcMNPV genome: Gene content, conservation, and function. *In* Baculovirus Molecular Biology, 2nd ed. National Library of Medicine (US), NCBI, Bethesda (MD).

Rohrmann, G. (2011b). Baculovirus late transcription. *In* Baculovirus Molecular Biology, 2nd ed. National Library of Medicine (US), NCBI, Bethesda (MD).

Rohrmann, G. (2011c). DNA Replication and genome processing. *In* Baculovirus Molecular Biology. G. Rohrmann, editor. National Library of Medicine (US), Bethesda (MD).

Rohrmann, G. (2011d). Early events in infection: Virus transcription. *In* Baculovirus Molecular Biology, 2nd ed. National Library of Medicine (US), NCBI, Bethesda (MD).

Rohrmann, G. (2011e). Structural proteins of baculovirus occlusion bodies and virions. *In* Baculovirus Molecular Biology. National Library of Medicine (US), NCBI, Bethesda (MD).

Rohrmann, G.F. (1986). Polyhedrin structure. *J Gen Virol.* 67:1499-1513.

Romanowski, V., and P.D. Ghiringhelli (2001). Biología molecular de baculovirus: Replicación y regulación de la expresión génica. *In* Los Baculovirus y sus Aplicaciones como Bioinsecticidas. M.L.-F.y.T.W. P. Caballero, editor, Phytoma-Universidad Pública de Navarra, Pamplona, España. 119-142.

Russell, R.L., and G.F. Rohrmann (1997). Characterization of P91, a protein associated with virions of an Orgyia pseudotsugata baculovirus. *Virology.* 233:210-223.

Singh, J., C.P. Singh, A. Bhavani, and J. Nagaraju (2010). Discovering microRNAs from Bombyx mori nucleopolyhedrosis virus. *Virology.* 407:120-128.

Slack, J., and B.M. Arif (2007). The baculoviruses occlusion-derived virus: virion structure and function. *Adv Virus Res.* 69:99-165.

Sparks, W.O., R.L. Harrison, and B.C. Bonning (2011). Autographa californica multiple nucleopolyhedrovirus ODV-E56 is a per os infectivity factor, but is not essential for binding and fusion of occlusion-derived virus to the host midgut. *Virology.* 409:69-76.

Sutherland, D., C. Samakovlis, and M.A. Krasnow (1996). branchless encodes a Drosophila FGF homolog that controls tracheal cell migration and the pattern of branching. *Cell.* 87:1091-1101.

Szewczyk, B., L. Hoyos-Carvajal, M. Paluszek, I. Skrzecz, and M. Lobo de Souza (2006). Baculoviruses - re-emerging biopesticides. *Biotechnol Adv.* 24:143-160.

Tang, X.D., Q. Xiao, X.C. Ma, Z.R. Zhu, and C.X. Zhang (2009). Morphology and genome of Euproctis pseudoconspersa nucleopolyhedrovirus. *Virus Genes.* 38:495-506.

Thiem, S.M. (1997). Prospects for altering host range for baculovirus bioinsecticides. *Curr Opin Biotechnol.* 8:317-322.

Thiem, S.M., X. Du, M.E. Quentin, and M.M. Berner (1996). Identification of baculovirus gene that promotes Autographa californica nuclear polyhedrosis virus replication in a nonpermissive insect cell line. *J Virol.* 70:2221-2229.

Thiem, S.M., and L.K. Miller (1989). Identification, sequence, and transcriptional mapping of the major capsid protein gene of the baculovirus Autographa californica nuclear polyhedrosis virus. *J Virol.* 63:2008-2018.

Titterington, J.S., T.K. Nun, and A.L. Passarelli (2003). Functional dissection of the baculovirus late expression factor-8 gene: sequence requirements for late gene promoter activation. *J Gen Virol.* 84:1817-1826.

Van Der Wilk, F., J.W.M. Van Lent, and J.M. Vlak (1987). Immunogold Detection of Polyhedrin, p10 and Virion Antigens in Autographa californica Nuclear Polyhedrosis Virus-infected Spodoptera frugiperda Cells. *J Gen Virol.* 68:2615-2623.

van Oers, M.M., M.H. Abma-Henkens, E.A. Herniou, J.C. de Groot, S. Peters, and J.M. Vlak (2005). Genome sequence of Chrysodeixis chalcites nucleopolyhedrovirus, a baculovirus with two DNA photolyase genes. *J Gen Virol.* 86:2069-2080.

Van Oers, M.M., J.T. Flipsen, C.B. Reusken, and J.M. Vlak (1994). Specificity of baculovirus p10 functions. *Virology.* 200:513-523.

van Oers, M.M., and J.M. Vlak (2007). Baculovirus genomics. *Curr Drug Targets.* 8:1051-1068.

Vanarsdall, A.L., K. Okano, and G.F. Rohrmann (2006). Characterization of the role of very late expression factor 1 in baculovirus capsid structure and DNA processing. *J Virol.* 80:1724-1733.

Vialard, J.E., L. Yuen, and C.D. Richardson (1990). Identification and characterization of a baculovirus occlusion body glycoprotein which resembles spheroidin, an entomopoxvirus protein. *J Virol.* 64:5804-5811.

von Sternberg, R., and J.A. Shapirob (2005). How repeated retroelements format genome function. *Cytogenet Genome Res.* 110:108-116.

Wang, M., E. Tuladhar, S. Shen, H. Wang, M.M. van Oers, J.M. Vlak, and M. Westenberg (2010a). Specificity of baculovirus P6.9 basic DNA-binding proteins and critical role of the C terminus in virion formation. *J Virol.* 84:8821-8828.

Wang, P., and R.R. Granados (1997). An intestinal mucin is the target substrate for a baculovirus enhancin. *Proc Natl Acad Sci U S A.* 94:6977-6982.

Wang, R., F. Deng, D. Hou, Y. Zhao, L. Guo, H. Wang, and Z. Hu (2010b). Proteomics of the Autographa californica nucleopolyhedrovirus budded virions. *J Virol.* 84:7233-7242.

Wang, X.-F., B.-Q. Zhang, H.-J. Xu, Y.-J. Cui, Y.-P. Xu, M.-J. Zhang, Y.S. Han, Y.S. Lee, Y.-Y. Bao, and C.-X. Zhang (2011). ODV-associated Proteins of the Pieris rapae Granulovirus. *J Proteome Res.* 10:2817-2827.

Wicker, T., F. Sabot, A. Hua-Van, J.L. Bennetzen, P. Capy, B. Chalhoub, A. Flavell, P. Leroy, M. Morgante, O. Panaud, E. Paux, P. SanMiguel, and A.H. Schulman (2007). A unified classification system for eukaryotic transposable elements. *Nat Rev Genet.* 8:973-982.

Williams, G.V., D.Z. Rohel, J. Kuzio, and P. Faulkner (1989). A cytopathological investigation of Autographa californica nuclear polyhedrosis virus p10 gene function using insertion/deletion mutants. *J Gen Virol.* 70:187-202.Willis, L.G., R. Seipp, T.M. Stewart, M.A. Erlandson, and D.A. Theilmann (2005). Sequence analysis of the complete genome of Trichoplusia ni single nucleopolyhedrovirus and the identification of a baculoviral photolyase gene. *Virology*. 338:209-226.

Wolff, J.L., F.H. Valicente, R. Martins, J.V. Oliveira, and P.M. Zanotto (2008). Analysis of the genome of Spodoptera frugiperda nucleopolyhedrovirus (SfMNPV-19) and of the high genomic heterogeneity in group II nucleopolyhedroviruses. *J Gen Virol*. 89:1202-1211.

Wormleaton, S., J. Kuzio, and D. Winstanley (2003). The complete sequence of the Adoxophyes orana granulovirus genome. *Virology*. 311:350-365.

Wu, W., T. Lin, L. Pan, M. Yu, Z. Li, Y. Pang, and K. Yang (2006). Autographa californica multiple nucleopolyhedrovirus nucleocapsid assembly is interrupted upon deletion of the 38K gene. *J Virol*. 80:11475-11485.

Wu, W., and A.L. Passarelli (2010). Autographa californica multiple nucleopolyhedrovirus Ac92 (ORF92, P33) is required for budded virus production and multiply enveloped occlusion-derived virus formation. *J Virol*. 84:12351-12361.

Wu, X., and L.A. Guarino (2003). Autographa californica nucleopolyhedrovirus orf69 encodes an RNA cap (nucleoside-2'-O)-methyltransferase. *J Virol*. 77:3430-3440.

Wu, Y., and E.B. Carstens (1996). Initiation of baculovirus DNA replication: early promoter regions can function as infection-dependent replicating sequences in a plasmid-based replication assay. *J Virol*. 70:6967-6972.

Xiang, X., L. Chen, A. Guo, S. Yu, R. Yang, and X. Wu (2011). The Bombyx mori nucleopolyhedrovirus (BmNPV) ODV-E56 envelope protein is also a per os infectivity factor. *Virus Res*. 155:69-75.

Xiao, H., and Y. Qi (2007). Genome sequence of Leucania seperata nucleopolyhedrovirus. *Virus Genes*. 35:845-856.

Xu, F., J.M. Vlak, and M.M. van Oers (2008). Conservation of DNA photolyase genes in group II nucleopolyhedroviruses infecting plusiine insects. *Virus Res*. 136:58-64.

Yang, S., and L.K. Miller (1999). Activation of baculovirus very late promoters by interaction with very late factor 1. *J Virol*. 73:3404-3409.

Zanotto, P.M., B.D. Kessing, and J.E. Maruniak (1993). Phylogenetic interrelationships among baculoviruses: evolutionary rates and host associations. *J Invertebr Pathol*. 62:147-164.

Zemskov, E.A., W. Kang, and S. Maeda (2000). Evidence for nucleic acid binding ability and nucleosome association of Bombyx mori nucleopolyhedrovirus BRO proteins. *J Virol*. 74:6784-6789.

Zhang, C.X., X.C. Ma, and Z.J. Guo (2005). Comparison of the complete genome sequence between C1 and G4 isolates of the Helicoverpa armigera single nucleocapsid nucleopolyhedrovirus. *Virology*. 333:190-199.

Zhao, Y., D.A. Chapman, and I.M. Jones (2003). Improving baculovirus recombination. *Nucleic Acids Res*. 31:E6-6.

Zhu, S.Y., J.P. Yi, W.D. Shen, L.Q. Wang, H.G. He, Y. Wang, B. Li, and W.B. Wang (2009). Genomic sequence, organization and characteristics of a new nucleopolyhedrovirus isolated from Clanis bilineata larva. *BMC Genomics*. 10:91.

3

Foot and Mouth Disease Virus Genome

Consuelo Carrillo
APHIS-NVSL-FADDL
USA

1. Introduction

Foot-and-Mouth Disease Virus (FMDV) is a member of the Picornaviridae family of viruses, which includes viruses that cause a number of high consequence human and animal diseases in addition to Foot-and-Mouth Disease (FMD), such as hand-foot-and-mouth disease, herpangina, polio, and encephalomyocarditis. FMDV infects domestic and wild cloven-hoofed animals, including bovine, caprine, ovine and swine species that are vital to the livestock industry. Depending on host and virus characteristics, FMD exhibits a broad range of clinical presentations resulting in significant morbidity. Generally, FMD produces fever and soreness, excessive salivation, loss of appetite and large vesicles on the feet, nose and tongue 24 to 72 hours post-infection (hpi). In some cases, complete prostration accompanied by the loss of hooves occurs (for review see refs. 1-4). Although mortality rates are typically low and usually associated with young or immunocompromised animals, the economic consequences of an FMD outbreak, such as dramatic decreases in livestock productivity and banning of the export of animals and animal products, are so significant that FMD is one of the most threatening diseases of domestic animals in the world. In many developing regions of Asia, Africa and South America, FMD is enzootic. Global economic activities and transboundary movement of people and animals presents a significant risk of accidental introduction of FMDV into previously FMDV-free countries. Additionally, terrorist groups may intentionally introduce FMDV into a country that is FMD-free and does not vaccinate against the virus (refs.5-8 and http://www.oie.int; http://iah.bbsrc.ac.uk/virus/Picornaviridae/Aphtovirus/fmd.htm).

2. Foot-and-mouth disease virus

Like all picornaviruses, FMDV is a non-enveloped virus with icosahedral symmetry and contains a single-stranded, positive-sense RNA molecule of approximately 8500 nucleotides (nt). The viral particle is small in size (approximately 30 nm diameter) and is composed of 60 copies each of proteins 1A, 1B, 1C and 1D (also known as VP4, VP2, VP3 and VP1, respectively) assembled in groups of increasing complexity. A single cluster of structural proteins, known as the *protomer*, involves one copy each of 1A, 1B, 1C and 1D. Five protomers assemble together to form a *pentamer*, which are then assembled into groups of twelve to form the complete viral capsid (9, 10).

FMDV RNA possesses genetic characteristics, such as positive polarity and a polyadenylated 3′ end, that allows it to act as messenger RNA (mRNA) *in vitro* and *in vivo*, and therefore should be considered a potentially infectious agent. RNA extracted from field samples has been used to produce infectious viruses using *in vitro* cell electroporation techniques. The rescued viruses were highly pathogenic in natural hosts and could be characterized using ELISA and full genome sequencing. This feature of the FMDV genome allows for manipulation of cDNA copies to study the genetics of pathogenesis and observe the phenotypic effects of mutations and other genomic alterations. This application of "reverse genetics" to construct chimeric and recombinant FMDVs has led to the discovery of several determinants of viral replication, host recognition and virulence (3, 4, 11). Continued use of reverse genetics will enhance current models of FMDV pathogenesis and further progress towards therapies and vaccines.

3. FMDV genomics publications

Analyses have been performed on the complete P1 polyprotein, the genomic region encoding all four structural proteins that compose the viral capsid (1A, 1B, 1C and 1D). However, most of the work published regarding FMDV genomics is limited to the coding region of capsid protein 1D (also known as VP1). This information has been used to analyze variability, selective pressures and immunogenicity of FMDV. Phylogenetic analysis employing 1D sequences and a 15% nucleotide difference as a cut-off organizes FMDV strains into major groups or genotypes. Interestingly, the genotypic information obtained with approximately 636 nucleotides of VP1 completely matches the phylogenetic results obtained when 2208 nucleotides of the complete P1 polyprotein are used (12, 13). Genetic lineages subsequently are geographically bound and are described as "topotypes". The viral capsid, notably the 1D protein, harbors immunogenic epitopes that are critical for neutralization of the virus. Hence, the topotype classification system has extraordinary value for vaccine selection.

The development of high-throughput sequencing techniques has allowed for complete genome sequencing of FMDV, significantly improving our understanding of infection, host range, and transmission. The use of complete FMDV genome sequences in phylogenetic studies has revealed much more complex epidemiological relationships between isolates than previously thought. Full genome comparisons suggest that the epidemiology of FMDV is heavily impacted by recombination; it also has led to the discovery of novel genetic lineages containing genomic sequences that appear equally distant from SAT and Euroasiatic lineages of the virus. Furthermore, complete genome sequencing has enhanced the discovery of FMDV variability, sequence conservation and universal genetic motifs that affect its virulence and transmission. Complete genome sequence analysis of FMDV isolates collected during the August 2007 outbreaks in England identified the initial and intermediate sources of the outbreaks, demonstrating the value of complete sequence analysis when examining virus phylogeny, an accomplishment previously impossible using partial genomic sequences (14, 15, 16).

4. Genome structure and proteins

4.1 FMDV genome

Upon entrance into the host cell the virus particle dissociates and the RNA is released into the cytoplasm. The genome of picornaviruses functions as a messenger (+) RNA, polyadenylated at its 3'-end and covalently linked to a small protein (VPg) at its 5'-end.

Translation occurs as a single polypeptide precursor (ORF) that is cleaved into functional proteins, mostly by virally encoded proteases (3, 10, 17).

Primary processing of the FMDV ORF results in three large intermediate polyproteins (L/P1, P2 and P3). Protease cleavage by FMDV proteins L, 3C and 2A produces smaller sub-products and 12 final mature proteins: L, 1A, 1B, 1C, 1D, 2A, 2B, 2C, 3A, 3B, 3C and 3D (Figs. 1 and 2). As mentioned above, 1A, 1B, 1C and 1D are the structural proteins that form the viral capsid. FMDV 2B and 2C proteins localize to ER-derived vesicles, the site of viral replication, but their functions remain poorly understood. 3A is thought to be a multifunctional integral membrane protein that enhances viral RNA synthesis and exhibits host-related markers. Unique to FMDV are three non-identical copies of the genome-linked 3B, a protein required for viral RNA replication. Finally, 3D encodes the viral RNA-dependent RNA polymerase, and along with 3A, co-localizes with ER membrane-associated replication complexes.

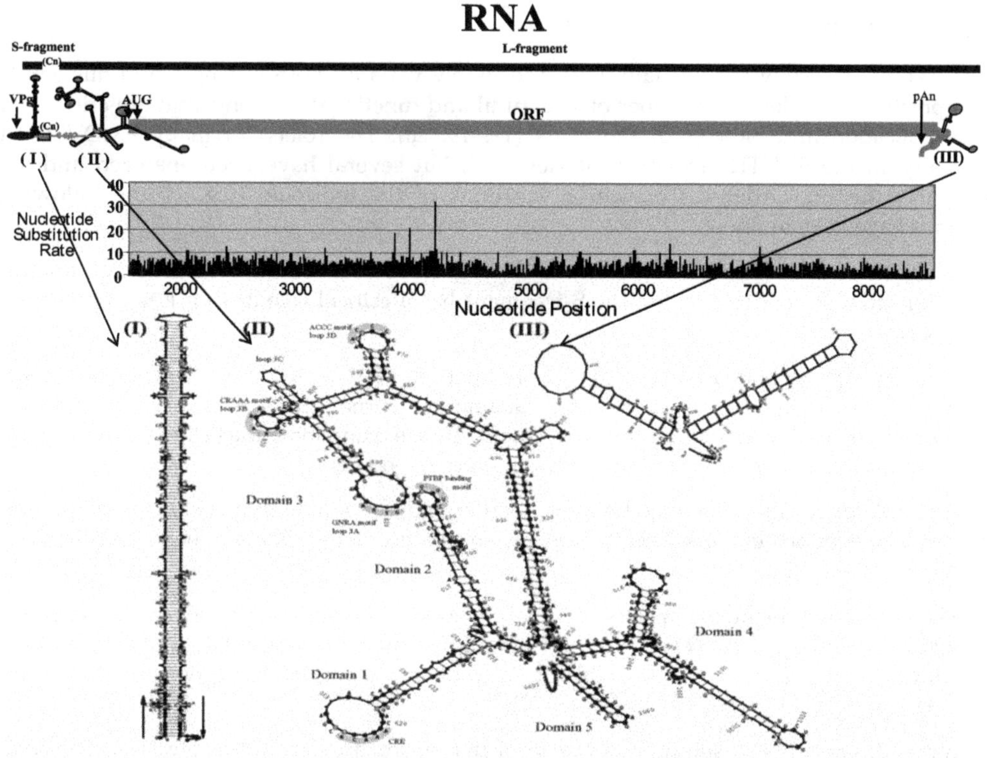

Fig. 1. FMDV RNA genome representation with detailed description of the UTR elements and its predicted secondary structure: (I) S-fragment; (II) IRES; (III) 3′ NCR. Also includes a graphic representation of the distribution of variability within the complete coding region (ORF), expressed as rate of substitutions per nucleotide site of a Clustal W multiple alignment of all serotype FMDV full length genomes (ref 14 for more detailed view).

The nomenclature used for FMDV proteins is similar amongst all picornaviruses and is based on their position in the viral RNA genome (18). However, it does not imply a conserved function of the proteins across all genera. In fact, there is increasing evidence that despite sharing similar genome organization and protein names there can be significant differences in functionality. For instance, FMDV encodes a 3A protein that is 50% longer than the equivalent protein encoded by poliovirus; FMDV also harbors three copies of 3B in contrast to a single copy in poliovirus. Additionally, the role of the poliovirus 3A protein in immune evasion and persistent infection seems to be played by FMDV 2BC .

Traditionally, the FMDV genome is classified into coding (ORF) and non-coding regions (NCRs) that distribute along three defined genomic intervals: (i) the 5′ untranslated region (5′-UTR), which contains non-coding nucleic acids that carry many regulatory elements; (ii) the protein coding region (ORF), which includes both structural and non-structural proteins; and (iii) the 3′ UTR or non-coding region, which also carries regulatory functions and a poly(A tail.

4.2 FMDV 5' UTR

The 5′-UTR of FMDV is unusually long and highly variable, both in length and nucleotide composition. It includes a number of structural and functional elements that are critical for the replication and biology of the virus (17, 19, 20). The role(s) of many FMDV RNA domains in the 5′ UTR are poorly understood, but several have been analyzed and are described herein. Ordered from the 5′-terminus of the molecule (Fig. 1), the following regions have been defined:

A highly structured small fragment or "S-fragment" of about 370 residues of unclear function. Evidence suggests that the S-fragment has a critical role during RNA replication (Fig. 1-I).

An internal polyribocytidylic, or poly(C), tract of 100 to 400 nucleotides, comprised predominantly of cytosine residues. Unusual among picornaviruses, it has been described as an element related to virulence, but subsequent studies using infectious clones (ICs) suggest the contrary. Its biological function is not well understood (21).

A pseudoknot region, also of unknown function and variable length. This region contains significant deletions in some FMDV isolates, and its presence has never been linked to any specific biological function.

A cis-Replicating Element (cre) or 3B-uridylylation site (bus), conforms of a stem-loop structure containing a conserved AAACA motif, functions as a template for addition of U residues to the protein primer 3B. It is critical during transcription in order for circularization of the viral RNA to occur (22).

A type-2 IRES of about 440 nt, comprised of five major domains, H-L (Fig 1-II). The IRES facilitates the internal initiation of protein synthesis in a CAP-independent fashion, allowing it to mediate ribosome recruitment to an internal site within the viral RNA (23). This process is facilitated by eukaryotic translation initiation factors (eIFs). Initiation of translation by the IRES begins with specific binding of the central domain initiation factor, eIF4G, to the J-K domains, which is stimulated by eIF4A. Then, these initiation factors induce a restructuration of the region and promote recruitment of ribosomal pre-initiation complexes. PTB and ITAF45 trans-acting factors are also required to stabilize the active conformation. Both binding of eIF4G to the IRES and IRES-directed translation are significantly impaired by mutations that impact the integrity of the double-stranded secondary structure. In fact, the primary sequence within the IRES of different FMDV isolates can be up to 50% variable and still retain similar overall secondary structure using compensatory base changes in helical elements.

4.3 FMDV coding region (ORF)

Protein synthesis starts at one of two functional in-frame AUG codons, separated from each other by an indispensable but highly variable tract of approximately 80 to 84 nucleotides. The long ORF that follows the AUG codon encodes a polyprotein of about 2330 amino acids (aa), although length and composition among natural isolates and even among passaged viruses can be variable. Although the polyprotein intermediaries of processing are biologically important, the current discussion will concentrate only on the twelve final protein products. For a complete review of cleavage sites, biologically critical residues, and variability/conservation between serotypes, see ref. 3, 4, 18.

Components of the polyprotein, from 5' to 3' (Fig. 2):

Polyprotein (ORF)

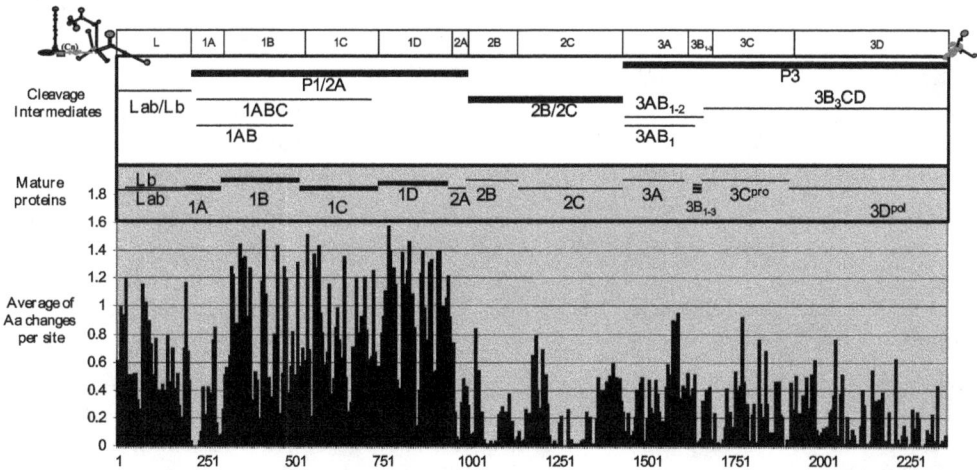

Fig. 2. Upper panel: schematic representation of FMDV genome and poly-protein coding region including protein-encoding regions, cleavage intermediates and mature protein products. Lower panel: schematic representation of the non-synonymous (amino acid) substitutions per site within the whole poly-protein (ORF).

4.3.1 Leader protease (Lpro)

Two in-frame AUG codons allow for two different initiation events within Lpro, producing two forms of the protein, named Lab and Lb. Both proteins catalyze their own cleavage at their C-terminus from the rest of the polyprotein as well as cleavage of the initiation factor eIF4G (p220) of the CAP-binding complex eIF4F, contributing to the shut-off of host cell protein synthesis. Approximately one-third of cell ribosomes initiate at the first AUG (AUG1, of Lab) site, while the majority of ribosomes start translation at the second AUG codon (AUG2, of Lb). It is unknown why ribosomal preference for the second AUG exists. Both proteins also limit the host innate immune response via inhibition of interferon beta (IFN-β) mRNA expression. *L(pro)* localizes to the nucleus of infected cells and disrupts the integrity of the nuclear factor NF-κβ using mechanisms that antagonize the cellular innate immune and inflammatory responses to FMDV infection.

4.3.2 P1 coding region

The P1 polypeptide sequence begins immediately downstream of the Lpro protein. Included within P1 are the four capsid proteins: *1A, 1B, 1C,* and *1D* (also known as *VP4, VP2, VP3,* and *VP1*, respectively). With the exception of *1A*, which is excluded from the virion surface, capsid proteins (*1B, 1C* and *1D*) are involved in antigenicity and binding to a subset of RGD-dependent integrins and heparin sulfate proteoglycan receptors on the cell surface (reviewed in Ref. 24).

FMDV structural proteins are involved in capsid assembly and stability, virus binding and antigenicity. Despite these essential characteristics, there is a high degree of flexibility in the primary sequence of most of these proteins. The structural proteins exhibit the highest rates of nucleotide and amino acid (aa) variability among all viral proteins, likely a response to intense selective pressures. *VP4* is an exception to this observation, since 73% to 84% of its nucleotide sequence is conserved among all FMDV isolates.

The only internal capsid protein, 1A, carries a swine-specific immunodominant and heterotypic T-cell antigenic site that is capable of providing help to a B-cell epitope when in tandem. Within the amino half of 1A, a conserved myristate binding site exists.

Structural protein *1B*, or *VP2*, plays a critical role in capsid stability and particle maturation, supported by the observation that 47% of its amino acids are invariant between and among different FMDV serotypes. At least 3 T-cell epitopes have been identified within *1B*, exemplifying its immunogenicity. A number of important conformational neutralizing epitopes and one T-cell epitope have also been identified in protein *1C* (*VP3*).

The best known FMDV protein is *1D*, also known as *VP1*. *VP1* is the most variable region of the FMDV genome; only 26% of its aa are universally conserved between serotypes. Many of the residues known or suspected to be critical for cleavage or other functions are located within invariant sequence motifs, indicating that the critical function of those residues may be contextual and require other specific residues. Protein *1D* is responsible for virus attachment and entry, protective immunity and serotype specificity. A major, non-essential immunodominant site is located within the so-called G–H loop of *VP1*. This loop appears highly disordered in X-ray diffraction patterns of crystallized virions, but it is known to protrude from the capsid surface when the capsid is bound to an antigen-binding fragment (Fab). After binding of a cellular integrin receptor to the RGD motif in *1D*, FMDV utilizes endocytosis to infect the cell. Viral replication commences when the viral capsid dissolves, allowing the release of RNA into the cell cytoplasm. Viruses that have sequences similar to the RGD motif can infect cells via different integrin receptors and can induce disease and transmission to susceptible animals. Other critical aa residues have been identified, such as the methionine (Met) at position 54, whose change to isoleucine (Ile) affects processing of precursor P1, decreasing production of *VP1* and accumulation of *VP1* precursor proteins. Although not in direct contact with the *VP1-VP3* cleavage site, residue M54 of *VP1* is exposed at the virion surface and it is close to an antigenic site within the B-C loop. Residues within the 1AB cleavage pocket and a 1C histidine (His-142-alpha-helix charge-dipole interaction at the twofold axes of symmetry between *pentamers*) play a role in acid-induced disassembly of the capsid.

4.3.3 P2 coding region

Most of the non-structural proteins (NSPs) are found within the P2 and P3 coding regions. These polypeptides and mature proteins are involved in RNA replication and viral maturation, although their specific roles remain to be elucidated. FMDV 2A is an 18 aa peptide that induces P1/2A polypeptide release from the rest of the genome through modification of the cellular translation apparatus. Generating the C-terminus of 2A and the N-terminus of 2B does not involve a protease, but rather cleavage of the ester linkage of peptidyl-tRNA within the peptidyl-transferase center of the ribosome during translation, a phenomenon termed 'StopGo'. The functional motif of 2A resides in a highly conserved aa sequence in its carboxy-terminal portion. This co-translational dissociation of the polyprotein and immediate recovery has been widely applied to develop research tools and gene therapies. The 2A protein is released from P1 by cleavage with the 3C viral protease in a later stage of processing. However, its function as an independent protein is not yet understood.

Little is known about the function of the FMDV 2B protein. It is a small hydrophobic protein that, upon individual expression, is localized to the endoplasmic reticulum (ER) and the Golgi complex. Differing from other picornaviral 2B proteins, FMDV 2B has minimal effects on Ca(2+) homeostasis and intracellular protein trafficking. However, it does cause accumulation of ER proteins in large vesicular structures around the nucleus. A transmembrane domain has been predicted between aa positions 120 and 140, supporting its involvement in vesicles and membrane-related stages of viral infection. Its expression in cells enhances membrane permeability and has been implicated in cytopathic effect. Only 37 of the 154 aa that compose the 2B protein are variable, and even those are restricted to one or two possible aa's, illustrating the great constriction of 2B protein variability.

Protein 2C is a highly conserved peptide with ATPase and RNA binding activity that is 318 aa in length. It has been assigned to the SF3 helicase family of AAA+ ATPases. In infected cells 2C is involved in the formation of membrane vesicles where it co-localizes with viral RNA replication complexes. Its 18 nt ATP-GTP binding motif is highly conserved in all serotypes. This sequence motif is generally referred to as the "A" consensus sequence or the P-loop, and is found in many protein families, such as thymidine kinases, ATP-binding proteins involved in active transport, DNA and RNA helicases, etc.. Protein 2C is involved in RNA synthesis and is the site of mutations that confer resistance to guanidine hydrochloride. Substitution of arginine 55 to tryptophan (R55W) within 2C mediates an increase in the extracellular release of viral RNA without a detectable increase of total viral RNA.

4.3.4 P3 coding region

FMDV protein 3A is a membrane-associated protein that localizes to a reticular structure. Some studies suggest that 3A influences host-range; for example, the amino acid substitution glutamine 44 to arginine (Q44R) in 3A, either alone or in combination with the replacement of isoleucine 248 with threonine (I248T) in 2C, was sufficient to give FMDV the ability to produce lesions in guinea pigs. Also, a 10 aa deletion and a series of substitutions (accumulated over the following 29 years) that surround the deletion were described to be a primary determinant of restricted growth of O/Taiwan 97 on bovine cells *in vitro* and as a contributor to bovine attenuation of O/Taiwan 97 *in vivo*. Subsequent experiments demonstrated that this deletion on its own does not contribute to porcine tropism of the virus, but that genome-wide changes (in addition to the deletion) produce the porcinophilic phenotype of current Asian viruses within this lineage.

The 3B region codes for 3 non-identical copies of the VPg protein. Covalent linkage of VPg occurs via a tyrosine (Y) residue to the 5′ end of the positive-sense RNA viral genome and to the nascent viral plus- and minus-stranded RNAs. This protein serves as a primer for the initiation of RNA replication and plays a role as an encapsidation signal. This priming step requires uridylation of the VPg peptide by the viral polymerase 3D and other viral or host cofactors. It remains to be elucidated why FMDV is the only picornavirus that encodes 3 tandem repeats of the VPg protein within the 3B coding region. Limited studies suggest it may be a critical component of the viral replication complex, enhancing transcription efficiency of the viral genome. Engineered FMDV infectious clones with either severe domain disruption or deletion of individual 3B proteins do not exhibit decreased infectivity *in vitro*, nor do they alter clinical disease in cattle or swine. Only clones encoding a single copy of VPg seem impaired in replicative competence. Interestingly, these mutant FMDVs did not produce plaques in BHK-21 tissue culture but produced a mild disease in swine and cattle.

The viral proteinase, 3Cpro, is a serine protease that catalyzes ten of the thirteen cleavages necessary to complete FMDV polyprotein processing. Its protease activity also affects host cell transcription since 3Cpro is responsible for the cleavage of the cellular histone H3 as well as the elongation factors eIF4G and eIF4A, resulting in cessation of host cell transcription. Crystal structure analyses indicate that FMDV 3Cpro adopts a chymotrypsin-like fold in the characteristic fingers, palm and thumb subdomains, with the presence of an NH2-terminal segment encircling the active site. The necessity of this tridimensional structure imposes serious restrictions on amino acid variability.

Replication of the RNA genome of the virus, via negative strand intermediates, involves an RNA-dependent RNA polymerase, 3Dpol. Several specific aa's have been determined as essential for maintaining the functional integrity of the polymerase. A NTP-binding motif and hydrophobic antigenic regions have also been described within 3Dpol.

4.4 FMDV 3' UTR

The 3′ end of foot-and-mouth disease virus is required for viral infectivity and stimulates IRES activity. It is composed of two distinct elements: a 90 nt untranslated region (3′-NCR), and a poly(A) tract. The 3′-NCR has a highly conserved structure composed of two stem-loops (SL-1 and SL-2) that interact with viral and host proteins during RNA replication. The poly(A) tract is generally heterogeneous in length and has an important structural role during replication. Independent deletions within the two predicted stem-loop structures of the 3'-NCR have provided information about potential functions. Deletion of SL2 was lethal for viral infectivity *in vitro*, while removal of SL1 negatively impacted viral growth kinetics and impaired negative-strand RNA synthesis, down-regulating genome replication. Studies examining the *in vivo* phenotype of these mutant viruses in pigs suggest that deletion of SL1 may contribute to FMDV attenuation, supporting the potential of RNA technology for the design of new FMDV vaccines. The 3' end of FMDV RNA establishes two distinct strand-specific, long-range RNA-RNA interactions: one with the S region and another with the IRES element. The S region interacts with each of the stem-loops, and such interaction is dependent of the poly(A) conformation.

5. FMDV cell entry and genome replication

FMDV initiates infection by binding to integrin receptors via an Arg-Gly-Asp (RGD) sequence found in the G-H loop of the structural protein VP1. The particle dissociates into pentamers at mildly acidic pH and the RNA is liberated into the cytoplasm of the infected cell. FMDV uses standard picornavirus cell entry mechanisms, forming 'altered' particle intermediates thought to induce membrane pores through which the genome can be transferred across the endosome membrane. Induction of viral RNA translation and cessation of cellular RNA translation occurs concurrently. The synthesis of cellular proteins is prevented by viral proteases that cleave cellular elongation factors, inhibiting CAP-dependent translation. The viral proteins required for replication are immediately obtained from translation of the positive-sense viral RNA. These proteins also synthesize negative-sense transcripts based on the positive-sense RNA template. The negative-sense RNA then becomes the template used to synthesize *di novo* positive-sense viral genomes (24). Anti-sense RNA is found at a one hundred-fold less concentration than sense strands in infected cells, suggesting that each negative-sense strand may serve as template for the synthesis of many positive-sense strands. Genome copying occurs via a complementary negative-sense RNA template and the formation of two replicating positive-sense strands. Partially double-stranded replicative intermediates may also be involved (review in ref. 22-24).

FMDV RNA replication is initiated by the covalent attachment of an uracil monophospate (UMP) molecule to the hydroxyl group of a tyrosine within the terminal VPg protein. This reaction is catalyzed by the virally encoded RNA-dependent RNA polymerase, 3D. The enzyme performs this operation, together with other viral and probably host proteins, in the cytoplasm of the host cell. Cytoplasmic RNA Helicase A (RHA) plays an essential role during replication of FMDV, interacting with the S fragment and the viral 2C and 3A proteins, as well as with cellular PABP, promoting the assembly of ribo-nucleoprotein replication complexes at the 5' end of the genome. Eukaryotic initiation factors (eIFs) are required for internal translation initiation at the internal ribosome entry site (IRES), an action common to all picornaviruses. The eIF4B is an RNA-binding protein that stimulates the ATPase and helicase activities of eIF4A and strengthens the mRNA-rRNA-tRNA interactions at the initiation codon. The eIF4A is an ATP-dependent RNA helicase; it is believed to unwind RNA secondary structure and is the prototypic member of the DEAD box family of helicases. The eIF4B is present both as part of the eIF4F complex bound to eIF4G, and also free in the cytoplasm. The helicase activity of eIF4A is strongly stimulated by eIF4B. The cleavage of eIF4G releases the N-terminal domain that contains the binding sites for eIF4E and the poly(A) binding protein. The residual portion of eIF4G, which is sufficient for IRES-directed translation, retains two binding sites for eIF4A and binding sites for eIF3 and Mnk-1. Thus eIF4G is a bridge between the mRNA and small ribosomal subunit. The eIF4B is incorporated into ribosomal 48S initiation complexes via the FMDV IRES. In contrast to the weak interaction of eIF4B with capped cellular mRNAs and its release upon entry of the ribosomal 60S subunit, eIF4B remains tightly associated with the FMDV IRES during formation of complete 80S ribosomes. Binding of eIF4B to the IRES is energy dependent, and binding of the small ribosomal subunit to the IRES requires the previous energy-dependent association of initiation factors with the IRES. The interaction of eIF4B with the IRES in 48S and 80S complexes is independent of the location of the initiator AUG and thus independent of the mechanism by which the small ribosomal subunit is placed at the actual start codon, either by direct internal ribosomal entry or by scanning.

Final assembly of the viral capsid and encapsidation of the viral RNA occur by mechanisms that are still obscure. Two hypotheses describe potential mechanisms of pentamer assembly into pro-virions. One idea postulates that the RNA is inserted into the virion after assembly of the capsid, while the other theory proposes that the viral RNA interacts directly with the pentamers to form the pro-virion prior to capsid formation.

6. Genetic variability of FMDV RNA

Due to the absence of proofreading-repair activity by the viral replicase, FMDV RNA genome replication is highly error-prone. The high mutation rates result in populations that consist of genetically related but non-identical viruses known as *quasispecies*. The *quasispecies* concept maintains that a viral population consists of a 'swarm' of genetic and phenotypic variants in perpetual renewal as genome replication proceeds in an infected host. The consensus nucleotide sequence of FMDV isolated from a clinical sample derives from a multimodal population of similar but distinct viruses; often the exact consensus sequence obtained does not exist within the population, but is a reflection of many co-existing *quasispecies*. The existence of quasispecies populations may explain the dramatic genetic plasticity observed in disease-causing RNA viruses, supporting pathogenic adaptations that expand their host repertoire and virulence profile (Review in Ref. 26-29).

Several *in vivo* experiments report the generation of highly variable FMD viruses from single animals during infection studies. These observations may have been influenced by molecular host factors and/or selective pressures indirectly incurred from lab methodologies (see ref. 28-32). However, additional controlled experimental infections in pigs confirmed these observations in every passage of *in vivo* infection. Interestingly, the location and nature of the genetic variation was not the same as *in vitro*-acquired differences (see ref. 31, 32, 33, 34, 37), including the estimated number of substitutions per nucleotide. Recently, a study conducted during the United Kingdom 2001 epidemic demonstrated that nucleotide changes occur throughout the genome at a rate of 2.26×10^{-5} nucleotide substitutions per site per day (95% confidence interval [CI], 1.75×10^{-5} to 2.80×10^{-5}) and nucleotide substitutions accumulate in the consensus nucleotide sequence at an average rate of 1.5 substitutions per farm infection. Data obtained from outbreaks like the 2001 epidemic support the experimental observations, demonstrating the role of host-related selective pressures on the variability and evolution of FMDV (35).

Comparative genomics studies using full-length sequences representative of all seven serotypes have identified highly conserved genomic regions, indicating functional constraints for variability as well as as-yet undefined motifs with likely biological significance (14, 34). At least 64% of all nt sites within the FMDV genome are susceptible to substitution, including compensatory substitutions. It is important to clarify that most of the "variant" or substitutable residues within the FMDV genome mutate in response to detrimental effects produced by mutations elsewhere in the genome. But most importantly, it indicates that at least 46% of the nucleotides are indispensable for FMDV survival; replacement of any of the known critical residues renders non-viable progeny.

In support of the comparative genomics analyses, sequence studies demonstrate that the most distantly related pair of FMDV isolates to-date do not differ more than 22% from each other. Within one serotype the differences are less than 15%. Although sequences have been intensely analyzed in terms of similarity and divergence, the genetic bases of most biological traits of FMDV remain to be discovered. Such analysis would benefit from initial studies examining the conserved regions within the ORF; notably, regions 2B and 3C exhibit the highest percentage of invariant nt (61% and 59%, respectively) and amino acids (76%) in the genome.

The most variable parts of the translated FMDV genome, like Lpro, 3A, 3B and the structural proteins (1B, 1C, and 1D) suggest that these proteins are subjected to strong selective pressures. Additional studies and characterization will reveal important molecular markers and signatures of epidemiological and forensic value. This information can be used to facilitate development of novel vaccines containing molecular markers that allow for differentiation of vaccinated and infected animals.

7. FMDV RNA recombination and evolution

Genomic comparisons of full-length sequences have been very useful to the understanding of FMDV evolution. Computer programs have been developed and used to estimate various parameters of evolution. The CODEML program is one of most popular, widely used to obtain the ω parameter ($\omega = dN/dS$), an indicator of selective pressure that considers and compares several models of evolution. For FMDV, CODEML ω rates obtained from different substitution models indicated that a few clusters of codons in Lpro, 1D, 3A and 3B may undergo diversifying selection. Evidence of positive selection has been identified in complete capsid sequences from all serotypes. Results suggest that novel antigenic variants benefit from a selective advantage in their interaction with the immune system, possibly throughout the course of an infection and/or during transmission to individuals with previous exposure to antigen (see ref. 14, 36-38).

Analysis of amino acid usage at sites under positive selection indicates that this selective advantage can be conferred by amino acid substitutions that share physical and chemical properties. Besides genetic drift, there is increasing evidence that recombination is an important mechanism of FMDV evolution. Using different recombination detection methods among the publicly available FMDV complete genome sequences, the large number of recombinant isolates suggests that horizontal recombination of sequences is common and probably advantageous in terms of fitness costs. Interestingly, the distribution of recombination breakpoints was found to be largely nonrandom (37, 38). Results suggest that genome regions encoding the structural proteins are functionally interchangeable modules, as can be deduced from evidence that the structural and nonstructural coding regions of picornaviruses evolve largely independently of one another (see ref. 14, 37-40).

Recombinant viruses may derive from an animal that is co-infected with different virus variants while also harboring viruses from a previous persistent or sub-clinical infection. We are still not very knowledgeable about subclinical and persistent infection. Indications of differential susceptibility for developing a subclinical course of disease has been observed in many instances: buffaloes and cattle present with different disease manifestations, with breed affecting severity (see ref. 41); sub-clinical symptoms in sheep variable and goats make detection of FMD difficult in those species (see ref. 42, 43, 44); and establishment of persistent infection in ruminants has been demonstrated (see ref. 45). Most interestingly, pigs experimentally infected with a highly virulent porcine-tropic strain (OTw97) exhibited gradual loss of virulence and the establishment of a subclinical infection upon serial passage in the absence of FMDV-specific antibodies (38). However, experimental evidence of possible mechanisms of transmission and its effects on the FMDV genome are sparse; additional controlled experimentation in this field is required (32, 33, 38).

The initial size of a virus population strongly influences evolution and replication fitness. While *in vitro* large population passages often result in fitness gains, repeated plaque-to-plaque transfers result in average fitness losses, known as Muller's Ratchet effect. On the contrary, experimental infection of a natural host with FMDV seems to require a relatively low number of FMDV particles to produce clinical disease. *In vivo* studies suggest that horizontal transmission of FMDV may be achieved with a low number of infectious particles. In this scenario, recombination events may rescue defective genomes (Muller's Ratchet) yielding a significant number of virulent viruses that spread into new hosts. In this manner, FMDV perpetuates within and between natural hosts and reservoirs, recovering the replicative capability of previously defective particles.

Additionally, recombination within the non-structural genome regions, potentially modifying the virulence of the virus, may be involved in the success of the new sub-lineage to regain infectiousness. It can also explain why phylogenetic analyses restricted to VP1 sequences appear to represent evolutionary cul-de-sacs and why they often reveal re-emergence of previously extinguished VP1 genetic lineages.

Recombination might be a decisive factor in the production of escape variants. Strong immunity includes multiple B-cell and T-cell epitopes that produce efficient humoral and cellular immune responses. Such an ample response is obtained after recovery of natural infection but is difficult to obtain as the immune response evoked by vaccination. Both B and T-cell epitopes have been identified in structural and non-structural proteins of FMDV. Some are highly conserved, but others are highly variable. Therefore, it is possible that re-arrangement of the antigenic display is one of the mechanisms to escape host immune response for FMDV, and that recombination is one major player in such re-arrangement of antigenic markers.

8. Final remarks

i. A new vaccine generation marked and targeting systemic and mucosal immunity is urgently needed for FMDV global eradication plans. Control of FMD is based on two major strategies: the slaughtering of affected and contact animals (the so called 'stamping out' procedure) or the regular vaccination of the major host species for FMDV. Unfortunately, classical vaccines cannot prevent the establishment of persistent FMDV infection in cattle. As an alternative to the conventional inactivated vaccines, the use of attenuated antigenically marked virus able to induce a solid and durable immunity through replication in the animal is highly desirable. FMDV escapes from vaccine production plants or diagnostic and research facilities, like what happen in UK 2007 (O1BFS 1860/UK/67) highlight the need for an alternative to the handling of large amounts of virus because of the danger of virus escaping from vaccine factories. Also, classical, inactivated whole-virus vaccines may be at the origin of outbreaks if inactivation prior to vaccine formulation was not complete. There is good evidence that some FMD outbreaks probably had a vaccine origin. All these are powerful arguments to design vaccines that do not require infectious virus at any stage of their preparation. However, to achieve this goal a deep understanding of the molecular bases that govern biological and immunological properties of FMDV is necessary. The prediction of viral cross-protection remains an important unsolved problem, transcripts that can be blocked at some steps of virus replication or assembly, genetic complementation and molecular basis of virulence factors should all be deeply explored from the genomic knowledge.

ii. Co-circulation of different types of FMDV is a reality in most parts of the endemic regions which represents a serious complication in the epidemiology of FMDV. Global epidemiological analysis is vital for implementing progressive regional foot-and-mouth disease control programs, but better knowledge of variability, recombination and evolution of FMDV is necessary. Development of spatial epidemic models to simulate transmission or to assess biosecurity planning and emergency-response preparedness requires better knowledge of FMDV evolution. Thus, to really understand FMD field epidemiology and how to contain the spreading of new outbreaks, wider molecular epidemiology analyses using full length genome information are necessary.

9. References

[1] Bachrach, H. L. 1968. Foot-and-mouth disease. Annu. Rev. Microbiol. 22: 201–244.

[2] Donaldson, A. I., and R. F. Sellers. 2000. Foot-and-mouth disease, p. 254–258. In W. B. Martin and I. D. Aitken (ed.), Diseases of sheep. Blackwell Science, Oxford, United Kingdom.

[3] Grubman M.J and Baxt B.. 2004. Foot-and-Mouth Disease. Clin Microbiol Rev 17 (2): 465-493

[4] Domingo E., Baranowski E., Escarmis C., Sobrino F. 2002. Foot-and-mouth disease virus. Comp Immun Microbiol and Infect Diseases, 25: 297-308.

[5] Samuel AR, Knowles NJ. Foot-and-mouth disease virus: cause of the recent crisis for the UK livestock industry. Trends Genet 2001; 17:421–4.

[6] Thompson, D., P. Muriel, D. Russell, P. Osborne, A. Bromley, M. Rowland, S. Creigh-Tyte, and C. Brown. 2002. Economic costs of the foot-and-mouth disease outbreak in the United Kingdom in 2001. Rev. Sci. Tech. Off. Int. Epizoot. 21:675–687.

[7] Yang, P. C., R. M. Chu, W. B. Chung, and H. T. Sung. 1999. Epidemiological characteristics and financial costs of the 1997 foot-and-mouth disease epidemic in Taiwan. Vet. Rec. 145:731–734.

[8] Pluimers, F. H., A. M. Akkerman, P. van der Wal, A. Dekker, and A. Bianchi. 2002. Lessons from the foot and mouth disease outbreak in the Netherlands in 2001. Rev. Sci. Tech. Off. Int. Epizoot. 21:711–721.

[9] Forss, S., K. Strebel, E. Beck, and H. Schaller. 1984. Nucleotide sequence and genome organization of foot-and-mouth disease virus. Nucleic Acids Res. 12:6587–6601

[10] Palmenberg, A. C. 1990. Proteolytic processing of picornaviral polyprotein. Annu. Rev. Microbiol. 44:603–623.

[11] Belsham GJ, Jamal SM, Tjørnehøj K, Bøtner A. 2011. Rescue of foot-and-mouth disease viruses that are pathogenic for cattle from preserved viral RNA samples. PLoS One; 6(1):e14621.

[12] Samuel AR, Knowles NJ. 2001. Foot-and-mouth disease type O viruses exhibit genetically and geographically distinct evolutionary lineages (topotypes). J Gen Virol. 82:609-21.

[13] Knowles, N. J., and A. R. Samuel. 2003. Molecular epidemiology of foot-and-mouth disease virus. Virus Res. 91:65–80.

[14] Carrillo C, Tulman ER, Delhon G, Lu Z, Carreno A, Vagnozzi A, Kutish GF, Rock DL. 2005. Comparative genomics of foot-and-mouth disease virus. J Virol. 79:6487-504.

[15] Klein J. 2009. Understanding the molecular epidemiology of foot-and-mouth-disease virus. Infect Genet Evol. 9(2):153-61.

[16] Cottam EM, Wadsworth J, Shaw AE, Rowlands RJ, Goatley L, Maan S, Maan NS, Mertens PP, Ebert K, Li Y, Ryan ED, Juleff N, Ferris NP, Wilesmith JW, Haydon DT, King DP, Paton DJ, Knowles NJ. 2008. Transmission pathways of foot-and-mouth disease virus in the United Kingdom in 2007. PLoS Pathog. 18;4(4):e1000050.

[17] Rueckert, R. R. 1996. Picornaviridae: the viruses and their replication, p. 609–654. In B. N. Fields, D. M. Knipe, and P. H. Howley (ed.), Fields virology, 3rd ed. Lippincott-Raven, Philadelphia, Pa.

[18] Rueckert, R. R., and E. Wimmer. 1984. Systematic nomenclature of picornavirus proteins. J. Virol. 50:957–959.

[19] Agol, V. I., A. V. Paul, and E. Wimmer. 1999. Paradoxes of the replication of picornaviral genomes. Virus Res. 62:129–147.

[20] Paul, A. V. 2002. Possible unifying mechanism of picornavirus genome replication, p. 227–246. In B. L. Semler and E. Wimmer (ed.), Molecular biology of picornaviruses. ASM Press, Washington, D.C.

[21] Rieder, E., T. Bunch, F. Brown, and P. W. Mason. 1993. Genetically engineered foot-and-mouth disease viruses with poly(C) tracts of two nucleotides are virulent in mice. J. Virol. 67:5139–5145.

[22] Mason, P. W., S. V. Bezborodova, and T. M. Henry. 2002. Identification and characterization of a cis-acting replication element (cre) adjacent to the internal ribosome entry site of foot-and-mouth disease virus. J. Virol. 76: 9686–9694.

[23] Pacheco A, Reigadas S, Martínez-Salas E. 2008. Riboproteomic analysis of polypeptides interacting with the internal ribosome-entry site element of foot-and-mouth disease viral RNA. Proteomics. 2008 Nov;8(22):4782-90.

[24] Baxt, B., S. Neff, E. Rieder, and P. W. Mason. 2002. Foot-and-mouth disease virus-receptor interactions: role in pathogenesis and tissue culture adaption, p. 115–123. In B. L. Semler and E. Wimmer (ed.), Molecular biology of picornaviruses. ASM Press, Washington, D.C.

[25] Gamarnik, A. V., and R. Andino. 1998. Switch from translation to RNA replication in a positive-stranded RNA virus. Genes Dev. 12:2293–2304.

[26] Domingo, E., and J. J. Holland. 1988. High error rates, population equilibrium and evolution of RNA replication systems, p. 3–36. In E. Domingo, J. J. Holland, and P. Ahlquist (ed.), RNA genetics, vol. III. Variability of RNA Genomes. CRC Press, Boca Raton, Fla.

[27] Domingo, E., M. G. Mateu, M. A. Matnez, J. Dopazo, A. Moya, and F. Sobrino. 1990. Genetic variabililty and antigenic diversity of foot-andmouth disease virus, p. 233–266. In R. G. M. E. Kurstak, F. A. Murphy, and M. H. V. Regenmortel (ed.), Virus variability, epidemiology and control, vol. 2. Plenum Publishing Corp., New York, N.Y.

[28] Drake, J. W., and J. J. Holland. 1999. Mutation rates among RNA viruses. Proc. Natl. Acad. Sci. USA 96:13910–13913.

[29] Eigen, M. 1971. Self-organization of matter and the evolution of biological macromolecules. Naturwissenschaften 58:465–523.

[30] Martinez MA, Carrillo C, Plana J, Mascarella R, Bergada J, Palma EL, Domingo E, Sobrino F. 1988. Genetic and immunogenic variations among closely related isolates of foot-and-mouth disease virus. Gene.;62(1):75-84.

[31] Martínez MA, Carrillo C, González-Candelas F, Moya A, Domingo E, Sobrino F. 1991. Fitness alteration of foot-and-mouth disease virus mutants: measurement of adaptability of viral quasispecies. J Virol; 65(7):3954-7.

[32] Carrillo C, Borca M, Moore DM, Morgan DO, Sobrino F. 1998. In vivo analysis of the stability and fitness of variants recovered from foot-and-mouth disease virus quasispecies. J Gen Virol.; 79 (Pt 7):1699-706.

[33] Carrillo C, Plana J, Mascarella R, Bergadá J, Sobrino F. 1990. Genetic and phenotypic variability during replication of foot-and-mouth disease virus in swine. Virology.;179(2):890-2.

[34] Carrillo C, Tulman ER, Delhon G, Lu Z, Carreno A, Vagnozzi A, Kutish GF, Rock DL. 2006. High throughput sequencing and comparative genomics of foot-and-mouth disease virus. Dev Biol (Basel); 126:23-30; discussion 323.

[35] Cottam EM, Haydon DT, Paton DJ, Gloster J, Wilesmith JW, Ferris NP, Hutchings GH,King DP. 2006. Molecular epidemiology of the foot-and-mouth disease virus outbreak in the United Kingdom in 2001. J Virol. 80(22):11274-82.

[36] Fares, M. A., A. Moya, C. Escarmis, E. Baranowski, E. Domingo, and E.Barrio. 2001. Evidence for positive selection in the capsid protein-coding region of the foot-and-mouth disease virus (FMDV) subjected to experimental passage regimens. Mol. Biol. Evol. 18:10–21.

[37] Haydon, D. T., A. D. Bastos, N. J. Knowles, and A. R. Samuel. 2001. Evidence for positive selection in foot-and-mouth disease virus capsid genes from field isolates. Genetics 157:7–15.

[38] Carrillo C, Lu Z, Borca MV, Vagnozzi A, Kutish GF, Rock DL. 2007. Genetic and phenotypic variation of foot-and-mouth disease virus during serial passages in a natural host. J Virol. 81(20):11341-51.

[39] Jackson AL, O'Neill H, Maree F, Blignaut B, Carrillo C, Rodriguez L, Haydon DT. 2007. Mosaic structure of foot-and-mouth disease virus genomes. J Gen Virol. 88(Pt 2):487-92.

[40] Wilson, V., P. Taylor, and U. Desselberger. 1988. Crossover regions in foot-and-mouth disease virus (FMDV) recombinants correspond to regions of high local secondary structure. Arch. Virol. 102:131–139.

[41] Kitching, P., and S. Alexandersen. 2002. Clinical variation in foot-and-mouth disease: pigs. Rev. Sci. Tech. Off. Int. Epizoot. 21:513–518.

[42] Geering, W. A. 1967. Foot and mouth disease in sheep. Aust. Vet. J. 43:485–489.

[43] Kitching, P., and G. H. Hughes. 2002. Clinical variation in foot-and-mouth disease: sheep and goats. Rev. Sci. Tech. Off. Int. Epizoot. 21:505–512.

[44] Hughes, G. J., V. Mioulet, R. P. Kitching, M. E. Woolhouse, S. Alexandersen, and A. I. Donaldson. 2002. Foot-and-mouth disease virus infection of sheep: implications for diagnosis and control. Vet. Rec. 150:724–727.

[45] Alexandersen, S., Z. Zhang, and A. I. Donaldson. 2002. Aspects of the persistence of foot-and-mouth disease virus in animals—the carrier problem. Microbes Infect. 4:1099–1110.

4

Ophioviruses: State of the Art

Maria Laura Garcia

Instituto de Biotecnología y Biología Molecular,
Facultad de Ciencias Exactas, Universidad
Nacional de La Plata, CONICET
Argentina

1. Introduction

As has happened in the last century with many plant diseases, the nature of the causal agents, particularly viruses, was not determined and studied until a few decades ago. Thus, an old disease named citrus psorosis was first described in 1891, but it was almost a century later when the viral agent was observed by immune electron microscopy as a novel spiral-filamentous particle (Derrick et al., 1988). In 1994, the real morphology of Citrus psorosis virus (CPsV) was observed by Robert G. Milne using negative staining electron microscopy; describing circular particles of different configurations resemble that of the tenuiviruses and the nucleocapsids of members of the family *Bunyaviridae* (Garcia et al., 1994). Due to the shape of the particles, they were called Ophiovirus, derived from the Greek "ophios", a serpent, referring to the snaky appearance of the virions (Figure 1, a).

Subsequently, in Japan, another ophiovirus is recognized in tulip, *Tulip mild mottle mosaic virus* (TMMMV) (Morikawa et al., 1995), and later, Robert G. Milne, a "virus hunter" (as once he called himself), found particles with similar morphology in diseased plants of lettuce and ranunculus (Milne, 2000). Thus began the study of the ophioviruses *Ranunculus white mottle virus* (RWMV) (Vaira et al., 1997), *Mirafiori lettuce big-vein virus* (MiLBVV) (Roggero et al., 2000), *Lettuce ring necrosis virus* (LRNV) (Torok et al., 2002, 2003) and *Freesia sneak virus* (FreSV) (Vaira et al., 2006). Most, if not all these viruses have been found around the world (Roistacher 1993; Navarro et al., 2004; Martin et al., 2006; Ghazal et al., 2008; Vaira et al., 2007, 2009; Plesko et al., 2009; Barcala Tabarrozzi et al., 2010).

1.1 Old diseases affecting major crops

In citrus: Citrus psorosis virus, the type member of the family

The first observation of symptoms of citrus psorosis disease was reported in 1891 (Swingle and Webber, 1896), and the first experimental evidence about that infectious disease transmitted by grafting in citrus trees was published in 1933 by H.S. Fawcett. Psorosis disease development is slow; it may take several years to manifest symptoms. Typical psorosis symptoms are bark-scaling of trunk and main branches, and more severe as rampant bark-scaling even on small limbs and twigs. Gum may accumulate below the bark scales and may impregnate the xylem producing wood staining and vessel occlusion. These symptoms have been used for field diagnosis of Psorosis (Roistacher, 1993). Chlorotic flecks and spots on young leaves can be observed in spring time in the field, and in infected seedlings in the greenhouse. (For symptoms and diagnosis of psorosis disease, see an excellent review of Alioto et al., 2007). The disease has been reported from many citrus-growing areas all over the world (Roistacher, 1993). Trees affected with psorosis have been less productive causing damage to citrus industry in the Mediterranean basin, and in some areas of South America. In the '80s, in Argentina and Uruguay it was a serious disease causing annual losses of about 5% of trees (Larocca 1985, Danós 1990) and the disease is still present as recently reported by Zanek et al (2006). There are reports of naturally spreading of psorosis in Argentina (Pujol and Beñatena, 1965), Uruguay (Campiglia et al., 1976), and in Texas, USA (Timmer and Garnsey, 1980). The suspected vector is unknown although the pattern of spread suggests an aerial vector (Beñatena and Portillo, 1984; Diamante et al., 1984). On the other hand, other ophioviruses are soil-transmitted by a root-infecting fungus from the *Olpidium* genus (see later in this section). Citrus psorosis virus has probably been vegetatively propagated for centuries around the world from citrus to citrus, and it could have lost any putative original capacity to be transmitted by *Olpidium*, and at the same time acquiring the ability of transmission by an aerial vector. Therefore, further studies are necessary to clarify this matter and to identify the natural vector of CPsV.

The first ophiovirus described was discovered in citrus, but most of them has been found in ornamental plants as ranunculus (dicotyledonous), freesia, tulips and lachenalia (monocotyledonous), and lettuce (dicotyledonous).

In ornamental plants: Tulip mild mottle mosaic virus, Ranunculus white mottle virus and Freesia sneak virus

Since 1979 the occurrence of mild mottle mosaic disease in tulip is described, and in 1989 it is reported for the first time in Japan by Yamamoto et al. (1989) as a virus-like disease of tulip, and as a soil-borne disease by Morikawa et al., (1993). In 1995, Morikawa and co-workers found a new virus recognized on tulip (*Tulipa gesneriana L.*, hybrids, *Liliaceae*) producing symptoms of venial chlorotic mottle mosaic on leaves and color-removing mottle on flower buds. They mechanically transmitted the virus in tulip and species as *Chenopodium quinoa, Tetragonia expansa, Nicotiana tabacum* and *Nicotiana benthamiana*, but it could not be back-inoculated from *C. quinoa* to tulip (Morikawa et al., 1995). The authors found that the disease spreads through bulbs of tulip and might be soil-borne. In 1998 Drs. T. Natsuaki and T. Morikawa (Utsunomiya University, Japan) have indicated that the vector of TMMMV is *Olpidium brassicae* (pers. comm.).

Other ornamental disease caused by an ophiovirus was found in ranunculus and anemone. In 1996 A.M. Vaira and co-workers described a new virus found in a plant of *Ranunculus* hyb. (cv. Grazia) collected in Liguria, Northern of Italy, in 1990. The symptoms described in ranunculus plants were mosaic, mottle and distortion of leaves and stems, giving the name *Ranunculus white mottle virus*. For years the virus was consistently isolated from plants and found in mixed infection with potyvirus (Vaira et al., 1997, 2009). They could mechanically transmit the virus to several herbaceous hosts (*N. benthamiana* and *N. clevelandii*), and by EM, in negative stain the particle morphology appeared similar to CPsV, *Tenuivirus* and *Bunyaviridae* (Vaira et al., 1996, 1997). So far, there are no reports about a vector for RWMV.

A severe disease called freesia leaf necrosis (FLN) has been known in freesia cultures for forty years in Europe (Verbeek and Meekes, 2005) but its causal agent was not identified until 2006 by Vaira et al., (2006). The authors found freesias (*Freesia refracta* hybrids, *Iridaceae*) with symptoms of FLN in the area around Sanremo (Italy), and later, in lachenalia cultivars (*Lachenalia* hyb., *Hyacinthaceae*) in South Africa (Vaira et. al, 2007). By electron microscopy the authors found an ophiovirus which is associated to this disease, which presents chlorotic inter-veinal lesions on the leaves, later coalescing and becoming sunken and necrotic. FLN is soil-transmitted as mild mottle mosaic disease in tulip (van Dorst, 1975; Vaira et al., 2006).

In lettuce: Mirafiori lettuce big-vein virus and Lettuce ring necrosis virus

Lettuce (*Lactuca sativa*) is other natural host for ophioviruses. In 1934 lettuce big-vein disease (BV) was described as possibly caused by a virus (Jagger et al., 1934). Big-vein is one of the most important diseases of lettuce crops worldwide. The symptoms, as the name refers are vein enlargement with chlorotic regions around the vascular tissue, making the plant no suitable for the market and producing important looses. The virus named Lettuce big-vein virus (LBVV), the type species of the genus *Varicosavirus* (van Regenmortel et al., 2000), was initially associated with big-vein disease (Kuwata et al., 1984). LBVV a rod-shaped virion transmitted by *Olpidium brassicae* (Kuwata et al., 1984; Vetten et al., 1987; Huijberts et al., 1990), but this varicosavirus had not been isolated or rigorously demonstrated to cause the disease. In big-vein affected lettuce the presence of unsuspected second virus with particles morphologically resembled those of ophioviruses was discovered by R. G. Milne and co-workers (Roggero et al., 2000). That ophiovirus was named *Mirafiori lettuce virus* (MiLV) since it was detected in Mirafiori, Turin (Italy). In 2002, Lot and co-workers demonstrated that the lettuce infected with MiLV alone consistently developed big-vein symptoms regardless of the presence or absence of LBVV (Lot et al., 2002). This important evidence showed that MiLV but not LBVV is the true causal agent of this disease, although both viruses are present in the diseased lettuce-plants. Later these viruses were renamed as *Mirafiori lettuce big-vein virus* (MiLBVV) and *Lettuce big-vein associated virus* (LBVaV) by the International Committee on Taxonomy of Viruses (ICTV). Recently, it has been determinate that both viruses are transmitted by *Olpidium virulentus*, a noncrucifer strain of *Olpidium brassicae* (Sasaya and Koganezawa, 2006).

Lettuce ring necrosis is still a serious disease producing coalescent necrotic rings and ring-like patterns on middle leaves of plants observed in greenhouses during winter and transmitted by the zoospores of *O. brassicae* (Bos et al., 1996). The disease was first described in The Netherlands and in Belgium as "kring necrosis" and observed in France where it was called "maladie des taches orangées".

As happened with LBVaV, a rod-shaped non-enveloped virus was tentatively named lettuce ring necrosis virus (LRNV) and both were closely associated to the diseases (Huijberts et al., 1990). In 2002, Torok et al. associated for the first time an ophiovirus with lettuce ring necrosis disease, and in 2003, the same authors published the molecular characterization of a this new ophiovirus (Torok et al., 2003). Later, the genome of LRNV was sequenced but no further analysis has been published so far.

1.2 Morphology of the ophiovirus particles – *In vitro* stability

Robert G. Milne described the particles as circles of at least two different contour lengths, the shortest length about 760 nm, and the largest about four times longer with 3 nm in diameter when appear in a circular form (Figure 1,a) (Garcia et al., 1994). The circles can collapse to form pseudolinear duplex structures, coiled filamentous about 9-10 nm in diameter. The presence of this pseudolinear form seems to be associated with long incubation (one to several days) in vitro (Milne et al., 1996). Figure 1, b shows a model of different configurations the particles can adopt, resemble that of the tenuiviruses and the nucleocapsids of members of the family *Bunyaviridae* (Garcia et al., 1994, Milne et al., 1996) (see Vaira et al., (1997) for EM photos of different RWMV forms). Thin sections of *N. clevelandii* leaf tissue infected with RWMV were observed by Vaira et al., (1997) using EM immunogold against RWMV coat protein, but no inclusions neither enveloped particles were found. The label was abundant in the cytoplasm of parenchyma cells, but the nuclei, chloroplast, mitochondria and microbodies were unlabelled (Vaira et al., 1997). So far, that has been the unique observation of any ophiovirus in thin sections. Attempts have been done to see CPsV particles in different tissues but they resulted unsuccessful (R.G. Milne, Peña E. and Kitajima E., pers. communications).

Ophiovirus particles are unstable in CsCl and in phosphotungstate but not in 2% aqueous uranyl acetate. Besides, the particle structure remains intact in cesium sulphate (D. Alioto, E. Luisoni and R.G. Milne, unpublished data). In order to purify and separate the smaller from the larger particles, virions of CPsV can be ultracentrifuged in sucrose or cesium sulphate density gradients (Derrick et al., 1988; Garcia et al., 1991, Sanchez de la Torre et al., 1998). The buoyant density in cesium sulfate is 1.22 g/cm3 for RWMV and MiLBVV (Vaira et al., 1997; Roggero et al. 2000). The particles have limited stability at pH below 8 (Garcia et al., 1991), and the infectivity does not survive in crude sap held at room temperature for more than 2 hr or 12-24 hr at 4 °C in the case of CPsV (Garcia et al., 1991) and TMMMV (Morikawa et al., 1995). Particle structure survives limited treatment with organic solvents and nonionic or zwitterionic detergents (Garcia et al., 1991; Roggero et al., 2000).

a   b

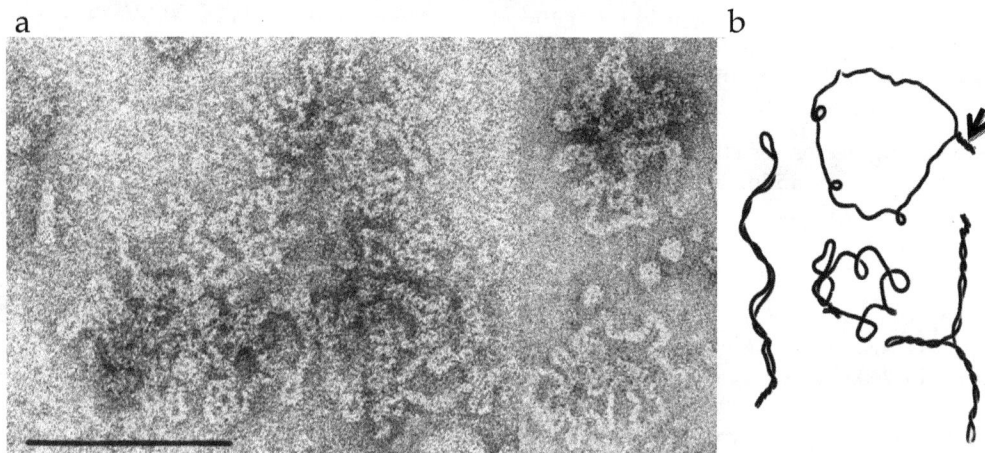

Fig. 1. a. Ophiovirus morphology: naked filamentous nucleocapsids. Circles of at least two different contour lengths, negatively stained in 1% uranyl acetate. Large (left) and small (right) particles. Bar = 100 nm. b. Wire models (not to scale) of possible forms for ophiovirus particles, representing larger and smaller particles in the circular and pseudolinear form. The putative "panhandle" structure is indicated by arrow (Milne et al., 1996 with modifications).

2. Genome organization, sequence analysis and putative proteins

Ophioviruses genome is divided into three or four individually encapsidated segments (Figure 2). CPsV, RWMV and FreSV have 3 RNAs (named as RNA 1, 2 and 3) and for MiLBVV and LRNV a fourth RNA has been reported (named as RNA 4).

The available information about ophiovirus genes and putative proteins are based on the sequences of CPsV, MiLBVV and LRNV, which are the ophioviruses completely sequenced so far. Partial sequences of RNA 3 of all ophiovirus species and the RNA-dependent RNA polymerase (RdRp) module of the RWMV RNA 1 are also available in database. Using ophiovirus-specific primers based on a highly conserved sequence of RNA 1, Vaira and co-workers (2003) amplified a 136 bp fragment detecting all ophiovirus species, making this RT-PCR the selected method to find new ophioviruses. All 136bp-fragment sequences are available in database.

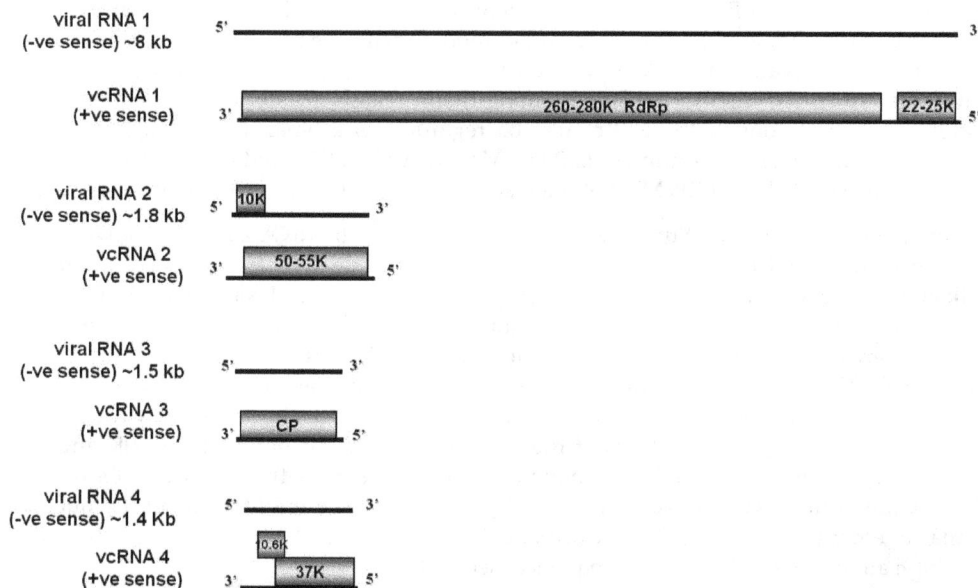

Fig. 2. Genome organization of ophiovirus. The length of the RNA segments and the predicted sizes of the ORF products are indicated. –ve sense: negative stranded RNA (viral RNA, vRNA), +ve sense: positive stranded RNA (viral complementary RNA, vcRNA). CPsV and RWMV have 3 RNAs. MiLBVV and LRNV contain 4 RNAs. The +ve sense of RNA 4 belongs to MiLBVV; LRNV contains only the 38K ORF (see the text). CP: coat protein. RdRp: RNA dependent-RNA polymerase.

In purified virus preparations the negative strand RNAs are the more abundant. The positive strands of all RNAs are also encapsidated although in much less amount. The size of RNA 1 is 8.2 kb for CPsV, 7.8 kb for MiLBVV, 7.6 kb for LRNV and 7.5 kb for RWMV (Naum et al. 2003; van der Wilk et al., 2002; Vaira et al., 1997; Torok et al., 2003). RNA 2 is about 1.8 kb for RWMV, MiLBVV and LRNV, 1.7 for FreSV and 1.6 kb for CPsV, and the RNA 3 is 1.3-1.5 kb for all ophioviruses (Vaira et al., 1997; van der Wilk et al., 2002; Torok et al., 2003; Sanchez de la Torre et al., 2002). The fourth genomic RNA reported for MiLBVV and LRNV is about 1.4 kb (van der Wilk et al., 2002; Torok et al., 2003).

In the RNA 1, a protein of 22-25 kDa of unknown function is encoded in the 5′ region of the positive strand. Separated by an intergenic region of hundred nucleotides a large ORF of RNA 1 is encoded (see Figure 2). The 109-nt intergenic region observed for CPsV (isolate CPV 4 from Florida, USA) is rich in A-U (88.3%) and contains 18-nt sequence (UUAAAA)$_3$ that could form a hairpin loop. Near the end of the intergenic region, a typical AAUAAA polyadenylation signal is found 12 nt upstream of the putative CA start polyadenylation site (Naum et al., 2003). However, these sequences were not found for the CPsV Spanish isolate P-121 (Martín et al., 2005), neither for MiLBVV (intergenic region of 147 nt, 66% of A+U), and LRNV (intergenic region of 80 nt, 65 % of A+U), making it unlikely that were involved in conserved functions among ophioviruses.

The largest ORF of the RNA 1 encode a protein of 261K for LRNV, 263K for MiLBVV and 280K for CPsV containing the core polymerase module with the five conserved motifs of the RdRp active site (van der Wilk et al., 2002; Naum et al., 2003; Vaira et al., 2003). The ophiovirus RdRp are highly conserved among them, mainly in the module sequence (see section 4). Two regions of the RdRp may be regarded as a bipartite nuclear localization signal (NLS) in the CPsV (Naum et al., 2003; Martín et al., 2005), and at least one NLS was also found in MiLBVV and RWMV polymerases (van der Wilk et al., 2002; Vaira et al., 2003).

A protein about 50-55 kDa of unknown function is encoded by RNA 2 in the positive strand of the ophioviruses CPsV, MiLBVV and LRNV (Sanchez de la Torre et al., 2002; van der Wilk et al., 2002; Torok et al., 2003). The 54K protein of CPsV has been detected in infected tissue confirming its size and coding assignment (Peña E. J. unpublished results). It is probably involved in virion movement and suppression of post transcriptional gene silencing (PTGS), the antiviral defence mechanism of the plant (Robles Luna and Peña personal communication). These two mentioned functions seem to be shared with the 24K protein of CPsV. Sequence analyses of the 54K protein and the homologous 50K and 55K proteins from LRNV and MiLBVV contain a conserved NLS, as the RdRp of ophioviruses CPsV, MiLBVV and RWMV, suggesting that part of the cycle might occur in the nucleus. Similarity among 50-55K and 22-25K proteins is lower than that found among CPs, and is very high among the RdRp module sequence (see section 4).

In the viral complementary RNA 2 of MiLBVV an additional minor ORF encodes a putative protein of about 10kDa (see Figure 2), but its function is unknown. In the RNA 2 of CPsV and LRNV this small protein is absent, and so far, it is unknown whether this putative protein is present in RWMV and FreSV genomes. In case the 10kDa polypeptide were not present in these viruses, MiLBVV would be the unique ophiovirus with ambisense RNA 2.

Negative and positive stranded RNAs of ophioviruses are encapsidated in a single coat protein of 43-50 kDa (Garcia et al., 1991; Vaira et al., 1997; Barthe et al., 1998). Using antibody obtained against purified virion particles and expressing RNA 2 and RNA 3 coded proteins in *E. coli*, Sanchez de la Torre et al., (1998) demonstrated that the coat protein of CPsV is encoded in the RNA 3. The protein encoded by MiLBVV RNA3 has similar molecular mass and high sequence similarity (44.6%) with the coat protein of CPsV, thus, it is presumed that coat protein of MiLBVV is also encoded by RNA 3 (van der Wilk et al., 2002). The CPs of MiLBVV and TMMMV are closer with 80% homology. Attempts to find similarities, with the exception of CPsV, some serological relationship between TMMMV and MiLBVV, and between RWMV and MiLBVV, have been found indicating that some epitopes in the capsid proteins among most of the ophiovirus are conserved (Roggero et al., 2000). In general, different isolates of the same ophiovirus specie showed highly conserved amino acid sequences in the coat protein as showed for CPsV (Martín et al, 2006) and MiLBV (Navarro et al., 2004), and less conserved among the different ophiovirus species.

Other than structural function of the coat protein can be assumed. Transmission facilitated by fungus zoospores has been reported for *Tombusviridae* family, involving coat protein (McLean et al., 1994), and even oligosaccharides as shown by Kakani et al, (2003). Rochon et al., (2004) proposed a model for the tombusvirus *Cucumber necrosis virus* transmission in which it binds to *Olpidium bornovanus* zoospores, showing that specific sites on the capsid as well as on the zoospore are involved. They also remark that the mechanism of the tombusvirus coat protein binding to the fungus is similar to poliovirus/host cell interactions and related viruses such as influenza, suggesting evolutionary conservation of functional features of plant and animal virus capsids.

MiLBVV and LRNV present a fourth RNA of negative polarity (van der Wilk et al., 2002; Torok et al., 2003, Torok et al., 2010). The RNA 4 of LRNV encodes a potential protein of 38 kDa, and the RNA 4 of MiLBVV one of 37kDa (p37). MiLBVV has an additional ORF of 10.6 kDa with a 38 nt overlapping sequence with the p37 (see Figure 2). This second ORF is proposed to be expressed by a +1 translational frameshift of p37, but lacks an initiation codon (van der Wilk et al., 2002). So far, the functions of these putative proteins are unknown.

3. The question about circular structures – A "panhandle" structure?

As mentioned before, virions of the ophioviruses appear circularized. The same morphology is observed for tenui- and phleboviruses particles, suggesting that ophiovirus can adopt a panhandle structure formed by the pairing of the conserved 5' and 3' ends of each genomic RNA (see Figure 1.b). Looking for this structure the 3´ and 5´ terminal sequences of CPsV, MiLBVV and LRNV have been checked (Figure 3). In CPsV, the first 12 nt of 5´ end of vcRNAs were found almost identical in the three RNAs, but unexpectedly no identity among the three RNAs at their 3´ ends was found (Figure 3.a), and were not able to form self-complementary panhandle structures between the 3´ and 5´ends of each RNA (Naum et al., 2003). Figure 3. b shows the alignment of the 3´ and 5´ends sequences of the four RNAs of LRNV presenting higher identities among the RNAs 1, 2 and 3, and less with the RNA 4. In the case of MiLBVV, both 5´and 3´ ends are conserved among the four viral RNAs. Both MiLBVV and LRNV do not anneal to perfect panhandle structures (Figure 3.c). Instead, van der Wilk et al., (2002) found that MiLBVV RNAs ends are able to fold into structures faintly resembling the "corkscrew" conformation of *Orthomyxoviridae* RNA termini. In the case of LRNV partial pairing of the conserved 5´ and 3´ ends of genomic RNAs can be found and a "corkscrew" conformation can also be inferred. However, since this structure was not found for CPsV, alternative explanations for the circular structure of CPsV particles are required.

```
RNA1   5´GAUACUUUUUUUCAAGGA-AAAGUAAUCAU
RNA2   5´GAUACUUUUUUUUGUGAUUAAAGC-AUCAC
RNA3   5´GAUACUUUUUUUUGUGGAAAAAGC-AUCAU
       ************    *   **    ****


RNA1   CGACAAGACGAUAAUCCUAAUUAGGAUACC 3´
RNA2   CACCCGUGGAAAUGUCACCAGGAUGACACC 3´
RNA3   UUUUAAAUAAACAAAAGUUCCCAAAUUCGG 3´
                   *              *
```

(a) CPsV

```
RNA1   5´UGAUAUAUUUCUAAAAUAUAUUACUGUUUC
RNA2   5´UGAUAAAUUUCUAUAAAAUUUAUCUGGCUU
RNA3   5´UGAUAGUUUUAUUUAAAGCUAUCAUGUUCA
RNA4   5´--AAAGUUGACUUUUUUAAAUAAAGUAUCA-
         *  *   *    *              *


RNA1   GAAACAGUAAUAUAUUUUAGAAAUA--UAUCA 3´
RNA2   -AAGCC-AGAUAAAUUUUAUAGAAAUUUAUCA 3´
RNA3   UGAACA-UGAUAG-CUUUAAAUAAAACUAUCA 3´
RNA4   -GAACU-UAAUAAUUUUUAUAAAAAAUUAUCA 3´
        * *    ***    **** * *  *****
```

(b) LRNV

```
RNA1   5´GAUUAUUUU UUAAAAAUAUAACAAGUUCUC
RNA2   5´GAUUAUUUUAAUAUAUGCAAAACAGUUUUC-
RNA3   5´GAUUAUUUU-UUAAAAAUAUAACAAGCUCAU
RNA4   5´GAUUAUUUU-UUAAAAAAUAAACAAGUAUCU
         *********  ** *      ****


RNA1   UUAAUCAAAAUAUAUUU--UAAAAAUAUAUCA 3´
RNA2   UCAACAACAAUAUUUUU--AGAAAAAUAUCA 3´
RNA3   UGAGCU--GAUAGUUUUAAUAAAAAACUAUCA 3´
RNA4   UGAACA--GAUAUUUUUAAUAAAAAAAUAUCA 3´
        * *    ***  ***    ****  *****
```

(c) MiLBVV

Fig. 3. Alignment of the 3´ and 5´ ends of positive stranded RNAs of CPsV, LRNV and
MiLBVV. Thirty terminal nucleotides of 5´ and 3´ ends of vcRNAs (a) CPsV (b) LRNV and
(c) MiLBVV. Identical nucleotides are denoted with asterisks. The conserved 5´ terminal
sequences among the three ophioviruses and 3´end of MiLBVV and LRNV are underlined.

Comparing among RNA termini of these ophioviruses it is noted that 5´ terminal sequences
GAUWNWUUUW (where N is any nucleotide and W is A or U) and 3´end of MiLBVV and
LRNV (UAUCA 3´) are quite conserved (see Figure 3, underlined sequences).

4. The analysis of the putative RdRp – Taxonomical relationship with the negative-stranded RNA viruses

The aa sequence of the large ORF of RNA 1 ophioviruses was aligned with the RdRp aa
sequences of members of the *Paramyxoviridae*, *Rhabdoviridae*, *Bornaviridae* and *Filoviridae*

families and *Varicosavirus* as reported by Naum et al, 2003, indicating that this protein is the putative RNA polymerase (Figure 4.a). The predicted 260-280K protein contains the core polymerase module with the five conserved motifs proposed to be part of the RdRp active site (Poch et al., 1989; Muller et al., 1994), and conserved residues recognized in all compared negative-stranded RNA viruses. Among the ophioviruses CPsV, MiLBVV, RWMV and LRNV, the aa sequences of the polymerase module is highly conserved (Figure 4.b). However, instead of the GDNQ of most of the non-segmented viruses present, the four ophioviruses have the SDD sequence in motif C, which is a signature for segmented negative-stranded RNA virus families *Orthomyxoviridae*, *Arenaviridae* and *Bunyaviridae*.

premotif A

(a)

```
CDV      535 KEKEIKEVGRLFAKMTYKMRACQVIAENL <95>
MEV      535 QEKEIKETGRLFAKMTYKMRACQVIAENL <95>
NIV      542 KEKETKQAGRLFAKMTYKMRACQVIAEAL <147>
SEV      543 KEKEIKQEGRLFAKMTYKMRAVQVLAETL <87>
MUV      553 KEKEIKATGRIFAKMTKRMRSCQVIAESL <83>
NDV      543 KEKEVKVNGRIFAKLTKKLRNCQVMAEGI <65>
HRSV     619 KERELS-VGRMFAMQPGMFRQIQILAEKM <49>
TRTV     554 KERELS-VGRMFAMQPGKQRQVQILAEKL <49>
MARV     556 KEKELN-IGRTFGKLPYRVRNVQTLAEAL <47>
ZEBOV    553 KEKELN-VGRTFGKLPYPTRNVQTLCEAL <47>
RYSV     537 KERELKIMARFFALLSFKMRLYFTATEEL <41>
SYNV     567 KEREMKTKARFFSLMSYKLRMYVTSTEEL <41>
NCMV     491 KEREMNPVARMFALMTLKMRSYVVITENM <42>
LBVaV    518 KEREIKVAARMYSLMTERMRYYFVLTEGL <39>
VSIV     530 KERELKLAGRFFSLMSWKLREYFVITEYL <42>
VSNJV    530 KERELKIAGRFFSLMSWRLREYFVITEYL <42>
BEFV     556 KERELKEEGRFFSLMSYELRDYFVSTEYL <42>
RABV     543 KERELKIEGRFFALMSWNLRLYFVITEKL <42>
BDV      326 KEKELKVKGRFFSKQTLAIRIYQVVAEAA <38>
IHNV     494 KEMELKIKGRGFGLMTFMPRLLQVLRES- <41>
MiLBVV   575 KEKEQKTAARFYGIASFKLKLWISSTMEM <37>
RWMV       1 KEREQKIAARFYGIASFKLKLWISSTMEM <37>
CPsV     635 KEKELKTEGRFYGVASFKLKIYISIIMEM <37>
```

```
         Motif A                    Motif B                      Motif C             Motif D

CDV      FITTDLKKYCLNWR <57> IFIKYPMGGVEGYCQKLWTISTIPYL <12> SLVQGDNQTI <60> YDGMLISQSLKSI
MEV      FITTDLKKYCLNWR <57> IFIKYPMGGIEGYCQKLWTISTIPYL <12> SLVQGDNQTI <60> YDGLLVSQSLKSI
NIV      FLTTDLKKFCLNWR <57> IFIHYPKGGIEGYSQKTWTIATIPFL <12> AIVQGDNESI <60> YDGKILPQCLKAL
SEV      FLTTDLKKYCLNWR <57> IFIHNPRGGIEGYCQKLWTLISMSAI <12> AMVQGDNQAI <60> YDGKILPQCLKAL
MUV      FLTTDLTKYCLNWR <57> IFIVSPRGGIEGLCQKLWTMISISTI <12> SMVQGDNQAI <60> YKGRILTQALKNV
NDV      FITTDLQKYCLNWR <57> IYIVSARGGIEGLCQKLWTMISIAAI <12> CMVQGDNQVI <59> KDGAILSQVLKNS
HRSV     SIITDLSKFNQAFR <58> GLYRYHMGGIEGWCQKLWTIEAISLL <12> ALINGDNQSI <59> HNGVYYPASIKKV
TRTV     SIVTDLSKFNQAFR <57> GLYRFHMGGIEGWCQKMWTMEAISLL <12> SLLNGDNQSI <59> SEGVMYPAAIKKV
MARV     SFVTDLEKYNLAFR <57> NAYHYHLGGIEGLQQKLWTCISCAQI <12> SSVMGDNQCI <60> LNGVQLPQSLKTM
ZEBOV    SFVTDLEKYNLAFR <57> SSYRGHMGGIEGLQQKLWTSISCAQI <12> SAVMGDNQCI <60> LNGVQLPQSLKTA
RYSV     VINMDFVKWNQQMR <55> VCWIDDGAGKEGIRQKAWTIMTVCDI <12> LVGGGDNQVL <64> YKGVPLRSPLKQV
SYNV     SMNIDFSKWNQNMR <54> WSRTGDESGKEGLRQKGWTITTVCDI <12> LIGGGDNQVL <63> YSGVPLRGRLKVI
NCMV     CINMDFEKWNLNMR <56> KSYEGHIRGFEGLRQKGWTVFTVVLI <12> LMGQGDNQVL <64> YKGVPLCSSLKRI
LBVaV    NINIDFSKWNTNMR <54> MCYRGHLGGFEGLRQKGWTVATVCLL <12> LMGQGDNQII <64> LDGRQLPQWYKKT
VSIV     ANHIDYEKWNNHQR <56> VCWQGQEGGLEGLRQKGWTILNLLVI <12> VLAQGDNQVI <63> FRGVVIRGLETKRW
VSNJV    ANHIDYEQWNNHQR <56> VCWNGQOKGGLEGLRQKGWSIVNLLVI <12> VLAQGDNQVI <63> FRGVIRGLETKRW
BEFV     ANNIDYEKWNNYQR <56> VCWEGQKGGLEGLRQKGWSILNYLMI <12> ILAQGDNQTI <63> IEGTIKGLPTKRW
RABV     AFHLDYEKWNNHQR <58> TCWNGQDGGLEGLRQKGWSLVSLLMI <12> VLAQGDNQVL <63> FRGNILVPESKRW
BDV      VINLDYSSWCNGFR <54> TCAVGTKTMGEGMRQKLWTILTSCWE <12> ILGQGDNQTI <51> FRGVPVPGCLKQL
IHNV     NKSLDINKFCTSQR <67> GVFSGLKGGIEGLCQYVWTICLLLRV <12> ILAQGDNVII <63> CP-QHLTLAIKKA
MiLBVV   SLFLDYSGHNTSQR <49> YISQGQGLGAIEGWLGSLWGIQSQLMI <12> IGTTYSDDSC <50> YRGKPMDMSIKKM
RWMV     SLFLDYSGHNTSQR <49> YYSKGQGLGAIEGWLGGLWGIQSQLML <12> IGTTYSDDSC <50> YKGIPIEMTLKKI
CPsV     TLFLDYSGHNTSQR <49> VWSKRQKGAIEGWFGPLWGIQSQLML <12> IGTTYSDDSC <50> YFDKPIDTSYKRI
```

(b)

<pre>
                              Premotif A
                     _____

MiLBVV <575> KEKEQKTAARFYGIASFKLKLWISSTMEMIKRAMKLLPGQMMTMTDDERRLIMFKMSEKL
RWMV   KEREQKIAARFYGIASFKLKLWISSTMEMVKRAMKLLPGQMMTMTDDERREIMFKMSEKL
CPsV   <635> KEKELKTEGRFYGVASFKLKIYISIIMEMIKRAMKLLPGQMMTMTEDERRTVMHRMSVML
LRNV   <475> KEREQKTSARFYGIASFKLKLWISSVMEMIKRAMKLLPGQMMTMTDDERREIMFKMSEKL
              :* *   .**:*******::**   ***:******:********:**** :*.:** *

                              Motif A

MiLBVV LSKNSYSLFLDYSGHNTSQRPENTNFILEEIANMYGYYEGTPEFNELTSLSYVFSNINII
RWMV   LEEDTYSLFLDYSGHNTSQRPENTNFIMEEIANMYGYFEGSLEYEEFTSLPYVFSNIHLI
CPsV   EEKDAYTLFLDYSGHNTSQRPENNLFILEEIADMYGFEENSIERKRLIQVVYLFNELEIL
LRNV   MEKDAYSLFLDYSGHNTSQRPENCVFLLEEIANMYGFFEGSDQYREIVSLAYVFANIHVI
        .:::*:**************** *:::****:***: *.: : ..: .: *:* ::.::

                                   Motif B   Motif C
                              _____   _____

MiLBVV VEDSWSDYVYISQGQLGAIEGWLGSLWGIQSQLMIEDMFMQLGMNDYIGTTYSDDSCGVF
RWMV   VEDTWSDMIYYSKGQLGAIEGWLGGLWGIQSQLMLEDMFMQLGVKDFIGTTYSDDSCGVF
CPsV   YEHLFSDMVVWSKRQKGAIEGWFGPLWGIQSQLMLDDMMVSMGIEKYIGTTYSDDSCGVF
LRNV   VEDNWSDIVYYSHGQLGAIEGWLGGLWGIQSQLMLEDMFLQLGVKDYIGTTYSDDSCGVF
        *. :** : *: * ******:* *********::**::.:*::.:************

                                                                  Motif D
                                                           _____

MiLBVV TQSSLDVHKLNGIIKNVQRYGEDMGLIVKLSQTQVTNGRCSMLKEHYYRGKPMDMSIKKM
RWMV   TQKSMDVQKLNGIIKNVQRCGEDMGLIVKLSQTQVTNGRCSMLKEHYYKGIPIEMTLKKI
CPsV   IKKNLDQEELNRMIEYIQSYSLKMGLLVKLSQTQITNGRCSMLKNHYYFDKPIDTSYKRI
LRNV   TQQKLNVSKLNDIVRNVQKYGEDMGLIVKLSQTQVTNGRCSMLKEHYYGEPIDMTLKKM
        :...::   :** ::. :*   . .***:*******:**********:*** . *:: : *::
</pre>

Fig. 4. **a.** Alignment of the core RdRp modules of some representative members of families *Borna-, Filo-, Paramyxo-* and *Rhabdoviridae,* and of the members of family *Ophioviridae* (CPsV, MiLBVV and RWMV). Conserved residues recognized previously in premotif A and motifs A, B, C and D, are shown in bold letters and additional strictly conserved residues are underlined. Numbers on the left of premotif A indicate the starting position, and numbers within brackets refer to the intervening sequences between motifs not represented. **b.** Alignment of the complete RdRp polymerase module of CPsV, MiLBVV, RWMV and LRNV ophioviruses. Identical residues are denoted with asterisks and similar residues by colons and dots. Virus acronyms are indicated in the legend of figure 5. Naum et al., (2003) with modifications.

These data support grouping these four viruses in the same genus, *Ophiovirus,* previously advanced on the basis of their similar virion morphology including TMMMV (Milne et al., 2000), and later FreSV (Vaira et al., 2011).

The amino sequence of the RdRp active site was exploited to study the phylogenetic relationships among the ophioviruses and other representative negative-stranded RNA viruses of families *Borna-, Filo-, Paramyxo-, Rhabdo-, Orthomyxo-, Bunya-* and *Arenaviridae,* and *Tenuivirus.* Figure 5 shows that ophioviruses CPsV, MiLBVV and RWMV appear as a monophyletic group that is separated from the other negative-stranded RNA viruses, reinforcing the taxonomic relatedness of the group.

Fig. 5. Unrooted phylogenetic tree showing the relationship among representative negative-stranded RNA viruses, from the *Mononegaviales* order, *Tenuivirus, Arenavirus* genera and the ophiovirus CPsV, MiLBVV and RWMV, *Ophioviridae* family, based on their conserved RdRp modules (see Figure 4.a). Branch lengths are proportional to genetic distances between sequences. The tree was generated by the neighbor-joining method and bootstrap values (indicated for each branch node) were estimated using 100 replicas. Branch lengths are proportional to genetic distances between sequences and the scale bar represents substitutions per amino acid site. Borna disease virus (BDV), Marburg virus (MARV), Zaire Ebola virus (ZEBOV), Sendai virus (SeV), Mumps virus (MuV), Newcastle disease virus (NDV), Measles virus (MEV), Canine distemper virus (CDV), Nipah virus (NIV), Human respiratory syncytial virus (HRSV), Turkey rhinotracheitis virus (TRTV), Vesicular stomatitis Indiana virus (VSIV), Vesicular stomatitis New Jersey virus (VSNJV), Bovine ephemeral fever virus (BEFV), Infectious hematopoietic necrosis virus (IHNV), Lettuce necrotic yellows virus (LNYV), Northern cereal mosaic virus (NCMV), Sonchus yellow net virus (SYNV), Rice yellow stunt virus (RYSV), Lettuce big-vein associated virus (LBVaV),

Influenza A virus (FLUAV), Influenza B virus (FLUBV), Influenza C virus (FLUCV), Thogoto virus (THOV), Dhori virus (DHOV), Infectious salmon anemia virus (ISAV), Bunyamwera virus (BUNV), La Crosse virus (LACV), Hantaan virus (HTNV), Puumala virus (PUUV), Sin nombre virus (SNV), Seoul virus (SEOV), Dugbe virus (DUGV), Rift Valley fever virus (RVFV), Toscana virus (TOSV), Uukeniemi virus (UUKV), Tomato spotted wilt virus (TSWV), Rice stripe virus (RSV), Lymphocytic choriomeningitis virus (LCMV), and Tacaribe virus (TCRV). Naum et al., (2003) with modifications.

As mentioned, ophioviruses possess two ORFs in the same strand of RNA 1, with the RdRp located downstream the intergenic region (see figure 2), which is a distinct genomic structure of all segmented- and negative stranded RNA viruses.

Taken all these characteristics, the family *Ophioviridae* has been proposed to the ICTV Ninth Report (Vaira et al., 2011), with only one genus recognized, *Ophiovirus*, containing six species: CPsV, TMMMV, RWMV, FreSV and MiLBVV, without an order assigned. Moreover, the phylogenetic analysis done by Vaira et al., (2003) with the 45 aa strings derived from the 136nt fragment amplified from the available isolates of CPsV, RWMV, LRNV and FreSV supported the positions of these ophioviruses as distinct species, and a closer relationship between MiLBVV and TMMMV species.

5. Cultivars and transgenic lines resistant to ophiovirus

Since most of the ophiovirus are soil-transmitted, the cultivation of commercially important species addresses the challenge of the disease control searching for resistant cultivars economically important. In Japan, natural resistance has been found in tulip cultivars for tulip mild mottle mosaic disease. Bulb lots of 214 cultivars were tested, some of which were resistant to TMMMV, although resistance varies greatly among them (Morikawa et al., 2004). In Virginia, USA, Hansen el at., (2009) have been reported that in freesia cvs. 'Honeymoon'and 'Santana' mixed infection with the potyvirus Freesia mosaic virus, and the ophiovirus FreSV can be found and probably making more difficult to control the disease.

Big-vein diseased lettuce plants are infected with MiLBVV, and usually together with the varicosavirus LBVaV (Lot et al., 2002; Navarro et al., 2004; Plesko et al., 2009; Barcala Tabarrozzi et al., 2010). To control this disease resistant cultivars have been developed by conventional breeding method, like the cultivars Thompson and Pacific, using several resistant sources (Ryder 1981; Ryder and Robinson, 1991, 1995). However, although with some cultivars losses were reduced, do not exhibit high levels of resistance and do not eliminate the disease. More recently, partial big-vein resistance was identified in *Lactuca sativa* cultivars Great Lakes 65, Pavane, Margarita. In the same work, *Lactuca virosa*, which is not used in the market, was found ophio- and varicosavirus-free and big-vein symptomless (Hayes et al., 2006). Big-vein resistance breeding efforts using this line has been reported (Hayes et al., 2004) generating *L. virosa–L. sativa* hybrid but variation for the frequency of symptomatic plants was found.

In last 20 years, different strategies using transgenic plants have been developed successfully to gain virus resistance cultivars (Sudarshana et al. 2007; Prins et al., 2008). The most widely used have been protein-mediated (pathogen-derived resistance, PDR) and more recently RNAi-mediated resistance by post-transcriptional gene silencing (PTGS) mechanism (e.g. Jan et al., 2000; Shimizu et al., 2009; Fahim et al., 2010). Expression of the coat protein gene of several RNA viruses were shown to confer virus resistance in experimental and natural hosts, and later, other virus-derived sequences in sense or antisense constructs carrying the movement protein or RdRp genes were expressed also conferring resistance against virus. Transgenic plants carrying the coat protein gene have resulted successes, which could indicate that the CP is involved in early events of virus infection (Reyes et al., 2011). However, it is not predictable which viral genes is the best to confer resistance (Morroni et al., 2008; Kamo et al., 2010).

The first attempt to get resistance against ophioviruses was done on lettuce (*Lactuca sativa* L) by agro-transformation expressing the coat protein gene of LBVaV in sense or antisense orientations. Interestingly, some of the lines were susceptible to LBVaV, but line A-2 was resistant to MiLBVV without big-vein symptoms regardless of the presence or absence of LBVaV (Kawazu and Fujiyama, 2006). In this line the LBVaV coat protein-mRNA derived from the transgene was not detected, probably due to RNA silencing. However, the mechanism by which line A-2 was resistant to MiLBVV is not clear, since there is no significant sequence homology between the transgene (LBVaV coat protein gene) and the MiLBVV coat protein gene. Later, lettuce was transformed with inverted repeats of a coat protein gene fragment of MiLBVV and two lines resulted resistant to this virus (Kawazu et al. 2009). These lines showed resistance to big-vein symptom expression but were susceptible to LBVaV. Moreover, MiLBVV was detected in roots but not in leaves of one of the lines after inoculation, suggesting that resistance to MiLBVV is less effective in roots than in leaves. Furthermore, T3, T4, and T5 generations showed high resistance to this ophiovirus and big-vein symptoms expression indicating that high resistance to lettuce big-vein disease is stably inherited (Kawazu et al., 2010).

Citrus plants do not present natural resistant species including oranges, mandarins and grapefruits, as well as hybrids and citrus relatives used as rootstock (Roistacher, 1993), promoting the generation of new alternatives of control. Transformation of woody plants present disadvantages as the time consuming in transformation procedure and multiplication. Lines are propagated by bud grafting onto seedlings used as rootstocks generating replicates of each line. For challenge, transgenic scions are infected by grafting using infective tissue. For citrus psorosis disease the first work was done by Zanek et al., (2008) producing 21 independent lines of transgenic sweet orange (cp-lines) expressing low and variable amounts of CPsV coat protein (isolate CPV4). In these lines no correlation between copy number and transgene expression was found and no significant differences were observed in the response to virus challenge among the lines or among the replicates. Although two different viral loads were evaluated to challenge the transgenic plants, no resistance or tolerance was found in any line after one year of continuous observations. An inherent difficulty in this assay is that as long the rootstock is susceptible to CPsV, the virus could move to the rootstock and replicate. Thus, viruses could be delivered to the scion as a continuous challenge overcoming the protection. Applying PDR strategy but using different viral genes, sweet orange transgenic lines were generated expressing the *54k* and *24k* genes of CPV 4 isolate (Reyes et al., 2011a). In these assays fourteen lines were selected including new cp-lines. These plants were evaluated for their acquired resistance against two isolates, PsA (CPV 4) and PsB (CPsV 189-34), which differ in symptoms severity. These lines were susceptible to both isolates when graft-infected, although one of them carrying the cp gene (CP-96 line) containing two copies of the transgene and expressing a low level of the coat protein showed a delay in symptom expression when inoculated with the PsB isolate.

Therefore, other transgenic approach was applied developing CPsV-resistant sweet orange plants. In order to trigger the PTGS prior to CPsV infection, transgenic sweet orange plants producing intron-hairpin RNA transcripts corresponding to *cp, 54k* or *24k* genes were generated. Lines carrying the ihpRNA derived from the *cp* gene (ihpCP, lines 10 and 15) provided a high level of virus resistance, but ihp54K and ihp24K lines resulted variable or highly susceptible to CPsV respectively (Reyes et al., 2011b). The siRNAs accumulation level was not directly correlated to the degree of the triggered virus resistance among the different lines, and no significant difference was observed between inoculated and non-inoculated ihpCP resistant lines, indicating that in these plants the virus has been controlled probably immediately after the virus enter to the cells. Moreover, these results support the idea that not all regions of the viral genome yield the same level of resistance applying pathogen-derived-resistance strategy (Valkonen et al., 2002). For negative-stranded RNA viruses as tospovirus, replication is regulated by the CP concentration at the point of switching from mRNA production to replication of the genome (Storms, 1998). In the case of CPsV and MiLBVV the coat protein could be early involved in these functions, and its absence could impede viral replication. Moreover, all these results indicate that pre-activation of the RNA-silencing machinery against the *cp* gene seems to be one alternative to prevent other ophiovirus infections.

6. Concluding remarks

Ophioviruses are the causal agent of old diseases as citrus psorosis and big-vein affecting major crops (citrus, ornamental plants and lettuce). Most of the ophioviruses are soil-transmitted by a root-infecting fungus *Olpidium*, and in the case of CPsV an aerial vector is also suspected, although there are no evidences so far. The virions are circles of at least two different contour lengths, particles can form pseudolinear duplex structures, and the coiled filamentous are about 9-10 nm in diameter. Ophiovirus genome is divided into three or four individually encapsidated segments. CPsV, RWMV and FreSV have 3 RNAs and MiLBVV and LRNV have a fourth RNA. In the RNA 1, a protein of 22-25 kDa of unknown function is encoded in the 5´ region of the positive strand. Separated by an intergenic region of hundred nucleotides, the large ORF of RNA 1 encodes the putative RNA-dependent RNA polymerase of about 260K- 280K. Two regions of the RdRp of CPsV may be regarded as a bipartite nuclear localization signal (NLS), one of them conserved in MiLBVV and RWMV. A protein about 50-55 kDa is encoded by vcRNA 2 of CPsV, MiLBVV and LRNV, probably involved in virion movement and suppression of post transcriptional gene silencing (PTGS). A conserved NLS sequence of the RNA 2 of CPsV, LRNV and MiLBVV is present, suggesting that part of the cycle might occur in the nucleus. In the vcRNA 2 of MiLBVV an additional putative protein of unknown function is encoded. Viral and viral complementary RNAs of ophioviruses are encapsidated in a single coat protein of 43-50 kDa, which is encoded in the RNA 3. RNA 4 of MiLBVV and LRNV encode a potential protein of 37-38 kDa, and an additional ORF in MiLBVV overlapping with the sequence of the p37, both of unknown function. The 5´and 3´ terminal sequences of MiLBVV are conserved among the four viral RNAs, but they do not anneal to perfect panhandle structures, which is expected to form according to the circular morphology. Instead, a "corkscrew" conformation similar to the *Orthomyxoviridae* RNA termini has been suggested. Similarly, for LRNV partial pairing of the 5´ with the 3' end sequences of genomic RNAs can be found and a "corkscrew" conformation could be inferred, but this structure was not found in CPsV.

Moreover, comparing among RNA termini among ophiovirus genomes, it is noted that among 5´ terminal sequences of CPsV, MiLBVV and LRNV, and between 3´ ends of MiLBVV and LRNV, the sequences are highly conserved. The amino sequence of the RdRp active site was exploited to study the phylogenetic relationships among the ophioviruses and other representative negative-stranded RNA viruses, forming a new family *Ophioviridae*, without order assigned, including five species.

For some cultivars of tulip and lettuce a natural resistance source has been used in breeding programs to gain resistant plants. However, that resistance has been variable and not enough to control these diseases. In the case of CPsV and MiLBVV pre-activation of the RNA-silencing machinery against the *cp* gene seems a good alternative to prevent ophiovirus infection in citrus and lettuce, and probably applicable in ornamental plants.

7. Acknowledgements

My gratitude to Dr. Robert G. Milne, who was able to see for the first time these viruses even "in the middle of the jungle," as he used to say. I thank to Carina A. Reyes for critical reading, Gabriel Robles Luna and Eliana Ocolotobiche for their help preparing the figures.

Maria Laura Garcia belongs to the staff of Instituto de Biotecnología y Biología Molecular (IBBM) – CCT-La Plata-CONICET, Argentina, and Facultad de Ciencias Exactas, UNLP, and is a member of the research career of the National Research Council of Argentina (CONICET). This work was supported by grants from Ministerio de Ciencia, Tecnología e Innovación Productiva, Argentina, CONICET and UNLP.

8. References

Alioto D., Guerri J., Moreno P., Milne R.G. 2007. Citrus psorosis virus. In Characterization, diagnosis and management of plant viruses, Vol. 2. Rao, G.P., Myrta, A. and Ling, K.S. (eds.). Studium Press LL. C., Ho.

Barcala Tabarrozzi A.E., Peña E.J., Dal Bo E., Robles Luna G., Reyes C.A. and M.L. Garcia (2010). Identification of Mirafiori lettuce big-vein virus and Lettuce big-vein associated virus infecting Lactuca sativa with symptoms of lettuce big-vein disease in Argentina. New Disease Reports 20, 38

Barthe, G.A., Ceccardi, T.L., Manjunath, K.L. and Derrick, K.S. (1998). Citrus psorosis virus: nucleotide sequencing of the coat protein gene and detection by hybridization and RT-PCR. J. Gen. Virol., 79, 1531-1537

Beñatena, H.N. and Portillo, M.M. (1984). Natural spread of psorosis in sweet orange seedlings pp. 159-164. In: Proceedings of the 9th Conference of the International Organization of Citrus Virologists, IOCV (eds: S.M. Garnsey, L.W. Timmer and J.A. Doods). University of California, Riverside, CA, USA.

Bos, L., Huijberts, N. Lettuce ring necrosis, caused by a chytrid-borne agent distinct from lettuce big-vein virus in (1996). European journal of plant pathology 102: 867-873

Campiglia, H.G., Silveira, C.M. and Salibe, A.A. 1976. Psorosis transmission through seeds of trifoliate orange pp.132-134. In: Proceedings of the 7th Conference of the International Organization of Citrus Virologists, IOCV (eds: E.C. Calavan). University of California, Riverside, CA, USA.

Danós, E., 1990. La psorosis de los cítricos: la epidemia en curso en Argentina y el desafio de su control. In: International Foundation for Science (IFS) e Instituto Nacional de Tecnologia Agropecuaria (INTA) (Eds.), Revista de Investigaciones Agropecuarias 22, 265–277.

Derrick, K.S., Brlansky, R.H., Da Graça, J.V., Lee, R.F., Timmer. L.W. and Nguyen, T.K. (1988). Partial characterization of a virus associated with citrus ringspot. Phytopathology, 78, 1298-1301.

Diamante de Zubrzycki, Zubrzycki, H.M. and Correa, M. 1984. Determination of the distribution of psorosis in commercial plantings pp. 165-169. In: Proceedings of the 9th Conference of the International Organisation of Citrus Virologists, IOCV (eds: S.M. Garnsey, L.W. Timmer, J.A. Dodds). University of California, Riverside, CA, USA.

Fahim, M.,Ayala-Navarrete,L.,Millar,A.A.,Larkin,P.J.,(2010).Hairpin RNA derived from viral NIa gene confers immunity to wheat streak mosaic virus infection in transgenic wheat plants.PlantBiotechnol.J.8,1–14.

Falk BW (1997) Lettuce big-vein. In: Davis RM, Subbarao KV, Raid RN, Kurtz EA.(eds) Compendium of lettuce diseases. APS Press, St Paul, pp 41–42

Fawcett HS (1933). New symptoms of psorosis, indicating a virus disease of citrus. Phytopathology 23: 930.

Ghazal SA, El-Dougdoug KhA, Mousa AA, Fahmy H, Sofy AR Isolation and identification of citrus psorosis virus Egyptian isolate (CPsV-EG). Commun Agric Appl Biol Sci. 2008;73(2):285-95.

García M.L., O. Grau & A.N. Sarachu. (1991). Citrus Psorosis is probably caused by a bipartite ssRNA virus. Research in Virology 142, 303-311

Garcia ML, Dal Bo E, Grau O and Milne R (1994).The closely related citrus ringspot and citrus psorosis viruses have particles of novel filamentous morphology. Journal of General Virology 75: 3585 – 3590.

Hansen M. AVaira., A. M., Murphy C., Hammond J., Dienelt M., Bush E. and C. Sutula.(2009). Freesia sneak virus (FreSV) on freesia: a first detection for Virginia and the United States.

Hayes, R. J., Ryder, E., and Robinson, B. (2004). Introgression of Big Vein Tolerance from Lactuca virosa L. into Cultivated Lettuce (Lactuca sativa L.). HortScience 39:881.

Hayes, R. J., Wintermantel, W. M., Nicely, P. A., and Ryder, E. J. 2006. Host resistance to Mirafiori lettuce big-vein virus and Lettuce big-vein associated virus and virus sequence diversity and frequency in California. Plant Dis. 90:233-239

Huijberts N, Blystad D-R, Bos L, 1990. Lettuce big-vein virus: Mechanical transmission and relationships to tobacco stunt virus. Annals of applied Biology 116, 463-476.

Jan, F.-J., Fagoaga, C., Pang, S.-Z., and Gonsalves, D. (2000). A minimum length of N gene sequence in transgenic plants is required for RNA-mediated tospovirus resistance. J. Gen. Virol. 81, 235–242.

Jagger, I. C., and Chandler, N. 1934. Big vein, a disease of lettuce. Phytopathology 24:1253-1256

Kakani K, Robbins M, Rochon DM (2003) Evidence that Binding of Cucumber Necrosis Virus to Vector Zoospores Involves Recognition of Oligosaccharides. J Virol 77:3922-3928

Kamo K, Jordan R, Guaragna MA, Hsu HT, Ueng P. (2010). Resistance to Cucumber mosaic virus in Gladiolus plants transformed with either a defective replicase or coat protein subgroup II gene from Cucumber mosaic virus. Plant Cell Rep 29(7):695-704.

Kawazu Y and R Fujiyama (2006). A transgenic lettuce line with resistance to both lettuce big-vein associated virus and Mirafiori lettuce virus J. Amer. Soc. Hort. Sci. 131: 709-819.

Kawazu Y, Fujiyama R and Y Noguchi (2009). Transgenic resistance to Mirafiori lettuce virus in lettuce carrying inverted repeats of the viral coat protein gene. Transgenic Res. 18:113-120

Kawazu Y, Fujiyama R, Noguchi Y, Kubota M, Ito H, Fukuoka H (2010). Detailed characterization of Mirafiori lettuce virus-resistant transgenic lettuce. Transgenic Res. 19:211-20.

Kuwata, S., and Kubo, S. 1984. Properties of two Olpidium-transmitted viruses: Tobacco stunt and lettuce big vein. 6th Int. Congr. Virol. Sendai 1984:331.

Larocca, L.H., 1985. La citricultura en la provincia de Entre R´ios. In: Instituto Nacional de Tecnologia Agropecuaria (INTA) (Eds.), Actas de la XVI, Jornadas T´ecnicas de Citricultura, pp. 42–51.

Lot H, Campbell RN, Souche S, Milne RG, Roggero P (2002) Transmission by Olpidium brassicae of mirafiori lettuce virus and lettuce big-vein virus, and their roles in lettuce big-vein etiology. Phytopathology 92: 288–293

Martín, S., García, M.L., Troisi, A., Rubio, L., Legarreta, G., Grau, O., Alioto, D., Moreno, P., Guerri, J. 2006. Genetic variation of populations of Citrus psorosis virus. J. Gen. Virol. 87, 3097-3102.

Martín, S., López, C., García, M.L., Naum-Onganía, G., Grau, O., Flores, R., Moreno, P., Guerri, J. 2005. The complete nucleotide sequence of a Spanish isolate of Citrus psorosis virus: comparative analysis with other ophioviruses. Arch. Virol. 150, 167-176.

McLean JS, Campbell RN, Hamilton RI, Rochon DM (1994) Involvement of cucumber necrosis virus coat protein in the specificity of fungus transmission by Olpidium bornovanus. Virol 204:840–842

Milne, R.G. 2000. Role of electron microscopy in discovering and characterizing plant viruses: A few examples. Proceedings of the 7th Aschersleben Symposium "New Aspects of Resistance Research on Cultivated Plants", November 17-18, 1999, Aschersleben, Germany, Beitraege zur Zuechtungsforschung, 6 (3): 35-40.

Milne, R.G., Djelouah, K., Garcia, L.M., Dal Bo, E. and Grau, O. 1996. Structure of citrus-psorosis-associated virus particles: Implications for diagnosis and taxonomy pp. 189-197. In: Proocedings of the 14th Conference of the International Organization of Citrus Virologists, IOCV (eds: J.V. da Graça, P. Moreno and R.K. Yokomi). University of California, Riverside, CA, USA.

Milne, R. G., Garcia, M. L. & Grau, O. (2000). Genus Ophiovirus. In Virus Taxonomy. Seventh Report of the International Committee on Taxonomy of Viruses, pp. 627-631. Edited by M. H. V. van Regenmortel, C. M. Fauquet, D. H. L. Bishop, E. B. Carstens, M. K. Estes, S. M. Lemon, J. Maniloff, M. A. Mayo, D. J. McGeoch, C. R. Pringle & R. B. Wickner. San Diego: Academic Press.

Morikawa, T., Nomura, Y., Ohura K. and Yamamoto T. (1993). Some propoerties of soil-borne mild mottle moseaic disease of tulip. Annals of Phytopathological Society of Japan, 59: 65.

Morikawa, T., Nomura, Y., Yamamoto, T. and Natsuaki, T. (1995). Partial characterization of virus-like particles associated with tulip mild mottle mosaic. Annals of Phytopathological Society of Japan, 61: 578-581.

Morikawa T, Taga Y and t Morii. (2004). Resistance of tulip cultivars to mild mottle mosaic disease. ISHS Acta Horticulturae 673: IX International Symposium on Flower Bulbs.

Morroni M, Thompson JR, Tepfer M. (2008).Twenty years of transgenic plants resistant to Cucumber mosaic virus. Mol Plant Microbe Interact. Jun;21(6):675-84.

Naum-Onganıa, G., Gago-Zachert, S., Peñ a, E., Grau, O. & Garcıa, M. L. (2003). Citrus psorosis virus RNA 1 is of negative polarity and potentially encodes in its complementary strand a 24K protein of unknown function and 280K putative RNA dependent RNA polymerase. Virus Res 96, 49–61.

Navarro, J. A., Torok, V. A., Vetten, H. J., and Pallas, V. (2004). Genetic variability in the coat protein genes of lettuce big-vein associated virus and Mirafiori lettuce big-vein virus. Arch. Virol. 150:681-694.

Prins, M., Laimer, M., Noris, E., Schubert, J., Wassenegger, M., Tepfer, M., (2008). Strategies for antiviral resistance in transgenic plants. Mol. Plant Pathol. 9, 73–83.

Plesko Mavric, M. Virscek Marn and M. Zerjav. 2009. Identification of Lettuce big-vein associated virus and Mirafiori lettuce big-vein virus Associated with Lettuce Big-Vein Disease in Slovenia. Plant disease, vol. 93(5): 549

Pujol, A.R. and Beñatena, H.N. 1965. Study of psorosis in Concordia, Argentina pp. 170-179. In: Proceedings of the 3th Conference of the International Organization of Citrus Virologists, IOCV (ed. W.C. Price). University of California, Riverside, CA, USA.

Reyes CA, De Francesco A, Peña EJ, Costa N, Plata MI, Sendin L, Castagnaro AP, García ML (2011b). Resistance to Citrus psorosis virus in transgenic sweet orange plants is triggered by coat protein-RNA silencing. J Biotechnol 151(1):151-8.

Reyes CA, Zanek MC, Velázquez K, Costa N, Plata MI, García ML (2011a). Generation of Sweet Orange Transgenic Lines and Evaluation of Citrus psorosis virus-derived Resistance against Psorosis A and Psorosis B. Jounal of Phytopathology 159: 531–537.

Roggero P., Ciuffo M., Vaira A.M., Accotto G.P., Masenga V. and Milne R.G. (2000). An Ophiovirus isolated from lettuce with big-vein symptoms. Arch. Virol., 145, 2629-2624.

Roistacher, C.N., 1993. Psorosis—a review. In: Moreno, P., da Graca, J.V., Timmer, L.W. (Eds.), Proceedings of the 12th Conference of the International Organization of Citrus Virologists, IOCV. Riverside, CA, USA, pp. 139-154.

Rochon D, Kakani K, Robbins M, Reade R (2004) Molecular aspects of plant virus transmission by Olpidium and plasmodiophorid vectors. Annu Rev Phytopathol 42:211–41

Ryder E J (1981). Thompson lettuce. HortScience 16:687-688.

Ryder E J and B J Robinson (1991). Pacific Lettuce. HortScience 26:437-438

Ryder E J and B J Robinson (1995). Big-vein resistance in lettuce: Indentifying, selection, and testing resistant ciltivars and breeding lines. J.Amer. Soc. Hort. Sci. 120:741-746.

Sanchez de la Torre, M.E., Lopez, C., Grau, O. and Garcia, M.L. (2002). RNA 2 of Citrus psorosis virus is of negative polarity and has a single open reading frame in its complementary strand. J. Gen. Virol., 83, 1777-1781.

Sanchez de la Torre, M.E., Riva, O., Zandomeni, R., Grau, O. and Garcia, M.L. (1998). The top component of citrus psorosis virus contains two ssRNAs, the smaller encodes the coat protein. Mol., Plant Pathol. On-Line,
http://www.bspp.org.uk/mppol/1998/1019sanchez.

Sasaya, T., Ishikawa, K. and Koganezawa, H. (2001). Nucleotide sequence of the coat protein gene of Lettuce big-vein virus. J. Gen. Virol. 82:1509-1515.

Sasaya, T. and Koganezawa, H. (2006). Molecular analysis and virus transmission tests place Olpidium virulentus, a vector of Mirafiori lettuce big-vein virus and tobacco stunt virus, as a distinct species rather than a strain of Olpidium brassicae. J Gen Plant Pathol 72:20-25

Storms, M.M.H., 1998. The role of NSs during tomato spotted wilt virus infection. PhD Thesis, Agricultural University Wageningen, The Netherlands

Swingle, W.T. and Webber, H.J. 1896. The principal disease of citrus fruits in Florida. US Department of Agriculture Division of Vegetable Physiology and Patholology Bulletin. N. 8.

Shimizu, T., Yoshii, M., Wei, T., Hirochika, H., Omura, T.,(2009). Silencing by RNAi of the gene for Pns12, a viroplasm matrix protein of Rice dwarf virus, results in strong resistance of transgenic rice plants to the virus. Plant Biotechnol. J. 7, 24-32.

Sudarshana MR, Roy G, Falk BW. Methods for engineering resistance to plant viruses. Methods Mol Biol. 2007;354:183-95.

Timmer L.W. and Garnsey S.M. 1980. Natural spread of citrus ringspot virus in Texas and its association with psorosis-like diseases in Florida and Texas pp. 167-173. In: Proceedings of the 8th Conference of the International Organization of Citrus Virologists, IOCV (eds: E.C. Calavan, S.M. Garnsey, L.W. Timmer), University of California, Riverside, CA, USA.

Torok, V.A. and Vetten, H.J. (2002). Characterisation of an ophiovirus associated with lettuce ring necrosis. Joint Conf Int Working Groups on Legume and Vegetable Viruses, Bonn 4-9 August 2002. Abstract p. 4

Torok VA and Vetten J (2003). Identification and molecular characterisation of a new ophiovirus associated with lettuce ring necrosis disease Proceedings of Arbeitskreis Viruskrankheiten der Pflanzen, March 27-28, Heidelberg, Germany

Torok, V.A. and Vetten, H.I. (2010). Ophiovirus associated with lettuce ring necrosis. European Society for Virology Meeting, Cernobbio, Italy, April 7-11 2010. Abstract p. 282.

Vaira AM, Accotto GP, Borghi V., Masenga V, Luisoni E, Milne RG (1996). A new virus isolated from Ranunculus hyb., with particles resembling supercoiled filamentous nucleocapsids. Acta Horticult. 432:36-42.

Vaira, A.M., Milne, R.G., Accotto, G.P., Luisoni, E., Masenga, V. and Lisa, V. (1997). Partial characterization of a new virus from ranunculus with a divided RNA genome and circular supercoiled thread-like particles. Arch. Virol., 142, 2131-2146.

Vaira, A.M., Accotto, G.P., Costantini, A. and Milne R.G. (2003). The partial sequence of RNA1 of the ophiovirus Ranunculus white mottle virus indicates its relationship to rhabdoviruses and provides candidate primers for an ophiovirus-specific RT-PCR test. Arch. Virol., 148, 1037-1050.

Vaira, A.M., Lisa, V., Costantini, A., Masenga, V., Rapetti, S. and Milne, R.G. (2006). Ophioviruses infecting ornamentals and a probable new species associated with a severe disease in freesia. Acta Hort. (ISHS) 722:191-199.

Vaira, A.M., Kleynhans, R. and Hammond, J. (2007). First report of Freesia sneak virus infecting Lachenalia cultivars in South Africa. Plant Disease, 91:770

Vaira, A.M., Hansen, M.A., Murphy, C., Reinsel, M.D. and Hammond, J. (2009). First report of Freesia sneak virus in Freesia sp. in Virginia, USA. Plant Disease, 93:965.

Vaira AM, Gago-Zachert S, ML Garcia, J Guerri, J Hammond, RG Milne, P Moreno, T Morikawa, T Natsuaki, JA Navarro, V Pallas, V Torok, M Verbeek and HJ Vetten. (2011). Ophioviridae. In: 9th ICTV Report of the International Committee on Taxonomy of Viruses Andrew King, Mike Adams,Eric Carstens, and Elliot Lefkowitz (EDS) In press.

Valkonen, J.P., Rajamäki, M.L., Kekarainen, T., 2002. Mapping of viral genomic regions important in cross-protection between strains of a potyvirus. Mol. Plant Microbe Interact. 15, 683–692

van der Wilk, F., Dullemans, A.M., Verbeek, M. and van den Heuvel, J.F.J.M. (2002). Nucleotide sequence and genomic organization of an ophiovirus associated with lettuce big-vein disease. J. Gen. Virol., 83, 2869-2877.

van Dorst, H.J.M. (1975). Evidence for the soil-borne nature of freesia leaf necrosis virus. Neth. J. Plant Pathol. 81,45-48.

Verbeek M and Meekes E. (2005). New insights in freesia leaf necrosis disease. FlowerTech 8:14-15. www.HortiWorld.nl

van Regenmortel, M. H. V., Fauquet, C. M., Bishop, D. H. L., Carstens, E. B., Estes, M. K., Lemon, S. M., Maniloff, J., Mayo, M. A., McGeoch, D. J., Pringle, C. R., and Wickner, R. B. 2000. Virus Taxonomy: Classification and Nomenclature of Viruses. (7th Report ICTV) Academic Press, San Diego, CA.

Vetten H J, Lesemann DE, Dalchow J. (1987): Electron microscopical and serological detection of virus-like particles associated with lettuce big vein disease. J. Phytopathology 120, 53-59.

Yamamoto T., Morikawa T. Inagaki K, Matsumoto M and Nahata K. (1989) Ocurrence of virus-like diseases of tulip. The symptoms and transmissions. . Annals of Phytopathological Society of Japan, 55: 105.

Zanek, M.C.,Peña, E.,Reyes,C.A.,Figueroa,J.,Stein,B.,Grau,O.,Garcia,M.L.,2006. Detection of Citrus psorosis virus in the northwestern citrus production area of Argentina by using an improved TAS-ELISA. J. Virol. Methods137,245–251.

Zanek, M.C.,Reyes,C.A.,Cervera,M.,Peña, E.J.,Velázquez,K.,Costa,N.,Plata,M.I., Grau, O.,Peña, L.,García,M.L.(2008). Genetic transformation of sweet orange with the coat protein gene of Citrus psorosis virus and evaluation of resistance against the virus.Plant Cell Rep.27,57–66.

Part 2

Regulation of Viral Replication and Gene Expression

5

Cis–Acting RNA Elements
of Human Immunodeficiency Virus

Mario P.S. Chin
Department of Microbiology and Immunology,
Center for Substance Abuse Research,
Temple University School of Medicine
USA

1. Introduction

The World Health Organization has indicated that effective control of the HIV/AIDS pandemic is the world's most urgent public health challenge. The 2009 UNAIDS Global Facts and Figures report estimated that almost 60 million people have been infected with the virus and that 25 million AIDS-related deaths have occurred since the pandemic began in the early 1980s. In 2008, there were approximately 33.4 million people living with HIV, 2.7 million new infections and 2 million deaths from AIDS-related causes. The pathogen that causes this pandemic is the Major (M) group of HIV type 1 (HIV-1). Group M HIV-1 dominates the global pandemic with at least nine subtypes and multiple intersubtype recombinants have been identified to date (Leitner et al., 2005). Many of these recombinants are circulating in multiple geographical regions and are integral parts in the HIV-1 pandemic (McCutchan, 2006; Takebe et al., 2004).

HIV-1 has become the most studied virus in history. Our understanding of the replication mechanism of HIV-1 has allowed scientists to develop several classes of antiviral therapies targeting various steps of the virus life cycle (Gilliam et al., 2011; Liao et al., 2010; Perno et al., 2008). Current antiviral treatments target the functions of several HIV-1 proteins; available drugs include nucleoside and non-nucleoside reverse transcriptase inhibitors; protease inhibitors, which block the maturation of the nascent virus; and integrase inhibitors, which prevent the integration of viral DNA into the host genome. Fusion and entry inhibitors are newer classes of antiviral drugs and can prevent viral infection before the virus's entry into the cell.

In addition to the proteins encoded by the viral genome, RNA secondary structures play important roles in the replication of HIV-1 by acting *in cis* to regulate and facilitate different stages of viral replication. Indeed, these RNA secondary structures appear to be promising targets for next-generation antiviral drugs (Berkhout, 2009; Biswas et al., 2004; Daelemans et al., 2002; Haasnoot et al., 2007; Houghton et al., 2010; Reyes-Darias et al., 2008; Rossi et al., 2007). Here, we provide an overview of the functions of several important *cis*-acting RNA elements that are crucial to HIV-1 replication. We will also present our latest in-depth analysis of a multi-functional viral RNA element that participates in the dimerization of HIV genomic RNA and virion packaging in producer cells, as well as reverse transcription (RT) and the recombination of viral RNA in infected cells.

2. The HIV-1 replication cycle

A schematic of the HIV life cycle is shown in Figure 1. The life cycle includes binding, entry, reverse transcription, integration, viral protein synthesis, assembly and budding.

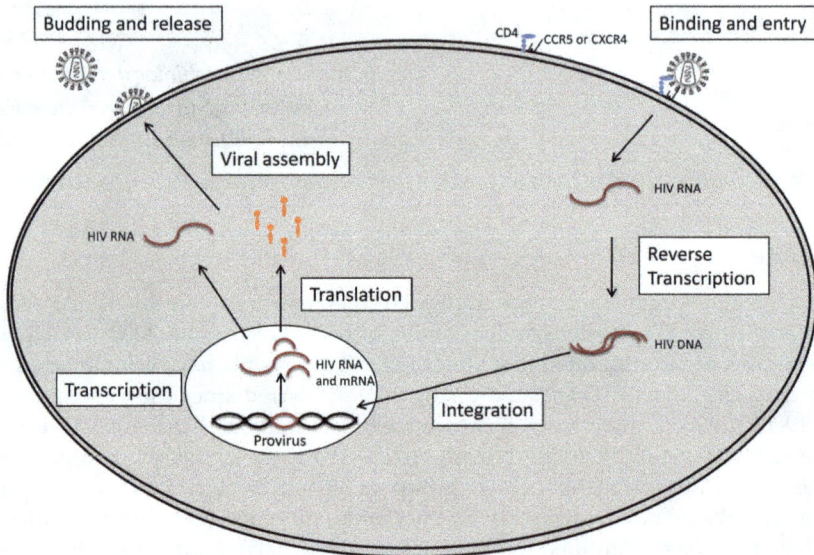

Fig. 1. Schematic representation of the life cycle of HIV-1.

2.1 Binding and entry

The HIV-1 replication cycle begins with a virion binding to a target cell. Both binding and entry depend upon the surface envelope proteins of the virus, which are trimeric glycoproteins composed of heterodimers of gp120 and gp41 (Checkley et al., 2011). Binding is mediated via the interaction between gp120 on the virion and CD4 on the T-lymphocyte (Yoon et al., 2010). Upon binding, the viral envelope glycoprotein undergoes a conformational change, exposing a specific domain capable of binding the CCR5 or CXCR4 chemokine receptors on the cell membrane (Trkola et al., 1996). The binding of gp120 to CD4 and one of the two chemokine receptors results in the fusion of gp41 on the viral envelope with the cellular membrane. After fusion of the viral envelope with the cell membrane, the virus core is released into the cytoplasm, and the viral RNA is uncoated from the viral core (Arhel, 2010). RT occurs in the cytoplasm, and the viral RNA is converted to a double-stranded cDNA by the polymerase and RNase H domains of the reverse transcriptase.

2.2 Reverse transcription and integration

Once HIV-1 genomic RNA is uncoated in the host cytoplasm, reverse transcriptase uses host tRNA as a primer for the viral primer binding site (PBS) to initiate minus-strand DNA synthesis (Figure 2) (Jiang et al., 1993; Mak et al., 1994). RT proceeds to the 5′ end of the genomic RNA, creating a DNA/RNA hybrid. The RNA component of the hybrid is degraded by the RNase H activity of reverse transcriptase, generating minus-strand strong-stop DNA. The direct repeat (R) sequence allows the minus-strand strong-stop DNA to anneal to the identical R at the 3′ end of the viral genome (first-strand transfer). Once first-strand transfer is completed, minus-strand DNA synthesis continues. The RNase H domain of the reverse transcriptase degrades the RNA template when DNA is synthesized, but the degradation is incomplete.

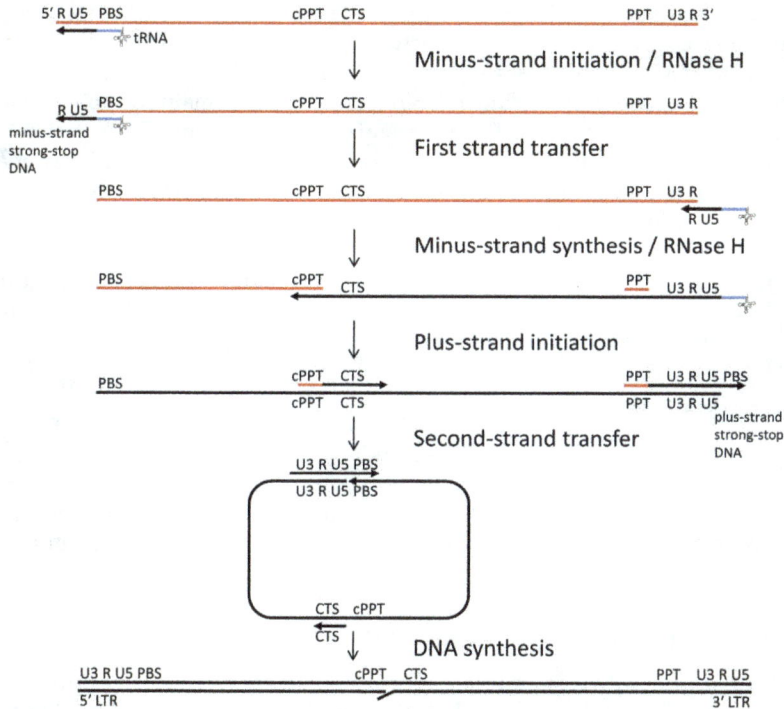

Fig. 2. HIV-1 reverse transcription. Reverse transcriptase uses host tRNA (blue line) bound to PBS as primer. Viral RNA is indicated as red lines. Black lines represent viral DNA. Approximate locations of *cis*-acting elements relevant to reverse transcription are shown. CTS, central termination signal.

The purine-rich region of the RNA genome is called the poly-purine tract (PPT) and central PPT (Charneau et al., 1992; Huber and Richardson, 1990). PPT acts as a primer for plus-strand DNA synthesis. The PPT and central PPT are relatively resistant to RNase H digestion and can prime plus-strand DNA synthesis. Plus-strand synthesis from the PPT continues to the 3′ end of the viral genome and the portion of the primer tRNA yielding plus-strand strong-stop DNA. RNase H removes the primer tRNA, allowing the PBS on the plus-strand strong-stop DNA to anneal to the upstream complementary PBS (second-strand transfer). DNA synthesis from the central PPT provides an additional primer for plus-strand synthesis. Plus- and minus-strand syntheses are then completed, with each strand of DNA serving as the template for the other. The resulting double-stranded HIV-1 cDNA is imported into the nucleus and integrated into the cell genome by integrase (Li et al., 2011). The virus then resides permanently in the genome as a provirus.

2.3 Viral protein synthesis, assembly and budding

When the host cell receives a signal to become active, cellular RNA polymerase uses the promoter and enhancer in the 5'LTR to initiate transcription of proviral DNA into viral RNA (Kingsman and Kingsman, 1996). The full-length unspliced viral RNA serves two purposes: it expresses Gag and Pol, and becomes incorporated into newly generated viral particles. Upon maturation, Gag forms the three structural proteins of the virion: the matrix, capsid and nucleocapsid (NC) (Freed, 1998). The protease, reverse transcriptase and integrase are encoded by the pol gene. Other viral mRNA encodes the remaining viral proteins. Gag expressed from the unspliced viral RNA recognizes viral genomic RNA that contains the major packaging signal and packages two copies of RNA into a virion (Clever et al., 1995). Virion assembly takes place at the cellular membrane, and the assembly process gives rise to immature viral particles (Adamson and Freed, 2007). The viral protease cleaves the Gag-Pol polyproteins into matrix, capsid, NC, reverse transcriptase and integrase proteins, producing mature and infectious virus particles. The mature particle buds through the infected cell membrane and acquires viral envelope glycoproteins that are encoded by the env gene and expressed on the cell membrane.

3. HIV-1 secondary RNA structure

The viral genome of HIV contains several secondary RNA structures that are important for the regulation of viral replication (Watts et al., 2009; Wilkinson et al., 2008). The known secondary RNA structures with well-defined functions are the trans-activation responsive (TAR) element, stem-loop (SL) 1 to SL4, ribosomal frameshift signal, PPT, central PPT, and Rev response element (RRE).

3.1 Trans-activation responsive element

The TAR element primarily resides in an approximately 45-nucleotide region of the 5' R of the HIV-1 genome. TAR RNA forms a hairpin stem-loop structure with a side bulge. The viral transactivator protein, Tat, binds to the bulge of the cis-acting TAR to activate transcription. The absence of Tat severely impairs viral replication, highlighting the importance of this protein in the viral life cycle. Transcription from the LTR is enhanced several hundred-fold in the presence of Tat. Upon binding to TAR, Tat promotes the binding of cellular proteins to form the ribonucleoprotein complex, a positive transcription elongation factor (P-TEFb) complex that contains Cyclin T1, cdk9 and Brd4 and ensures efficient transcription of the full-length HIV genome (Jang et al., 2005; Marshall and Price, 1992). The interactions of TAR with Tat and P-TEFb allow it to bind RNA polymerase II and increase its processivity (Isel and Karn, 1999; Parada and Roeder, 1996).

3.2 Stem-loops in the 5' untranslated region

In addition to TAR, HIV-1 possesses RNA secondary structures at the 5' end of the HIV-1 RNA in the untranslated region. This region forms a series of four SLs preceding and overlapping the Gag start codon that are important for the regulation of viral replication (Figure 3A) (Berkhout and van Wamel, 2000; Clever et al., 1995; Watts et al., 2009; Wilkinson et al., 2008). Despite some sequence variation, different subtypes of HIV-1 all have similar secondary structures in this region (Berkhout and van Wamel, 1996; Laughrea et al., 1997). SL1 contains the dimerization initiation sequence that controls partner selection during viral RNA dimerization in the cytoplasm (Figure 3B and see Section 4 below). In the absence of SL1, HIV-1 cannot replicate in human T cell lines, highlighting the crucial role of this element in HIV-1 replication. SL2 is the splice donor that directs the splicing of viral mRNA transcripts such as tat and rev. SL3 is the major packaging signal that allows Gag to recognize and package viral genomic RNA into the virion. HIV-1 RNA is encapsidated into virions through Gag–RNA interactions involving the recognition of SL3 by zinc finger motifs in the Gag NC. In addition, SL3 is present in unspliced genomic HIV-1 RNA but absent from spliced viral mRNAs, ensuring efficient packaging of the full length HIV-1 viral genome. Moreover, SL1, SL2 and SL4 are integral components of the packaging signal (Amarasinghe et al., 2000; Clever et al., 1995; Clever and Parslow, 1997; Damgaard et al., 1998; McBride and Panganiban, 1996, 1997; Sakaguchi et al., 1993). Biochemical analysis has indicated that short RNAs possessing HIV-1 SL2 or SL3 have the highest affinity for NC, whereas those with SL1 or SL4 have lower affinity for NC (Shubsda et al., 2002). Mutation analyses have shown that all of these structures are important for RNA packaging (Berkhout and van Wamel, 1996; Clever and Parslow, 1997; Laughrea et al., 1997; Shankarappa et al., 2001).

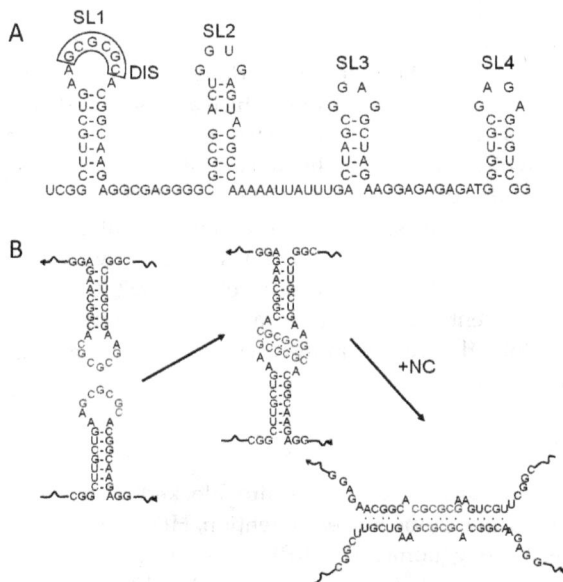

Fig. 3. Stem-loops of the HIV-1 5' untranslated region. (A) RNA structure of the four stem-loops. The dimerization initiation signal (DIS) sequence is shown in box. (B) Mechanism of viral RNA dimerization. Dimerization is initiated by base-pairing of the DIS forming a kissing-loop complex. Gag nucleocapsid (NC) promotes the formation of a more stable RNA dimer.

3.3 Ribosomal frameshift signal

The compactness of the HIV-1 genome makes it challenging for the virus to express multiple viral proteins. HIV-1 overcomes this problem by incorporating a ribosomal frameshift signal between the gag and pol transcripts (Jacks et al., 1988). The signal has a slippery sequence (UUUUUUA) that causes a frameshift, and the sequence immediately downstream forms a stem-loop structure (Dinman et al., 2002; Dulude et al., 2002; Jacks et al., 1988; Parkin et al., 1992). The stem-loop RNA structure of the signal is hypothesized to stall the ribosome, resulting in a switch from the zero reading frame to the minus-one frame in the 5' direction; translation continues in the new frame (Jacks et al., 1988). The signal is a translational control mechanism that is responsible for a minus-one ribosomal frameshift that, in turn, produces a specific ratio of Gag and Gag-Pol polyproteins from the overlapping Gag-Pol open reading frames.

3.4 The polypurine and central polypurine tracts

The RNA genome of HIV-1 contains two short PPTs that are involved in the initiation of plus-strand DNA synthesis (Figure 2). The 3′ PPT is a purine-rich sequence (AAAAGAAAAGGGGGGA) located just upstream of U3 (Huber and Richardson, 1990). The central copy of the PPT, which has an additional function in the nuclear import of HIV-1 cDNA, is an exact copy of the 3′ PPT (Charneau et al., 1992; Zennou et al., 2000). PPT acts as a primer for plus-strand DNA synthesis because it is relatively resistant to RNase H degradation. Downstream plus-strand synthesis is primed by the central PPT. Mutations in the central PPT significantly reduce viral replication as a result of reduced plus-strand initiation (Charneau et al., 1992). The 3′ end PPT primes the synthesis of the 3′ LTR, which is paused after the primer tRNA is degraded to produce a plus-strand strong-stop DNA. After strand transfer of the plus-strand strong-stop DNA, DNA synthesis continues to the center of the viral genome, which is defined by a central termination signal (CTS) located approximately 68 nucleotides downstream of the central PPT (Charneau and Clavel, 1991). The central initiation of the plus-strand DNA at the central PPT and the downstream termination at the CTS generate a linear DNA molecule with a three-stranded DNA structure called the central DNA flap (Charneau et al., 1992; Charneau and Clavel, 1991; Zennou et al., 2000). This central DNA flap promotes HIV-1 DNA nuclear import *in cis*. Absence of the central DNA flap results in the accumulation of unintegrated linear DNA in the cytoplasm of infected cells.

3.5 Rev response element

HIV-1 genomic RNA and unspliced mRNA are blocked from nuclear export and are retained in the nucleus. To overcome nuclear retention, HIV-1 expresses the Rev protein and harbors an RRE in its RNA genome. The RRE is an approximately 200-nucleotide RNA element located at the junction between the surface (gp120) and transmembrane (gp41) domains of the env gene. The RRE has multiple high-affinity binding sites for the Rev viral protein (Dillon et al., 1990; Zapp and Green, 1989). Rev contains the nuclear export signal and is expressed from a fully spliced HIV-1 mRNA that can be exported from the nucleus normally. After expression and nuclear entry, Rev binds to RRE and facilitates the nuclear export of viral genomic RNA and viral unspliced mRNA via the Crm1 nuclear export pathway (Fridell et al., 1996; Fukuda et al., 1997; Neville et al., 1997). The Rev-RRE interaction is an essential regulatory switch in the viral life cycle. At the beginning of viral replication, Rev concentration is low, and only fully spliced viral mRNA, e.g., Rev mRNA, is exported to the cytoplasm. Following expression and nuclear entry, Rev concentration increases, and the protein binds to and multimerizes with the RRE to recruit nuclear export complexes (Olsen et al., 1990; Zapp et al., 1991). This process results in the export of viral genomic and unspliced mRNA from the nucleus to the cytoplasm, which marks the late stage of viral replication.

4. Multiple functions of stem-loop 1

Unlike most cis-acting elements that participate in a defined step of viral replication, SL1 has multiple well-defined functions in the virus life cycle (Berkhout and van Wamel, 1996). Studies have shown that SL1 directs the dimerization of HIV genomic RNA and its packaging into the virion in producer cells, as well as RT and recombination of the viral RNA in infected cells (Chin et al., 2007; Chin et al., 2008; Chin et al., 2005; Moore et al., 2007; Moore and Hu, 2009).

4.1 Stem-loop 1 directs viral RNA dimerization and controls genetic recombination

HIV-1 virions contain two copies of the viral RNA genome. The genomic RNA is held together as a dimer by a noncovalent linkage at the 5′ end (Hoglund et al., 1997; Song et al., 2007). The dimerization process occurs in the cytoplasm, and the dimeric RNA is then packaged through Gag-RNA interactions, as described above (Chen et al., 2009; Moore et al., 2007; Moore et al., 2009). The viral element that directs the dimerization process is a 6-nucleotide palindromic sequence called the dimerization initiation signal (DIS), located at the SL1 loop in the 5′ untranslated region (Figure 3A) (Berkhout and van Wamel, 1996; Chin et al., 2005; Laughrea and Jette, 1994; Moore et al., 2007; Moore and Hu, 2009; Muriaux et al., 1995; Skripkin et al., 1994; Song et al., 2007). The DIS sequences of HIV-1 are either subtype B-like, i.e., GCGCGC, or subtype C-like, i.e., GTCGAC (Leitner et al., 2005). Once full-length HIV-1 genomic RNAs are exported into the cytoplasm, the DIS sequences of two viral RNAs interact through Watson-Crick base-pairing (Figure 3B) (Clever et al., 1996; Muriaux et al., 1996b; Paillart et al., 1996b). The dimerization process is then initiated and produces a kissing loop complex (Clever et al., 1996; Kieken et al., 2006; Laughrea and Jette, 1994; Muriaux et al., 1996b; Skripkin et al., 1994). The NC domain of Gag then promotes the conversion of the kissing loop complex to a more stable extended dimer (Feng et al., 1996; Muriaux et al., 1996a). Gag then packages the viral RNA dimer into the virion by interacting with the major packaging signal in SL2.

Studies have shown that the DIS-mediated base-pairing of two viral RNA molecules is a major determinant in the selection of the copackaged RNA partners (Chin et al., 2005; Moore et al., 2007; Moore and Hu, 2009). Using an assay that measures recombination rate as a proxy for the efficiency of packaging of genotypically distinct HIV-1 RNA molecules, studies have found that the copackaging of two subtype B or subtype C HIV-1 RNAs is very efficient (Chin et al., 2005; Rhodes et al., 2005). However, the copackaging of a subtype B RNA with a subtype-C RNA occurs with much lower efficiency (9-fold reduction) compared to the copackaging of homologous sequences. Therefore, HIV-1 copackaging of genotypically different genomic RNAs is restricted. The major element that restricts the copackaging of subtype B and subtype C RNAs was mapped to the DIS on SL1. Subtype B and subtype C HIV-1 possess different palindromic sequences in their DIS sequences. This sequence difference reduces the co-packaging of subtype B and subtype C viral RNA molecules. Although the frequency of template-switching or recombination by reverse transcriptase is unchanged, now only a small population of virions contains two different subtypes of RNA is present. Because genotypically distinct recombinants can only be generated from viruses containing two RNA molecules that encode different sequences (heterozygous virions) but not from viruses containing two identical RNAs (homozygous virions) (Hu and Temin, 1990), sequence differences in the DIS result in a drastic decrease in recombinant HIV-1 formation.

Based on the numbers and prevalence of circulating and unique recombinant forms of HIV-1, it is evident that recombination has played a significant role in generating the diversity of virus strains in the infected population. Recombination can occur during reverse transcription, generating DNA that contains genetic information from each co-packaged RNA (Coffin, 1979). The studies described above showed that the DIS sequence identity plays a pivotal role in determining the packaging efficiency of RNAs from different HIV-1 strains and thus governs the opportunities for recombination to occur. It has been suggested that the recombination potential between two HIV-1 subtypes can be predicted from their DIS sequences. One study explored this possibility by measuring the recombination rate between subtype B, subtype C and circulating recombinant forms 01_AE (AE) strains of HIV-1 (Chin et al., 2007). In that study, the recombination rate between AE and subtype B viruses, which have different DIS sequences, was four-fold lower than the rate between AE and subtype C viruses, which have identical DIS sequences. Moreover, the lower recombination rate between the AE and B viruses could be recovered by changing the subtype B DIS to a subtype C DIS. Therefore, mismatches that affect base-pairing within the DIS can severely disrupt recombination between HIV-1 subtypes. Although the intersubtype HIV-1 recombination rate is much lower than the intrasubtype rate, HIV-1 has exceedingly high recombination rates, approximately 10-fold higher than those of murine leukemia virus or spleen necrosis virus (Anderson et al., 1998; Hu and Temin, 1990). Therefore, even for different DIS sequences, intersubtype HIV-1 recombination still occurs at levels similar to gammaretrovirus recombination rates.

4.2 Stem-loop 1 maintains proper nucleic acid structures in the reverse transcription complex

Studies have suggested that the multi-functional SL1 of HIV-1 helps to facilitate RT. It was shown that SL1 deletion impairs plus-strand HIV-1 DNA transfer in RT (Paillart et al., 1996a; Shen et al., 2000). In addition, template-switching is restricted in a 2-kb region immediately downstream of SL1 mutations (Chin et al., 2008), which affects the efficiency of RT and the synthesis of full-length HIV-1 DNA (King et al., 2008). The second observation is intriguing because most HIV-1 RNA secondary structures that are thought to stall RT and thereby increase recombination are limited to a very short region (Derebail and DeStefano, 2004; Galetto et al., 2004; Moumen et al., 2001). Unlike these RNA structures, the SL1 mutations that cause improper base-pairing between two HIV-1 RNA molecules have a long-range effect on the template-switching tendency of reverse transcriptase (Chin et al., 2008). In that study, viruses that packaged two RNAs containing different DIS sequences were examined. The lack of perfect base-pairing between the two DIS regions caused an apparent recombination gradient with far fewer recombination events immediately downstream from the DIS compared to the pol region, which is more than 2 kb downstream from the DIS.

The long-range effect can be corrected when there is perfect base-pairing between the DIS of the two viral RNAs, indicating that the long-range effect is caused by the DIS rather than by other local sequences. These results suggest that the two RNA molecules in the RT complex are organized in a particular structure(s) and that the base-pairing of the DIS sequences has an important role in forming this structure. It is possible that the DIS serves as a nucleation point to allow proper arrangement of the dimeric RNA structures immediately downstream from it. Without this nucleation point, the 2 kb region immediately following the DIS is not structurally suitable for recombination. The effect of DIS base-pairing diminishes after approximately 2 kb; most of the pol regions had similar numbers of recombination events regardless of whether the DIS could base-pair perfectly. This observation suggests that the remainder of the RNA sequences adopt the proper dimer structure. This result is consistent with the conclusion generated by several studies that, despite the importance of the DIS, the base-pairing of DIS sequences is not absolutely essential for the generation of virion RNA dimers (Berkhout and van Wamel, 1996; Laughrea and Jette, 1996; Moore et al., 2007; Muriaux et al., 1996b). These findings reveal that the DIS plays a critical role in maintaining proper nucleic acid structure in the RT complex.

4.3 Stem-loop 1 regulates the packaging of spliced and unspliced viral RNA

The zinc finger motifs of NC recognize the major packaging signal within the SL3 in a full-length unspliced genomic HIV-1 RNA to promote packaging into virions. Partially spliced and completely spliced viral RNAs, which do not contain SL3, are largely excluded from packaging. An SL1 deletion mutant of HIV-1 is non-viable and has an abnormal packaging preference for full-length unspliced HIV-1 genomic RNA and singly and fully spliced viral mRNA (Clever and Parslow, 1997; Clever et al., 2000; Houzet et al., 2007; McBride and Panganiban, 1997; Ristic and Chin, 2010; Russell et al., 2003). The ΔSL1 mutant packaged genomic RNA two-fold less efficiently than the wildtype (Figure 4A) (Ristic and Chin, 2010). This result is not surprising because the SL1 has been suggested to have a role in binding Gag during packaging (Clever et al., 1995; Clever and Parslow, 1997; Shubsda et al., 2002). In contrast, three- to four-fold more spliced viral mRNA is packaged into the virion when SL1 is deleted (Figure 4B). The deletion of SL1 increased the amount of spliced viral mRNA relative to HIV-1 genomic RNA by seven- to nine-fold (Ristic and Chin, 2010). This aberrant packaging of genomic and spliced viral RNA is caused by an abnormal interaction between the RNA and Gag; three-fold less ΔSL1 genomic RNA co-immunoprecipitates with Gag compared to wildtype RNA. This result indicates that the decreased packaging efficiency of ΔSL1 genomic RNA is caused by a reduced association of Gag with the ΔSL1 RNA. In accordance with this observation, Gag showed an enhanced association with ΔSL1 spliced viral mRNA, immunoprecipitating approximately four-fold more singly spliced and fully spliced viral mRNA (Ristic and Chin, 2010).

A

B

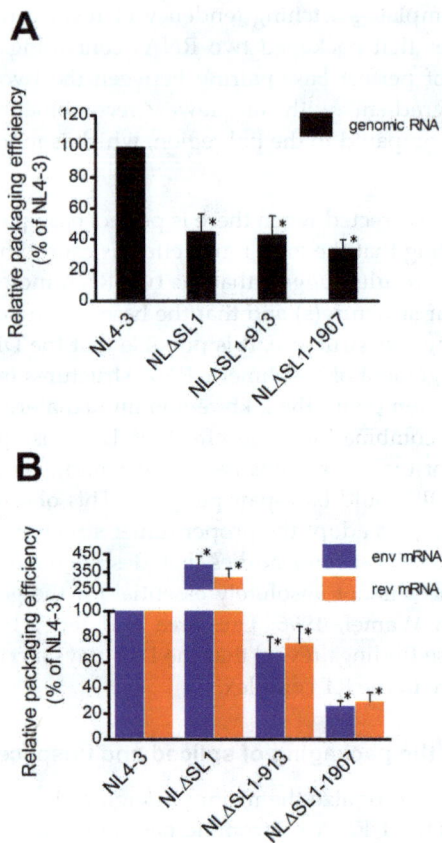

Fig. 4. Quantification of HIV-1 RNA content in the virion by real-time PCR. (A) Efficiency of HIV-1 genomic RNA packaging. NL4-3, wildtype HIV-1; NLΔSL1, SL1 deletion mutant; NLΔSL1-913, ΔSL1 with compensatory mutation in matrix; NLΔSL1-1907, ΔSL1 with compensatory mutation in SP1. The amount of NL4-3 genomic RNA was set at 100%. *, indicates $p < 10^{-4}$ and significant deviation from the wild-type copy number as determined by Student's t test. (B) Efficiency of spliced HIV-1 RNA (env and rev mRNA) packaging. The amount of NL4-3 spliced mRNA was set at 100%. *, indicates significant deviation from the wild-type copy number as determined by Student's t test; $p < 10^{-4}$, except for NLΔSL1-913, $p < 10^{-3}$.

4.4 Is stem-loop 1 a potential target for antiviral intervention?

Given the important role of SL1 in regulating multiple stages of the viral life cycle, it has been proposed as a target for RNA-based antiviral therapies including RNA interference and antisense approaches (Elmen et al., 2004; Ennifar et al., 2006; Sugiyama et al., 2009). However, mutation, sequence deletion and recombination are common mechanisms by which HIV-1 escapes antiviral intervention and continues to replicate in the host. Indeed, ΔSL1 HIV-1 replicated in human peripheral blood mononuclear cells (PBMCs), although it was >10-fold less infectious than the wildtype (Hill et al., 2003; Jones et al., 2008). Several studies have shown that HIV-1 can replicate without SL1 by acquiring changes in the genome (Liang et al., 1998; Liang et al., 1999; Ristic and Chin, 2010; Russell et al., 2003). We have demonstrated that, despite the absence of a vital element regulating RNA dimerization, the ΔSL1 mutant still dimerized, copackaged and subsequently recombined with a recombination-competent HIV-1 and was restored to infectivity similar to that of wildtype. We also showed that, in PM-1 cells infected with ΔSL1 HIV-1, syncytia were observed 14 days post-infection; wildtype infected cells showed syncytia by seven days post-infection. Virus production by the infected PM-1 cells was detected by ELISA in the culture supernatant three to four days before cytopathogenicity was observed (Ristic and Chin, 2010).

Sequence analysis of the near full-length genome of the ΔSL1 virus at 14 days post-infection in PM-1 showed that HIV-1 variants still harbored the SL1 deletion found in the ΔSL1 input virus. Two independent mutations were identified in the matrix domain and the SP1 domain of Gag (Ristic and Chin, 2010). When these two mutations were separately placed into a ΔSL1 HIV-1 backbone, both were able to enhance the infectivity of the deletion mutant by partially restoring the packaging specificity of viral RNA (Figures 4A and 4B). These compensatory mutations allow Gag to exclude spliced viral mRNA from packaging, thus reducing interference with the production of infectious virus in the ΔSL1 mutant. Flow cytometry analysis of infected PBMCs showed that ΔSL1 HIV-1 carrying these compensatory mutations depleted CD4+ cells more rapidly than the original ΔSL1 mutant. These data indicate that more than one pathway can compensate for the loss of SL1 secondary RNA structure. HIV-1 adapts quickly to the deletion of SL1 by compensatory mutations or recombination with a variant. These results highlight the ever-changing nature of HIV-1. In addition to SL1, future antiviral drug design should also target essential and highly conserved gene coding sequences.

5. Conclusion

Despite advances in antiviral therapy against HIV and greater understanding of the biology of the virus, the eradication of HIV/AIDS remains elusive. HIV continues to evade drug interventions and vaccines by mutation and recombination, which allow rapid diversification of HIV population. Facing this challenge, antiviral development has expanded to target HIV replication at the RNA level. Viral cis-acting RNA elements play crucial roles in regulating various steps of viral replication; in particular, the SL1 participates in multiple stages of the virus's lifecycle. Indeed, RNA interference-based antivirals targeting these elements have been tested in vitro but is far from success mainly because of the high variability of the virus. To this end, we have demonstrated that HIV adapts quickly to a severe defect at the SL1 region and regains wild type-like infectivity and pathogenicity. Therefore, in the light of this urgent public health problem, scientists need to continue the endeavor to elucidate new viral cis-acting elements and the mechanisms by which they regulate replication, thereby revealing new targets for antiviral intervention and to develop combination therapy.

6. Acknowledgment

We thank Natalia Ristic for excellent technical assistance in this study. This work was supported by the National Institutes of Health through grant DA026293.

7. References

Adamson, C.S. & Freed, E.O. (2007). Human immunodeficiency virus type 1 assembly, release, and maturation, *Adv Pharmacol* 55: 347-87.

Amarasinghe, G.K., De Guzman, R.N., Turner, R.B. & Summers, M.F. (2000). NMR structure of stem-loop SL2 of the HIV-1 psi RNA packaging signal reveals a novel A-U-A base-triple platform, *J Mol Biol* 299(1): 145-56.

Anderson, J.A., Bowman, E.H. & Hu, W.S. (1998). Retroviral recombination rates do not increase linearly with marker distance and are limited by the size of the recombining subpopulation, *J Virol* 72(2): 1195-202.

Arhel, N. (2010). Revisiting HIV-1 uncoating, *Retrovirology* 7: 96.

Berkhout, B. (2009). Toward a durable anti-HIV gene therapy based on RNA interference, *Ann N Y Acad Sci* 1175: 3-14.

Berkhout, B. & van Wamel, J.L. (1996). Role of the DIS hairpin in replication of human immunodeficiency virus type 1, *J Virol* 70(10): 6723-32.

Berkhout, B. & van Wamel, J.L. (2000). The leader of the HIV-1 RNA genome forms a compactly folded tertiary structure, *Rna* 6(2): 282-95.

Biswas, P., Jiang, X., Pacchia, A.L., Dougherty, J.P. & Peltz, S.W. (2004). The human immunodeficiency virus type 1 ribosomal frameshifting site is an invariant sequence determinant and an important target for antiviral therapy, *J Virol* 78(4): 2082-7.

Charneau, P., Alizon, M. & Clavel, F. (1992). A second origin of DNA plus-strand synthesis is required for optimal human immunodeficiency virus replication, *J Virol* 66(5): 2814-20.

Charneau, P. & Clavel, F. (1991). A single-stranded gap in human immunodeficiency virus unintegrated linear DNA defined by a central copy of the polypurine tract, *J Virol* 65(5): 2415-21.

Checkley, M.A., Luttge, B.G. & Freed, E.O. (2011). HIV-1 Envelope Glycoprotein Biosynthesis, Trafficking, and Incorporation, *J Mol Biol* 410(4): 582-608.

Chen, J., Nikolaitchik, O., Singh, J., Wright, A., Bencsics, C.E., Coffin, J.M., Ni, N., Lockett, S., Pathak, V.K. & Hu, W.S. (2009). High efficiency of HIV-1 genomic RNA packaging and heterozygote formation revealed by single virion analysis, *Proc Natl Acad Sci U S A* 106(32): 13535-40.

Chin, M.P., Chen, J., Nikolaitchik, O.A. & Hu, W.S. (2007). Molecular determinants of HIV-1 intersubtype recombination potential, *Virology* 363(2): 437-46.

Chin, M.P., Lee, S.K., Chen, J., Nikolaitchik, O.A., Powell, D.A., Fivash, M.J., Jr. & Hu, W.S. (2008). Long-range recombination gradient between HIV-1 subtypes B and C variants caused by sequence differences in the dimerization initiation signal region, *J Mol Biol* 377(5): 1324-33.

Chin, M.P., Rhodes, T.D., Chen, J., Fu, W. & Hu, W.S. (2005). Identification of a major restriction in HIV-1 intersubtype recombination, *Proc Natl Acad Sci U S A* 102(25): 9002-7.

Clever, J., Sassetti, C. & Parslow, T.G. (1995). RNA secondary structure and binding sites for gag gene products in the 5' packaging signal of human immunodeficiency virus type 1, *J Virol* 69(4): 2101-9.

Clever, J.L. & Parslow, T.G. (1997). Mutant human immunodeficiency virus type 1 genomes with defects in RNA dimerization or encapsidation, *J Virol* 71(5): 3407-14.

Clever, J.L., Taplitz, R.A., Lochrie, M.A., Polisky, B. & Parslow, T.G. (2000). A heterologous, high-affinity RNA ligand for human immunodeficiency virus Gag protein has RNA packaging activity, *J Virol* 74(1): 541-6.

Clever, J.L., Wong, M.L. & Parslow, T.G. (1996). Requirements for kissing-loop-mediated dimerization of human immunodeficiency virus RNA, *J Virol* 70(9): 5902-8.

Coffin, J.M. (1979). Structure, replication, and recombination of retrovirus genomes: some unifying hypotheses, *J Gen Virol* 42(1): 1-26.

Daelemans, D., Afonina, E., Nilsson, J., Werner, G., Kjems, J., De Clercq, E., Pavlakis, G.N. & Vandamme, A.M. (2002). A synthetic HIV-1 Rev inhibitor interfering with the CRM1-mediated nuclear export, *Proc Natl Acad Sci U S A* 99(22): 14440-5.

Damgaard, C.K., Dyhr-Mikkelsen, H. & Kjems, J. (1998). Mapping the RNA binding sites for human immunodeficiency virus type-1 gag and NC proteins within the complete HIV-1 and -2 untranslated leader regions, *Nucleic Acids Res* 26(16): 3667-76.

Derebail, S.S. & DeStefano, J.J. (2004). Mechanistic analysis of pause site-dependent and -independent recombinogenic strand transfer from structurally diverse regions of the HIV genome, *J Biol Chem* 279(46): 47446-54.

Dillon, P.J., Nelbock, P., Perkins, A. & Rosen, C.A. (1990). Function of the human immunodeficiency virus types 1 and 2 Rev proteins is dependent on their ability to interact with a structured region present in env gene mRNA, *J Virol* 64(9): 4428-37.

Dinman, J.D., Richter, S., Plant, E.P., Taylor, R.C., Hammell, A.B. & Rana, T.M. (2002). The frameshift signal of HIV-1 involves a potential intramolecular triplex RNA structure, *Proc Natl Acad Sci U S A* 99(8): 5331-6.

Dulude, D., Baril, M. & Brakier-Gingras, L. (2002). Characterization of the frameshift stimulatory signal controlling a programmed -1 ribosomal frameshift in the human immunodeficiency virus type 1, *Nucleic Acids Res* 30(23): 5094-102.

Elmen, J., Zhang, H.Y., Zuber, B., Ljungberg, K., Wahren, B., Wahlestedt, C. & Liang, Z. (2004). Locked nucleic acid containing antisense oligonucleotides enhance inhibition of HIV-1 genome dimerization and inhibit virus replication, *FEBS Lett* 578(3): 285-90.

Ennifar, E., Paillart, J.C., Bodlenner, A., Walter, P., Weibel, J.M., Aubertin, A.M., Pale, P., Dumas, P. & Marquet, R. (2006). Targeting the dimerization initiation site of HIV-1 RNA with aminoglycosides: from crystal to cell, *Nucleic Acids Res* 34(8): 2328-39.

Feng, Y.X., Copeland, T.D., Henderson, L.E., Gorelick, R.J., Bosche, W.J., Levin, J.G. & Rein, A. (1996). HIV-1 nucleocapsid protein induces "maturation" of dimeric retroviral RNA in vitro, *Proc Natl Acad Sci U S A* 93(15): 7577-81.

Freed, E.O. (1998). HIV-1 gag proteins: diverse functions in the virus life cycle, *Virology* 251(1): 1-15.

Fridell, R.A., Bogerd, H.P. & Cullen, B.R. (1996). Nuclear export of late HIV-1 mRNAs occurs via a cellular protein export pathway, *Proc Natl Acad Sci U S A* 93(9): 4421-4.

Fukuda, M., Asano, S., Nakamura, T., Adachi, M., Yoshida, M., Yanagida, M. & Nishida, E. (1997). CRM1 is responsible for intracellular transport mediated by the nuclear export signal, *Nature* 390(6657): 308-11.

Galetto, R., Moumen, A., Giacomoni, V., Veron, M., Charneau, P. & Negroni, M. (2004). The structure of HIV-1 genomic RNA in the gp120 gene determines a recombination hot spot in vivo, *J Biol Chem* 279(35): 36625-32.

Gilliam, B.L., Riedel, D.J. & Redfield, R.R. (2011). Clinical use of CCR5 inhibitors in HIV and beyond, *J Transl Med* 9 Suppl 1: S9.

Haasnoot, J., Westerhout, E.M. & Berkhout, B. (2007). RNA interference against viruses: strike and counterstrike, *Nat Biotechnol* 25(12): 1435-43.

Hill, M.K., Shehu-Xhilaga, M., Campbell, S.M., Poumbourios, P., Crowe, S.M. & Mak, J. (2003). The dimer initiation sequence stem-loop of human immunodeficiency virus type 1 is dispensable for viral replication in peripheral blood mononuclear cells, *J Virol* 77(15): 8329-35.

Hoglund, S., Ohagen, A., Goncalves, J., Panganiban, A.T. & Gabuzda, D. (1997). Ultrastructure of HIV-1 genomic RNA, *Virology* 233(2): 271-9.

Houghton, J.L., Green, K.D., Chen, W. & Garneau-Tsodikova, S. (2010). The future of aminoglycosides: the end or renaissance?, *Chembiochem* 11(7): 880-902.

Houzet, L., Paillart, J.C., Smagulova, F., Maurel, S., Morichaud, Z., Marquet, R. & Mougel, M. (2007). HIV controls the selective packaging of genomic, spliced viral and cellular RNAs into virions through different mechanisms, *Nucleic Acids Res* 35(8): 2695-704.

Hu, W.S. & Temin, H.M. (1990). Genetic consequences of packaging two RNA genomes in one retroviral particle: pseudodiploidy and high rate of genetic recombination, *Proc Natl Acad Sci U S A* 87(4): 1556-60.

Huber, H.E. & Richardson, C.C. (1990). Processing of the primer for plus strand DNA synthesis by human immunodeficiency virus 1 reverse transcriptase, *J Biol Chem* 265(18): 10565-73.

Isel, C. & Karn, J. (1999). Direct evidence that HIV-1 Tat stimulates RNA polymerase II carboxyl-terminal domain hyperphosphorylation during transcriptional elongation, *J Mol Biol* 290(5): 929-41.

Jacks, T., Power, M.D., Masiarz, F.R., Luciw, P.A., Barr, P.J. & Varmus, H.E. (1988). Characterization of ribosomal frameshifting in HIV-1 gag-pol expression, *Nature* 331(6153): 280-3.

Jang, M.K., Mochizuki, K., Zhou, M., Jeong, H.S., Brady, J.N. & Ozato, K. (2005). The bromodomain protein Brd4 is a positive regulatory component of P-TEFb and stimulates RNA polymerase II-dependent transcription, *Mol Cell* 19(4): 523-34.

Jiang, M., Mak, J., Ladha, A., Cohen, E., Klein, M., Rovinski, B. & Kleiman, L. (1993). Identification of tRNAs incorporated into wild-type and mutant human immunodeficiency virus type 1, *J Virol* 67(6): 3246-53.

Jones, K.L., Sonza, S. & Mak, J. (2008). Primary T-lymphocytes rescue the replication of HIV-1 DIS RNA mutants in part by facilitating reverse transcription, *Nucleic Acids Res* 36(5): 1578-88.

Kieken, F., Paquet, F., Brule, F., Paoletti, J. & Lancelot, G. (2006). A new NMR solution structure of the SL1 HIV-1Lai loop-loop dimer, *Nucleic Acids Res* 34(1): 343-52.

King, S.R., Duggal, N.K., Ndongmo, C.B., Pacut, C. & Telesnitsky, A. (2008). Pseudodiploid genome organization aids full-length human immunodeficiency virus type 1 DNA synthesis, *J Virol* 82(5): 2376-84.

Kingsman, S.M. & Kingsman, A.J. (1996). The regulation of human immunodeficiency virus type-1 gene expression, *Eur J Biochem* 240(3): 491-507.

Laughrea, M. & Jette, L. (1994). A 19-nucleotide sequence upstream of the 5' major splice donor is part of the dimerization domain of human immunodeficiency virus 1 genomic RNA, *Biochemistry* 33(45): 13464-74.

Laughrea, M. & Jette, L. (1996). HIV-1 genome dimerization: formation kinetics and thermal stability of dimeric HIV-1Lai RNAs are not improved by the 1-232 and 296-790 regions flanking the kissing-loop domain, *Biochemistry* 35(29): 9366-74.

Laughrea, M., Jette, L., Mak, J., Kleiman, L., Liang, C. & Wainberg, M.A. (1997). Mutations in the kissing-loop hairpin of human immunodeficiency virus type 1 reduce viral infectivity as well as genomic RNA packaging and dimerization, *J Virol* 71(5): 3397-406.

Leitner, T., Korber, B., Daniels, M., Calef, C. & Foley, B. (2005). HIV-1 Subtype and Circulating Recombinant Form (CRF) Reference Sequences, 2005, In: *HIV Sequence Compendium 2005*, Leitner, T., Foley, B., Hahn, B., Marx, P., McCutchan, F., Mellors, J.W., Wolinsky, S. & Korber, B. (Eds.), pp. 41-8. Theoretical Biology and Biophysics Group, Los Alamos National Laboratory, Los Alamos, NM.

Li, X., Krishnan, L., Cherepanov, P. & Engelman, A. (2011). Structural biology of retroviral DNA integration, *Virology* 411(2): 194-205.

Liang, C., Rong, L., Laughrea, M., Kleiman, L. & Wainberg, M.A. (1998). Compensatory point mutations in the human immunodeficiency virus type 1 Gag region that are distal from deletion mutations in the dimerization initiation site can restore viral replication, *J Virol* 72(8): 6629-36.

Liang, C., Rong, L., Quan, Y., Laughrea, M., Kleiman, L. & Wainberg, M.A. (1999). Mutations within four distinct gag proteins are required to restore replication of human immunodeficiency virus type 1 after deletion mutagenesis within the dimerization initiation site, *J Virol* 73(8): 7014-20.

Liao, C., Marchand, C., Burke, T.R., Jr., Pommier, Y. & Nicklaus, M.C. (2010). Authentic HIV-1 integrase inhibitors, *Future Med Chem* 2(7): 1107-22.

Mak, J., Jiang, M., Wainberg, M.A., Hammarskjold, M.L., Rekosh, D. & Kleiman, L. (1994). Role of Pr160gag-pol in mediating the selective incorporation of tRNA(Lys) into human immunodeficiency virus type 1 particles, *J Virol* 68(4): 2065-72.

Marshall, N.F. & Price, D.H. (1992). Control of formation of two distinct classes of RNA polymerase II elongation complexes, *Mol Cell Biol* 12(5): 2078-90.

McBride, M.S. & Panganiban, A.T. (1996). The human immunodeficiency virus type 1 encapsidation site is a multipartite RNA element composed of functional hairpin structures, *J Virol* 70(5): 2963-73.

McBride, M.S. & Panganiban, A.T. (1997). Position dependence of functional hairpins important for human immunodeficiency virus type 1 RNA encapsidation in vivo, *J Virol* 71(3): 2050-8.

McCutchan, F.E. (2006). Global epidemiology of HIV, *J Med Virol* 78 Suppl 1: S7-S12.

Moore, M.D., Fu, W., Nikolaitchik, O., Chen, J., Ptak, R.G. & Hu, W.S. (2007). Dimer initiation signal of human immunodeficiency virus type 1: its role in partner selection during RNA copackaging and its effects on recombination, *J Virol* 81(8): 4002-11.

Moore, M.D. & Hu, W.S. (2009). HIV-1 RNA dimerization: It takes two to tango, *AIDS Rev* 11(2): 91-102.

Moore, M.D., Nikolaitchik, O.A., Chen, J., Hammarskjold, M.L., Rekosh, D. & Hu, W.S. (2009). Probing the HIV-1 genomic RNA trafficking pathway and dimerization by genetic recombination and single virion analyses, *PLoS Pathog* 5(10): e1000627.

Moumen, A., Polomack, L., Roques, B., Buc, H. & Negroni, M. (2001). The HIV-1 repeated sequence R as a robust hot-spot for copy-choice recombination, *Nucleic Acids Res* 29(18): 3814-21.

Muriaux, D., De Rocquigny, H., Roques, B.P. & Paoletti, J. (1996a). NCp7 activates HIV-1Lai RNA dimerization by converting a transient loop-loop complex into a stable dimer, *J Biol Chem* 271(52): 33686-92.

Muriaux, D., Fosse, P. & Paoletti, J. (1996b). A kissing complex together with a stable dimer is involved in the HIV-1Lai RNA dimerization process in vitro, *Biochemistry* 35(15): 5075-82.

Muriaux, D., Girard, P.M., Bonnet-Mathoniere, B. & Paoletti, J. (1995). Dimerization of HIV-1Lai RNA at low ionic strength. An autocomplementary sequence in the 5' leader region is evidenced by an antisense oligonucleotide, *J Biol Chem* 270(14): 8209-16.

Neville, M., Stutz, F., Lee, L., Davis, L.I. & Rosbash, M. (1997). The importin-beta family member Crm1p bridges the interaction between Rev and the nuclear pore complex during nuclear export, *Curr Biol* 7(10): 767-75.

Olsen, H.S., Cochrane, A.W., Dillon, P.J., Nalin, C.M. & Rosen, C.A. (1990). Interaction of the human immunodeficiency virus type 1 Rev protein with a structured region in env mRNA is dependent on multimer formation mediated through a basic stretch of amino acids, *Genes Dev* 4(8): 1357-64.

Paillart, J.C., Berthoux, L., Ottmann, M., Darlix, J.L., Marquet, R., Ehresmann, B. & Ehresmann, C. (1996a). A dual role of the putative RNA dimerization initiation site of human immunodeficiency virus type 1 in genomic RNA packaging and proviral DNA synthesis, *J Virol* 70(12): 8348-54.

Paillart, J.C., Skripkin, E., Ehresmann, B., Ehresmann, C. & Marquet, R. (1996b). A loop-loop "kissing" complex is the essential part of the dimer linkage of genomic HIV-1 RNA, *Proc Natl Acad Sci U S A* 93(11): 5572-7.

Parada, C.A. & Roeder, R.G. (1996). Enhanced processivity of RNA polymerase II triggered by Tat-induced phosphorylation of its carboxy-terminal domain, *Nature* 384(6607): 375-8.

Parkin, N.T., Chamorro, M. & Varmus, H.E. (1992). Human immunodeficiency virus type 1 gag-pol frameshifting is dependent on downstream mRNA secondary structure: demonstration by expression in vivo, *J Virol* 66(8): 5147-51.

Perno, C.F., Moyle, G., Tsoukas, C., Ratanasuwan, W., Gatell, J. & Schechter, M. (2008). Overcoming resistance to existing therapies in HIV-infected patients: the role of new antiretroviral drugs, *J Med Virol* 80(4): 565-76.

Reyes-Darias, J.A., Sanchez-Luque, F.J. & Berzal-Herranz, A. (2008). Inhibition of HIV-1 replication by RNA-based strategies, *Curr HIV Res* 6(6): 500-14.

Rhodes, T.D., Nikolaitchik, O., Chen, J., Powell, D. & Hu, W.S. (2005). Genetic recombination of human immunodeficiency virus type 1 in one round of viral replication: effects of genetic distance, target cells, accessory genes, and lack of high negative interference in crossover events, *J Virol* 79(3): 1666-77.

Ristic, N. & Chin, M.P. (2010). Mutations in matrix and SP1 repair the packaging specificity of a Human Immunodeficiency Virus Type 1 mutant by reducing the association of Gag with spliced viral RNA, *Retrovirology* 7: 73.

Rossi, J.J., June, C.H. & Kohn, D.B. (2007). Genetic therapies against HIV, *Nat Biotechnol* 25(12): 1444-54.

Russell, R.S., Roldan, A., Detorio, M., Hu, J., Wainberg, M.A. & Liang, C. (2003). Effects of a single amino acid substitution within the p2 region of human immunodeficiency virus type 1 on packaging of spliced viral RNA, *J Virol* 77(24): 12986-95.

Sakaguchi, K., Zambrano, N., Baldwin, E.T., Shapiro, B.A., Erickson, J.W., Omichinski, J.G., Clore, G.M., Gronenborn, A.M. & Appella, E. (1993). Identification of a binding site for the human immunodeficiency virus type 1 nucleocapsid protein, *Proc Natl Acad Sci U S A* 90(11): 5219-23.

Shankarappa, R., Chatterjee, R., Learn, G.H., Neogi, D., Ding, M., Roy, P., Ghosh, A., Kingsley, L., Harrison, L., Mullins, J.I. & Gupta, P. (2001). Human immunodeficiency virus type 1 env sequences from Calcutta in eastern India: identification of features that distinguish subtype C sequences in India from other subtype C sequences, *J Virol* 75(21): 10479-87.

Shen, N., Jette, L., Liang, C., Wainberg, M.A. & Laughrea, M. (2000). Impact of human immunodeficiency virus type 1 RNA dimerization on viral infectivity and of stem-loop B on RNA dimerization and reverse transcription and dissociation of dimerization from packaging, *J Virol* 74(12): 5729-35.

Shubsda, M.F., Paoletti, A.C., Hudson, B.S. & Borer, P.N. (2002). Affinities of packaging domain loops in HIV-1 RNA for the nucleocapsid protein, *Biochemistry* 41(16): 5276-82.

Skripkin, E., Paillart, J.C., Marquet, R., Ehresmann, B. & Ehresmann, C. (1994). Identification of the primary site of the human immunodeficiency virus type 1 RNA dimerization in vitro, *Proc Natl Acad Sci U S A* 91(11): 4945-9.

Song, R., Kafaie, J., Yang, L. & Laughrea, M. (2007). HIV-1 viral RNA is selected in the form of monomers that dimerize in a three-step protease-dependent process; the DIS of stem-loop 1 initiates viral RNA dimerization, *J Mol Biol* 371(4): 1084-98.

Sugiyama, R., Habu, Y., Ohnari, A., Miyano-Kurosaki, N. & Takaku, H. (2009). RNA interference targeted to the conserved dimerization initiation site (DIS) of HIV-1 restricts virus escape mutation, *J Biochem* 146(4): 481-9.

Takebe, E.Y., Kusagawa, S. & Motomura, K. (2004). Molecular epidemiology of HIV: tracking AIDS pandemic, *Pediatr Int* 46(2): 236-44.

Trkola, A., Dragic, T., Arthos, J., Binley, J.M., Olson, W.C., Allaway, G.P., Cheng-Mayer, C., Robinson, J., Maddon, P.J. & Moore, J.P. (1996). CD4-dependent, antibody-sensitive interactions between HIV-1 and its co-receptor CCR-5, *Nature* 384(6605): 184-7.

Watts, J.M., Dang, K.K., Gorelick, R.J., Leonard, C.W., Bess, J.W., Jr., Swanstrom, R., Burch, C.L. & Weeks, K.M. (2009). Architecture and secondary structure of an entire HIV-1 RNA genome, *Nature* 460(7256): 711-6.

Wilkinson, K.A., Gorelick, R.J., Vasa, S.M., Guex, N., Rein, A., Mathews, D.H., Giddings, M.C. & Weeks, K.M. (2008). High-throughput SHAPE analysis reveals structures in HIV-1 genomic RNA strongly conserved across distinct biological states, *PLoS Biol* 6(4): e96.

Yoon, V., Fridkis-Hareli, M., Munisamy, S., Lee, J., Anastasiades, D. & Stevceva, L. (2010). The GP120 molecule of HIV-1 and its interaction with T cells, *Curr Med Chem* 17(8): 741-9.

Zapp, M.L. & Green, M.R. (1989). Sequence-specific RNA binding by the HIV-1 Rev protein, *Nature* 342(6250): 714-6.

Zapp, M.L., Hope, T.J., Parslow, T.G. & Green, M.R. (1991). Oligomerization and RNA binding domains of the type 1 human immunodeficiency virus Rev protein: a dual function for an arginine-rich binding motif, *Proc Natl Acad Sci U S A* 88(17): 7734-8.

Zennou, V., Petit, C., Guetard, D., Nerhbass, U., Montagnier, L. & Charneau, P. (2000). HIV-1 genome nuclear import is mediated by a central DNA flap, *Cell* 101(2): 173-85.

6

Ribosomal Frameshift Signals in Viral Genomes

Ewan P. Plant
Food and Drug Administration
USA

1. Introduction

Viral genomes contain fewer genes and are more compact than eukaryotic or prokaryotic genomes. Overlapping reading frames are often present in viral genomes and this enables multiple proteins to be made from a single transcript. This genomic layout can serve different purposes: more efficient use of the genome; fewer resources needed for making transcripts; and the abundance of each protein can be optimized by regulating translation from one transcript. Programmed ribosomal frameshifting is a mechanism by which viruses can express overlapping reading frames.

Viruses conscript the host translation machinery for the production of viral proteins. Many viruses disrupt transcription and translation of host mRNAs and optimize translation of viral messages. Different methods are used to subvert the fidelity of the translation machinery so that specific viral proteins are produced. These include leaky scanning, reinitiation, suppression of termination, internal ribosome entry sites and programmed ribosomal frameshifting. Normally ribosomes decode the codon triplets with high fidelity. At each step the mRNA passes through the ribosome three nucleotides at a time, from the initiating codon, along the length of the open reading frame, to the termination codon. However, some viruses manipulate the ribosome so that the mRNA does not proceed forward three nucleotides. This disruption is referred to as programmed ribosomal frameshifting.

Programmed ribosomal frameshifting allows for the production of two proteins with the same amino terminus yet differing carboxyl termini. The same initiation codon is used, but at a defined position in the message, the ribosome is stimulated such that it sometimes changes reading frame and continues translating in a new reading frame. The signal that stimulates this change in reading frame is encoded within the message, often comprised of a seven nucleotide slippery sequence followed by a stimulatory element. Here I describe viral frameshift signals with a particular emphasis on -1 frameshift signals. I review the discovery of these signals; provide an update on how the analysis of frameshift signals has broadened our understanding of virus protein translation, virus replication and ribosome fidelity; and describe how frameshift signals can be targeted by antiviral compounds.

2. Ribosome frameshift signals are found in diverse viral genomes

Frameshift signals have been found in the genomes of several double-stranded RNA and plus-strand RNA viruses. Although frameshift signals may be found in one member of a virus family, they are not necessarily ubiquitous among that family. Most of the frameshift signals are positioned upstream of the open reading frame (ORF) that encodes an RNA-dependent RNA polymerase (RdRP), but signals have been identified that direct ribosomes to translate proteins involved in other functions such as cell-to-cell movement. The sequences that comprise the frameshift signals in viral genomes are diverse and form tertiary structures ranging from simple stem-loops to complex pseudoknots. This section summarizes the discovery, distribution and function of frameshift signals in viral genomes.

Proteins that had the same amino sequence but differed at the carboxyl terminus were first identified by tryptic digests in the 1970's. This led to the development of new ideas about how proteins are translated including programmed ribosomal frameshifting. Frameshifting was first demonstrated in cells infected with Rous sarcoma virus. Analysis of other virus sequences resulted in the rapid identification of more frameshift signals in retroviruses and coronaviruses containing overlapping ORFs. Initially proteins were sequenced to demonstrate that frameshifting occurred. Quickly molecular tools were developed that allowed DNA sequences to be cloned, transcribed and translated so that the ratios of frameshifted and non-frameshifted products could be measured. These methods have continued to evolve and protein expression systems are currently the main method of confirming frameshift signal function. Although recent advances in mass spectrometry have brought protein sequence identification to the fore again.

Over the last two decades considerably more viral sequences have become available for analysis. Many putative frameshift signals have been identified because of homology to an existing virus with a well-described frameshift signal. This approach isn't straightforward because of the diversity of sequences and secondary structures that stimulate frameshifting. For example, there are three very different structures that have been shown to facilitate frameshifting in coronavirues (see Figure 1). Computational efforts have been extended to find new frameshift signals. Two general approaches have been pursued; one looking for potential frameshift signals, that is slippery sites followed by potential stimulatory elements. The second approach is to look for out of frame ORFs and then potential frameshift signals (Belew et al., 2008; Bekaert et al., 2010). Both approaches have resulted in the generation of databases containing numerous potential frameshift signals. Some of these frameshift signals have been validated by functional assays or mass spectrometry analysis. Additional virological studies have helped unravel the function of some of the frameshift proteins although much work remains to be done.

Frameshift signals have been identified in several viral families including *Astroviridae*, *Flaviviridae, Luteoviridae, Potyviridae, Retroviridae, Togaviridae, Tombusviridae, Totiviridae*, the three families in the Order Nidovirales and a *Paramyxovirus*. Frameshift signals have also been used to support the assignment of viruses to particular families (den Boon et al., 1991; Cowley et al., 2000; Snijder et al., 1990 for example). In addition, frameshift signals have been characterized in several retrotransposons. Recently programmed ribosomal frameshift (PRF) signals were characterized in several invertebrates (Baranov et al., 2011). Currently there is only one characterized example of a frameshift signal in a mammalian genome but it is likely that more will be discovered and described as interest in recoding signals spreads (Manktelow et al., 2005). The proteins produced via frameshifting have varied functions. For example: 1) a frameshift product has a role in the neuro-invasiveness of the Kunjin subtype of West Nile virus (Melian et al., 2010); 2) the ORF3 protein from the Southern Bean Mosaic virus is required for cell-to-cell movement (Sivakumaran et al., 1998) and 3); altering frameshifting efficiency has been shown to detrimentally affect viral viability for both double-stranded and single-stranded RNA viruses, including the retrovirus HIV (discussed below).

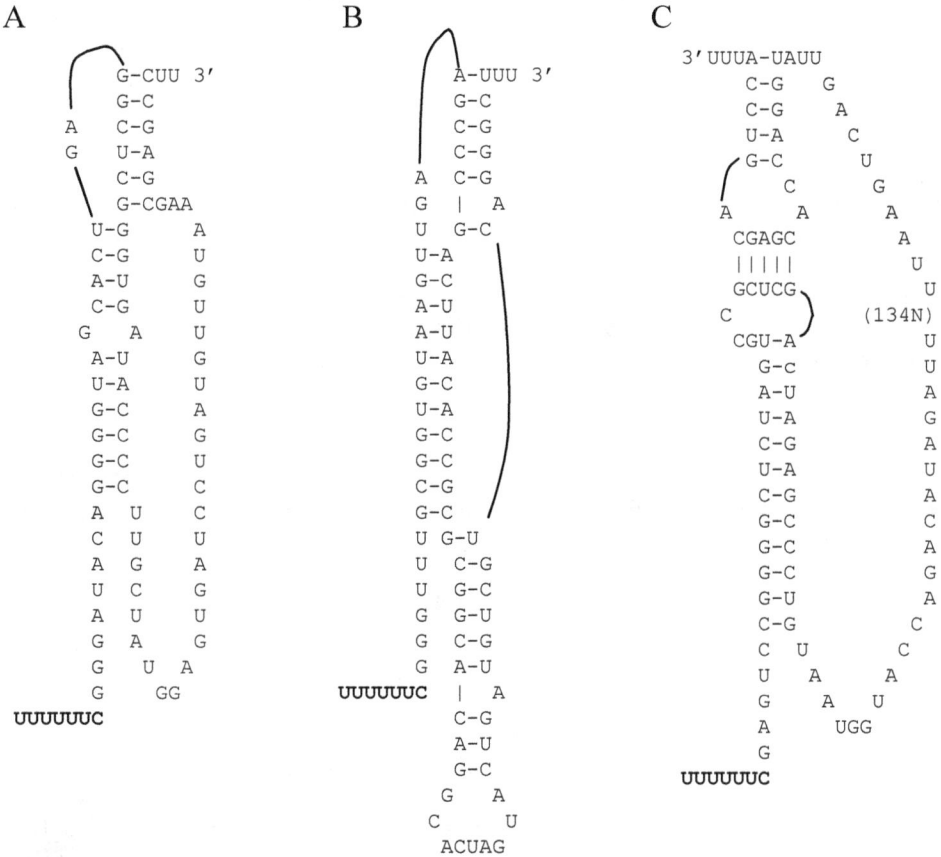

```
A                          B                          C

                                                       3' UUUA-UAUU
        G-CUU 3'                   A-UUU 3'                  C-G   G
        G-C                        G-C                       C-G   A
    A   C-G                        C-G                       U-A     C
    G   U-A                        C-G                      G-C       U
        C-G                   A    C-G                         C       G
        G-CGAA                G  |   A                    A    A       A
      U-G    A                U  G-C                       CGAGC       A
      C-G    U                U-A                          |||||       U
      A-U    G                G-C                          GCUCG       U
      C-G    U                A-U                      C        )    (134N)
      G    A  U               A-U                          CGU-A       U
      A-U    G                U-A                           G-c        U
      U-A    U                G-C                           A-U        A
      G-C    A                U-A                           U-A        G
      G-C    G                G-C                           C-G        A
      G-C    U                G-C                           U-A        U
      G-C    C                C-G                           C-G        A
      A  U   C                G-C                           G-C        C
      C  U   U                U  G-U                        G-C        A
      A  G   A                U  C-G                        G-C        G
      U  C   G                U  G-C                        G-U        A
      A  U   U                G  G-U                        C-G        C
      G  A   G                G  C-G                        C   U        C
      G   U A                 G  A-U                        U     A    A
      G    GG                 G  |   A                      G     A  U
   UUUUUUC             UUUUUUC   |   A                      A        UGG
                                 C-G                        G
                                 A-U                     UUUUUUC
                                 G-C
                                 G   A
                                 C     U
                                 ACUAG
```

Fig. 1. Diverse stimulatory structures from coronaviruses. Shown are diagrams of the sequence and base-pairing of three different coronavirus stimulatory elements. A) the avian infectious bronchitis virus two-stemmed pseudoknot (Brierley et al., 1989), B) the severe acute respiratory syndrome three-stemmed pseudoknot (Plant et al., 2005), and C) the kissing loops of the human coronavirus 229E stimulatory element (Herold & Siddell 1993). The heptameric slippery sites are shown in bold.

2.1 Frameshift signals in the *Totiviridae* family

The *Totiviridae* are double-stranded RNA viruses infecting arthropod, fungal and protozoan hosts. There are three genera; Totivirus, Giardiavirus and Leishmaniavirus. One of the early frameshift signals detected and well characterized was that of the *Saccharomyces cerevisiae* L-A totivirus (Dinman et al., 1991). It uses frameshifting to control the ratio of structural capsid (gag) protein to enzymatic polymerase (pol) protein. It has been shown that either increasing or decreasing frameshifting efficiency in the L-A genome alters the gag to gag-pol ratio and disrupts propagation of a satellite RNA (Dinman & Wickner, 1992). The yeast cells harboring the L-A virus are easily manipulated. This allowed many pivotal frameshifting studies to be performed in the Dinman laboratory. The ability to make specific mutants in yeast has permitted the analysis of many different genes that affect frameshifting and virus propagation. Additionally yeast cells are susceptible to many of the drugs that affect higher eukaryotes and this has facilitated the design of assay systems for the investigation of drugs that affect frameshifting (Rakauskaite et al., 2011). The insights from studying the L-A frameshift signal has helped other groups home in on specific regulatory genes, or drugs, that affect other viruses in cell culture systems using mammalian cells.

The Giardiavirus GLV has been shown to have a functional frameshift signal that, like L-A, modulates the gag to gag-pol ratio (Li et al., 2001a). A pseudoknot has been shown to stimulate frameshifting in the L-A virus (Dinman et al., 1991) while a stem-loop structure is used in the GLV virus (Li et al., 2001a). More recent analyses of *Totiviridae* genomes have identified a frameshift signal in the penaeid shrimp infectious myonecrosis virus (Nibert, 2007), Armigeres subalbatus totivirus (Zhai et al., 2010), Omono River virus (Isawa et al., 2011), helminthosporium victoriae virus and trichomonas vaginalis viruses II and 3 (Bekaert & Rousset, 2005). A +1 recoding mechanism for the Leishmania virus LRV1-4 has been reported (Kim et al., 2005). This frameshift also regulates the ratio of the putative polymerase protein to the structural protein. These findings suggest that a common viral problem, that of regulating the abundance of structural and enzymatic viral proteins, is solved in the *Totiviridae* family by manipulating translation fidelity, albeit by slightly different mechanisms.

2.2 Frameshift signals in the *Retroviridae* family

The *Retroviridae* are RNA viruses that produce a DNA copy of their genome that is integrated into the host cell genome. Like the *Totiviridae*, retroviruses use frameshift signals to modulate the ratio of gag protein to gag-pol fusion protein. The ratio of gag to pol proteins has been shown to be important for viral replication and infectivity in several retroviral systems. Altering the ratio has been shown to adversely affect replication, RNA dimerization and particle formation (Biswas et al., 2004; Chen and Montelaro, 2003; Gendron et al., 2005; Hung et al. 1998; Karacostas et al., 1993; Shehu-Xhilaga et al., 2001).

In 1985 it was noted that the 5' end of the polymerase ORF for the Deltaretrovirus bovine leukemia virus overlapped with the upstream ORF for the gag ORF (Rice et al., 1985). The authors postulated that a recoding event was responsible for the production of the polymerase. The first characterization of a viral frameshift signal was of the Alpharetrovirus Rous sarcoma virus (Jacks & Varmus, 1985). The region of the genome containing the frameshift signal was cloned into an expression vector, transcribed and translated. ^{35}S-labeled products were immunoprecipitated with antiserum against either the gag protein or the polymerase. The result clearly showed that a gag-pol polyprotein was produced from the same transcript as the gag protein (Jacks & Varmus, 1985). Frameshift signals have since been identified in Alpharetroviruses, Betaretroviruses, Deltaretroviruses and Lentiviruses.

Soon after Jacks and Varmus described the first frameshift signal, signals in the Betaretrovirus mouse mammary tumor virus (MMTV) and the Lentivirus human immunodeficiency virus (HIV-1) were characterized by the same group (Jacks et al., 1987, 1988a). The frameshift signal in the MMTV genome is located in the 5' end of the gag ORF, and when frameshifting occurs a protease encoded in the overlapping pro ORF is produced. The order of the amino acid motifs that identify the enzyme domains encoded in the frameshift-regulated ORF varies between different retroelements. So, although some commentaries indicate that the frameshift-regulated ORF encodes a protease, the ORF usually codes other enzymatic proteins, including the polymerase, that are produced as a polyprotein which is then cleaved by a protease. The frameshift signal from the Mason-Pfizer monkey Betaretrovirus, (simian retrovirus 1), has also been characterized (ten Dam et al., 1994).

Frameshift signals from the Deltaviruses human T-cell leukemia virus types I and II, have been characterized (Kollmus et al., 1994; Nam et al., 1993) and the putative signals in the primate T-lymphocyte virus 3 and simian T-lymphocyte viruses 1 and 2 have been described (Bekaert & Rousset, 2005; Bekaert et al., 2010). Additional Lentivirus frameshift signals from equine infectious anemia virus, feline immunodeficiency virus and simian immunodeficiency virus have been characterized (Bekaert & Rousset, 2005; Chen & Montelaro, 2003; Morikawa & Bishop, 1992). The HIV-2 genome layout suggests that, like HIV-1, there is a frameshift signal between the gag and pol ORFs. Putative signals in bovine immunodeficiency virus, caprine arthritis virus, ovine lentivirus, Jembrana disease virus and Visna virus have been described (Bekaert & Rousset, 2005; Bekaert et al., 2010). Additional frameshift signals in an Alpharetrovirus genome (avian leukosis virus), and a Deltavirus genome (bovine leukemia virus) have also been identified (Bekaert & Rousset, 2005; Bekaert et al., 2010). A putative signal from a Betaretrovirus, the cancer causing Jaagsiekte sheep retrovirus, has also been been described (Bekaert et al., 2010).

In several retrovirus genomes, the gag and pol genes are separated by a stop codon. It has been shown that readthrough of the stop codon can produce a fusion protein. A pseudoknot facilitates the readthrough of murine leukemia virus and pseudoknots are predicted to be present at the gag-pol junction of several retroviruses (Wills et al., 1994). Interestingly, two competing structures, a stem-loop structure and a pseudoknot, were described for the murine leukemia virus (Alam et al., 1999). Both pseudoknots and stem-loops have been proposed to stimulate frameshifting in HIV. This suggests that secondary structures are important for modulating ribosome fidelity during retroviral protein production in both readthrough and frameshifting, although exactly how they stimulate the ribosome remains a mystery.

2.3 Frameshift signals in the order Nidovirales

The order Nidovirales includes *Coronaviridae*, *Arteriviridae* and *Okavirus*, all of which are positive-stranded RNA viruses that infect animals. They have large single-strand genomes and express structural proteins from subgenomic RNAs transcribed from the 3' region of the genome. The nonstructural proteins are expressed from the genomic RNA. The coronavirus replicase genes are encoded in the 5' portion of the genome and a frameshift event is required for the production of several proteins including the RNA-dependent RNA polymerase (RdRP). The replicase proteins are translated as two polyproteins that are processed by self-encoded proteases. The smaller polyprotein has several domains including papainlike cysteine protease, chymotrypsin-like cysteine protease, metal binding and transmembrane motifs. The larger polyprotein, resulting from a frameshift event contains the domains of the smaller polyprotein and the RdRP, helicase, 3'-to-5' exonuclease and S-adenosylmethionine-dependent ribose 2'-O-methyltransferase domains (Ziebuhr, 2005). The complexities of coronaviral replication are still being unravelled but it is apparent that frameshifting is essential for the production of the proteins involved. Altering frameshifting efficiency has been shown to be detrimental for coronavirus replication (Ahn et al., 2011; McDonagh et al., 2011; Plant et al., 2010).

The first frameshift signal identified in a coronavirus was by Brierley et al. (1987). The infectious bronchitis virus (IBV) frameshift signal is now perhaps one of the most well characterized frameshift signals along with the HIV frameshift signal. However, although several luteovirus frameshift signals have been crystalized the size of the pseudoknot that promotes frameshifting in IBV has thwarted attempts at crystalization. Additionally there is an example of a kissing stem-loop that promotes coronavirus frameshifting (Herold and Siddell, 1993). A lot of nuclease mapping and NMR data is available for some coronavirus signals but this does not reflect the diversity of nidovirus frameshift signals (Dos Ramos et al., 2004; Napthine et al., 1999; Plant et al., 2005, 2010; Su et al., 2005).

Functional frameshift signals have been identified in some Arteriviridae including equine arteritis virus, lactate dehydrogenase-elevating virus and porcine reproductive and respiratory syndrome virus (Bekaert & Rousset, 2005; den Boon et al., 1991). A frameshift signal has been identified in the simian hemorrhagic fever virus (Bekaert & Rousset, 2005). A frameshift signal in the Gill-associated Okavirus has been shown to be functional (Cowley et al., 2000). Putative frameshift signals have been identified in several coronaviruses including bovine coronavirus, porcine epidemic diarrhia virus, transmissible gastroenteritis virus and human coronavirus OC43 (Bekaert & Rousset, 2005). The functionality of the IBV, human coronavirus 229E, Berne virus, murine hepatitis virus and SARS coronavirus frameshift signals have been demonstrated (Baranov et al., 2005; Brierley et al., 1987, 1989, 1991, 1992; Dos Ramos et al., 2004; Herold & Siddell, 1993; Plant et al., 2005, 2010; Su et al., 2005).

2.4 Frameshift signals in other positive-stranded RNA viruses

The *Astroviridae* are non-enveloped positive sense single-stranded RNA viruses. The genome has three ORFs, with the first two overlapping. A frameshift signal that separates the protease and the RdRP was identified in the Human astrovirus and subsequently shown to be functional (Jiang et al., 1993; Marczinke et al., 1994). More recently frameshift signals have been identified in the Chicken, Mink, Ovine and Turkey astroviruses (Bekaert & Rousset, 2005; Bekaert et al., 2010).

The *Flaviviridae* are positive sense single-stranded RNA viruses with a genome of approximately 10kb. Recently a frameshift signal was identified and characterized in the West Nile virus. Interestingly the frameshift is within an ORF encoding the non-structural protein NS1. The NS1′ frameshift protein plays a role in the neuro-invasiveness of the Kunjin subtype of West Nile virus (Melian et al., 2010). Potential frameshift signals have been identified in three other flaviviruses; Japanese encephalitis virus, Murray Valley encephalitis virus and Usutu virus (Firth & Atkins, 2009).

The *Luteoviridae* are positive sense single-stranded RNA viruses that infect plants. The RdRP ORF is downstream from a coat protein (CP) ORF in plant luteoviruses. The expression of the RdRP is regulated by readthrough of a stop codon or by -1 frameshifting. Production of a CP-RdRP fusion protein is required for aphid transmission (Demler & de Zoeten, 1991; Di et al., 1993). Frameshift signals have been found in Enamoviruses, Luteoviruses and Poleroviruses.

A well described example of a Luteoviral signal is upstream from the Barley yellow dwarf virus, PAV serotype polymerase (Di et al., 1993). However, the BYDV frameshift signal differs from other frameshift signals in that a sequence four kilobases downstream from the slippery site is required to stimulate frameshifting (Paul et al., 2001). A similar genome arrangement and frameshift signal has been found for the rose spring dwarf-associated virus (Salem et al., 2008). Interestingly, readthrough signals have been described for the BYDV PAV serotype coat protein which also requires an interaction with a distal sequence (Brown et al., 1996). Other luteoviruses with putative frameshift signals include the Bean leafroll virus and Soybean dwarf virus (Domier et al., 2002; Bekaert et al., 2010).

A frameshift signal has been identified in the Enamovirus pea enation mosaic virus (Demler & de Zoeten, 1991). Like the BYDV virus serotypes there are pea enation mosaic viruses that have been described that have a putative readthrough mechanism (Harrell et al., 2002). The structure of the pseudoknot from the pea enation mosaic virus-1 frameshift signal has been characterized (Giedroc et al., 2003; Nixon et al., 2002).

A number of frameshifting sequences have been identified in Polerovirus genomes including beet mild yellowing virus, cereal yellow dwarf viruses RPS and RPV and turnip yellows virus (Bekaert et al., 2010). The frameshift signals for beet chlorosis virus, beet western yellows virus, cucurbit aphid-borne yellows virus, potato leafroll virus have all been shown to be functional (Bekaert & Rousset, 2005; Kim et al., 2000; Kujawa et al., 1993; Prüfer et al., 1992) and the structure of several stimulatory elements have been elucidated (Cornish et al., 2005; Pallan et al., 2005; Su et al., 1999).

The *Togaviridae* are positive sense single-stranded RNA viruses that include rubella virus and the alphaviruses. The alphaviruses are transmitted by arthropods and many cause encephalitis. The discovery of a functional frameshift signal in the Semliki Forest virus 6K gene is expected to have wide ranging effects on the understanding of alphavirus lifecycle (Firth et al., 2008). The 6K protein is involved in envelope processing, membrane permeabilization, virus budding and virus assembly. Original observations of a 6K protein doublet are now in doubt as Firth et al., (2008) have confirmed the presence of the frameshifted product using amino acid sequencing. The additional protein is refered to as the transframe, or TF, protein. Mutant viruses lacking the ability to frameshift showed reduced growth. Sequence comparisons indicated that frameshift signals are present in other alphaviruses including Seal louse virus, Middleburg virus, Venezuelan equine encephalitis virus, Ndumu virus, Sinbis virus, Barmah Forest virus, Sleeping disease virus and Eastern equine encephalitis virus. The functionality of these frameshift signals has been confirmed (Chung et al., 2010).

The *Tombusviridae* are positive sense single-stranded RNA plant viruses. Some of these viruses are transmitted by fungal species, but either the virion or genetic material are infective. Putative frameshift signals have been identified in Dianthovirus RNA 1-like RNA, Pelargonium line pattern virus, carnation ringspot virus 1, sweet clover necrotic mosaic virus RNA-1, subterranean clover mottle virus, turnip rosette virus, carrot mottle mimic virus, groundnut rosette virus, pea enation mosaic virus-2 and tobacco bushy top virus (Bekaert & Rousset, 2005; Castano & Hernandez, 2005; Ge et al., 1993; Miranda et al., 2001). Frameshifting has been demonstrated from the cocksfoot mottle virus and red clover necrotic mosaic virus RNA-1 (RCNMV) signals (Kim & Lommel, 1994; Tamm et al., 2009; Xiong et al., 1993). It has been shown that the RdRP, which is produced via the frameshifting mechanism, is supplied in a cis-preferential manner for the synthesis of negative-strand

RCNMV RNA (Okamoto et al., 2008). Thus there appears to be some link between control of translation and replication of the virus. Like many other positive-strand RNA viruses the replication process results in a surplus of positive-strand RNA.

3. Programmed ribosomal frameshift signals

Programmed Ribosomal Frameshift (PRF) signals are sequences within a messager RNA (mRNA) that stimulate a portion of translating ribosomes to change reading frames. PRF signals are typically comprised of two features, a heptameric slippery site and a stimulatory element. The slippery site is a series of seven nucleotides in the mRNA from which the tRNAs in the translating ribosome can un-pair from the zero frame and re-pair to in the -1 frame. The stimulatory element has a dual function, it causes the translating ribosome to pause on the message when the slippery site is positioned within the ribosome and it stimulates ribosomal error.

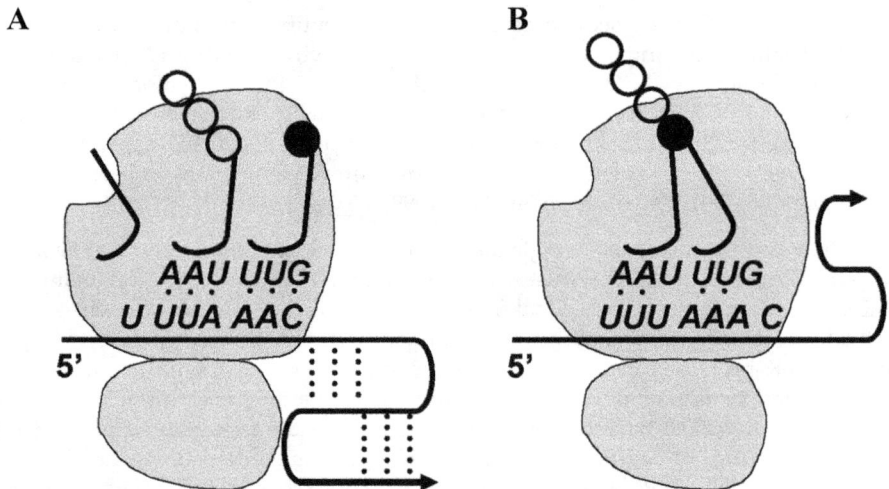

Fig. 2. Simultaneous slippage model of programmed ribosomal frameshifting. A) Cartoon of two tRNAs paired with the mRNA in the zero (AUG-initiated) reading frame before frameshifting occurs. B) Cartoon with the two tRNAs each paired to the mRNA at two of the three codon positions after frameshifting has occured. The 5′ end of the mRNA is indicated and dashed lines indicate secondary structure. Nucleotides involved in the codon:anticodon interaction are shown with dots indicating base pairing. The tRNA in the ribosomal P-site is attached to the elongating peptide chain (open circles) and the aminoacylated tRNA in the ribosomal A-site is shown with a filled circle.

3.1 Heptameric slippery sites

Evidence for the requirement of slippage of both the aminoacyl- and peptidyl-tRNAs has accumulated in the last few decades. The actual point of slippage has been confirmed by protein sequencing for mouse mammary tumor virus (MMTV) (Hizi et al., 1987); HIV (Jacks et al., 1988a); RSV (Jacks et al., 1988b); barley yellow dwarf virus (BYDV) (Di et al., 1993), human T-cell leukemia virus type 1 (HTLV-1) (Nam et al., 1993) and Semiliki Forest virus (Firth et al., 2008). A number of mutagenesis experiments have been performed that demonstrate that there are specific sequence requirements at both the A- and P-site codons in many viruses and even a eukaryotic gene (for example: HIV virus, Wilson et al., 1988; L-A virus, Dinman et al., 1991; potato leafroll virus (PLRV), Prüfer et al., 1992; red clover necrotic mosaic dianthovirus (RCNMV), Kim & Lommel, 1994; *Edr*, Manktelow et al., 2005). Rules defining the heptameric sequences on which tRNAs could slip were elucidated in part in yeast (Dinman et al., 1991) and more extensively in reticulocyte lysate (Brierley et al., 1992). In general the slippery site can be defined as N NNW WWH, where N is any three identical bases, W is A or U, and H is A, C or U (the frame of the initiator AUG is indicated by the spacing), although there are exceptions (see section 4). This sequence is often described as X XXY YYZ in the literature. The efficiency of frameshifting promoted by each heptameric slippery site varies depending on the system used to assay frameshifting.

3.2 Stimulatory elements 3' of the slippery site

The sequence 3' of the heptameric slippery site has been shown to be required for optimal frameshifting in a number of systems (for example: Brierley et al., 1987; Jacks et al., 1987, 1988a, 1988b). The stimulatory sequences were predicted to fold into stem loops until Brierley et al., (1989) demonstrated that the Avian Infectious Bronchitis virus (IBV) stimulatory element was a pseudoknot. Sequence comparisons indicated that some structures downstream from slippery sites in other viruses could also be pseudoknots (for example: Bredenbeek et al., 1990; Kujawa et al., 1993; Cowley et al., 2000). It has also been shown that the stimulatory structure could involve long range interactions (Herold & Siddell, 1993; Paul et al., 2001). It has also been postulated that stop codons stimulate frameshift events, presumably by pausing translation (Castano & Hernandez, 2005; Horsfield et al., 1995). Mutagenesis or deletion analyses have been used to demonstrate that the 3' stimulatory sequences are stem-loops, pseudoknots or other higher order structures (see Table 1). Nuclease mapping, NMR, crystallography and mass spectrometry has confirmed some of the 3' stimulatory structures but many remain unresolved at the atomic level due to difficulties purifying larger structures. Molecular modeling has been useful in elucidating some of these structures (Ahn et al., 2011). A list of stimulatory elements (with the exception of HIV which is discussed below) is provided in Table 1.

3.3 Stimulatory element for HIV

The structure of one 3' stimulatory element, the HIV structure, remained elusive for many years. Frameshifting efficiencies for many of the viruses listed in Table 1 were dramatically reduced when the 3' sequence was removed. However, removal or alterations in the HIV 3' sequence produced more subtle changes in frameshifting and, as a result, different groups reached different conclusions. Jacks et al., (1988a) proposed that a 3' stem-loop was required for efficient frameshifting. This was refuted later that year by Wilson et al., (1988) who performed experiments in both rabbit reticulocyte lysate and yeast cells in which they obtained efficient frameshifting (5-10% measuring labeled methionine incorporation) from a 26 nucleotide HIV sequence that lacked the proposed stem-loop sequence. Using luciferase reporter plasmids Moosmayer et al., (1991) obtained 2-4% frameshifting in BHK cells using the shorter sequence used by Wilson et al., but did not make comparisons to sequences containing the stem-loop. Parkin et al., (1992) made mutations to disrupt and reform the stem-loop to demonstrate that, when present, an intact stem-loop stimulated 4-9 fold higher frameshifting in avian and simian cells. This result was also obtained by Reil et al., (1993) in BHK-21 cells (3.2% with the stem loop versus 0.9% without). This difference in frameshifting efficiency was recapitulated in both mouse fibroblasts and human lymphoid cells (Cassan et al., 1994). Additionally it has been shown that HIV frameshifting efficiency is several fold higher in human T-cells than in a bacterial lysate (Plant & Dinman, 2006). Telenti et al. (2002) found that clinical isolates with variations in the stem-loop sequence predicted to disrupt the structure had lower levels of frameshifting. In 1998 Kang used UV absorbance melting and nuclease assays all under the same conditions to show the formation of a stem-loop structure.

Viruses with Stem Loops	Deletion Analysis or Mutagenesis	Nuclease Mapping	NMR, Crystallography or Mass Spectrometry
GLV	Li et al., 2001a	Li et al., 2001a	
HAst-1	Marczinke et al., 1994	Marczinke et al., 1994	
HTLV-II	Falk et al., 1993		
mIAP	Fehrmann et al., 1997		
PLRV-G	Prüfer et al., 1992		
RCNMV	Kim & Lommel, 1998		
Viruses with Pseudoknot			
BWYV	Sung & Kang, 1998		
BYDV	Paul et al., 2001		Su et al., 1999
EIAV	Chen and Montelaro, 2003		
FIV	Morikawa & Bishop, 1992		
IBV	Brierley et al., 1989; 1991	Napthine et al., 1999	Yu et al., 2005
L-A	Dinman et al., 1991		
	Tzeng et al., 1992		
MHV		Plant et al., 2010	
MMTV	Lee et al., 1995		
PEMV			Shen & Tinoco, 1995
			Giedroc et al., 2003
PLRV-P	Kujawa et al., 1993		Nixon et al., 2002
	Kim et al., 2000		Pallan et al., 2005
RSV	Marczinke et al., 1998	Marczinke et al., 1998	
SARS	Baranov et al., 2005	Dos Ramos et al., 2004	
	Plant et al., 2005	Plant et al., 2005	Plant et al., 2005
		Su et al., 2005	
ScYLV			
SRV	ten Dam et al., 1994	ten Dam et al., 1994	Cornish et al., 2005
	Sung & Kang, 1998		Du et al., 1997
TYMV			Michiels et al., 2001
VMV	Pennell et al., 2008	Pennell et al., 2008	Kolk et al., 1998

Table 1. Viral PRF Stimulatory Elements. References are listed that describe different methods used to characterize viral stimulatory elements. GLV, giardiavirus; HAst-1, human astrovirus; HTLV-II, human T-cell leukemia virus type 2; mIAP, mouse interstitial A-type particle; PLRV-G, potato leafroll virus; RCNMV, red clover necrotic mosaic dianthovirus; BWYV, beet western yellows virus; BYDV, barley yellow dwarf virus; EIAV, equine infectious anemia virus; FIV, feline immunodeficiency virus; L-A, *Saccharomyces cerevisiae* virus L-A; MHV, mouse hepatitis virus; MMTV, mouse mammary tumor virus; PEMV, pea enation mosaic virus; RSV, Rous sarcoma virus; SARS, severe acute respiratory syndrome virus; ScYLV, sugarcane yellow leaf virus; SRV, simian retrovirus; TYMV, turnip yellow mosaic virus; VMV, Visna-Maedi retrovirus.

These experiments and observations helped reinforce the idea that a stem-loop was the 3' stimulatory element in the HIV frameshift signal. However because so many other 3' stimulatory elements were pseudoknots different groups looked for potential pseudoknot folding downstream from the HIV slippery site. Taylor et al., (1994) analyzed the HIV genome for sequences with the potential to form pseudoknot structures and proposed that a few nucleotides of the sequence 3' of the stem-loop could fold back, and pair with the loop, creating a pseudoknot. Based on the structure of other known pseudoknots Du et al., (1996) proposed that a small pseudoknot involving the spacer region could form (Figure 3B). Dulude et al., (2002) proposed that the 3' region was not folding back to form a stem-loop, but was in fact extending stem 1 by pairing with the spacer region (Figure 3G). Dulude et al. supported this hypothesis by mutating each sequence individually and in tandem (to disrupt and reform pairing potential) and assayed frameshifting efficiencies in cultured cells. NMR experiments with the HIV 3' stimulatory sequence have shown that the extended stem loop proposed by Dulude et al., can form (Gaudin et al., 2005; Staple and Butcher, 2005). Contrasting experiments by Dinman et al. (2002), using comparative genomics, nuclease mapping and frameshifting assays on a variety of mutants, argued that a pseudoknot structure formed. However, instead of the second stem proposed by Taylor et al., (1994) it was suggested that the top portion of stem 1 could form triplex structure with loop 2, in effect creating stem 2. Baril et al., (2003) also used mutagenesis and nuclease mapping to argue that the HIV group O frameshift signal is a pseudoknot. The Brakier-Gingras group has shown that there are a variety of functional structures that stimulate –1 PRF in different HIV strains (Baril et al., 2003; Dulude et al., 2002). Thus it is apparent for HIV at least, that absolute conservation of one particular structure is not essential for frameshifting: there is sequence variation in the stimulatory element between strains, and efficient frameshifting has been demonstrated when proposed stimulatory elements are altered. Examples of these structures are shown in Figure 3.

It is possible that, as suggested by Taylor et al., (1994), the conversion between structures may be important for frameshifting. This notion is important when thinking about the structures proposed by Du et al., (1996) and Dulude et al., (2002) which incorporated the spacer region between the slippery site and stimulatory element. This spacer region needs to be un-paired to fit within the ribosome entry tunnel when the slippery site is correctly positioned inside the ribosome.

In addition to the possibility of different HIV 3' structures stimulating –1 PRF, there is evidence that the frameshifting efficiencies observed may be the sum of different frameshifting mechanisms. Jacks et al., (1988a) noted that, when they sequenced the frameshifted protein, there was some variation at the position where frameshifting occurs. This variation suggested that different aspects of ribosome fidelity were affected during frameshifting. The effect of complementary DNA sequences on HIV frameshifting efficiency has been analyzed (Vickers and Ecker, 1992). Oligonucleotides predicted to bind to the sequences 3' of the stem loop, which would disrupt both the proposed extended stem-loop and pseudoknot structures proposed above, actually enhanced frameshifting in a rabbit reticulocyte lysate, as did addition of a second stem-loop downstream from the first. Different groups spent the next two decades analyzing several frameshift signals from a variety of virus families with the goal of understanding frameshift mechanisms.

A
```
    C A
    A   A
    U-G
    C-G
    C-G
    U-A
    U-A
    C-G
    C-G
    G-C
    G-C
    U-A
    C-G
    A   G
    G     A
    A       3'
    A
    G
    G
    G
UUUUUUA
```

B
```
                    3'
                    A
                    G
            U-G
            C-G
        A   U-A
            G-C
            G-C
            C-G
            G-C   G
            A-U   A
            A-U   G
            G-C   G
            G-C   G
            G-U   A
        UUUUUUA  ACA
```

C
```
              G-C 3'
              G-C
              G-C
              G-C
              C-G
        G     A-U
        C     C-G
        C-G   A
        U-A   A
        C-G     C
        C-G     A
        G-C     A
        G-C     A
        U-A     G
        C-G     A
        A   G   C
        G   C   G
        A   A   U
        A   A   G
        G   U U
        G     UA
        G
    UUUUUUA
```

D
```
        3'
    C A-U
    A   A-U
    U   G-C
    C-G   U
    C-G   U
    U-A   U
    U-A   U
    C-G   A
    C-G   A
    G-C   G
    G-C   G
    U-A
    C-G
    U
    A
    G
    A
    A
    G
    G
    G
UUUUUUA
```

E
```
        3'
    C A-U
    A   A-U
    U   G-C
    C   G-U
    C   G-U
    U   A-U
    U   A-U
    C-G   A
    C-G   A
    G-C   G
    G-C   G
    U-A
    C-G
    U
    A
    G
    A
    A
    G
    G
    G
UUUUUUA
```

F
```
        3'
    C A-U
    A   A-U
    U   G-C
    C:G:U
    C:G:U
    U:A:U
    U:A:U
    C-G   A
    C-G   A
    G-C   G
    G-C   G
    U-A
    C-G
    U
    A
    G
    A
    A
    G
    G
    G
UUUUUUA
```

G
```
    C A
    A   A
    U-G
    C-G
    C-G
    U-A
    U-A
    C-G
    C-G
    G-C
    G-C
    U-A
    C-G G
    |   G
    U-A A
    A-U
    G-U
    A-U
    A-U
    G-C
    G-U
UUUUUUAAG  UCG3'
```

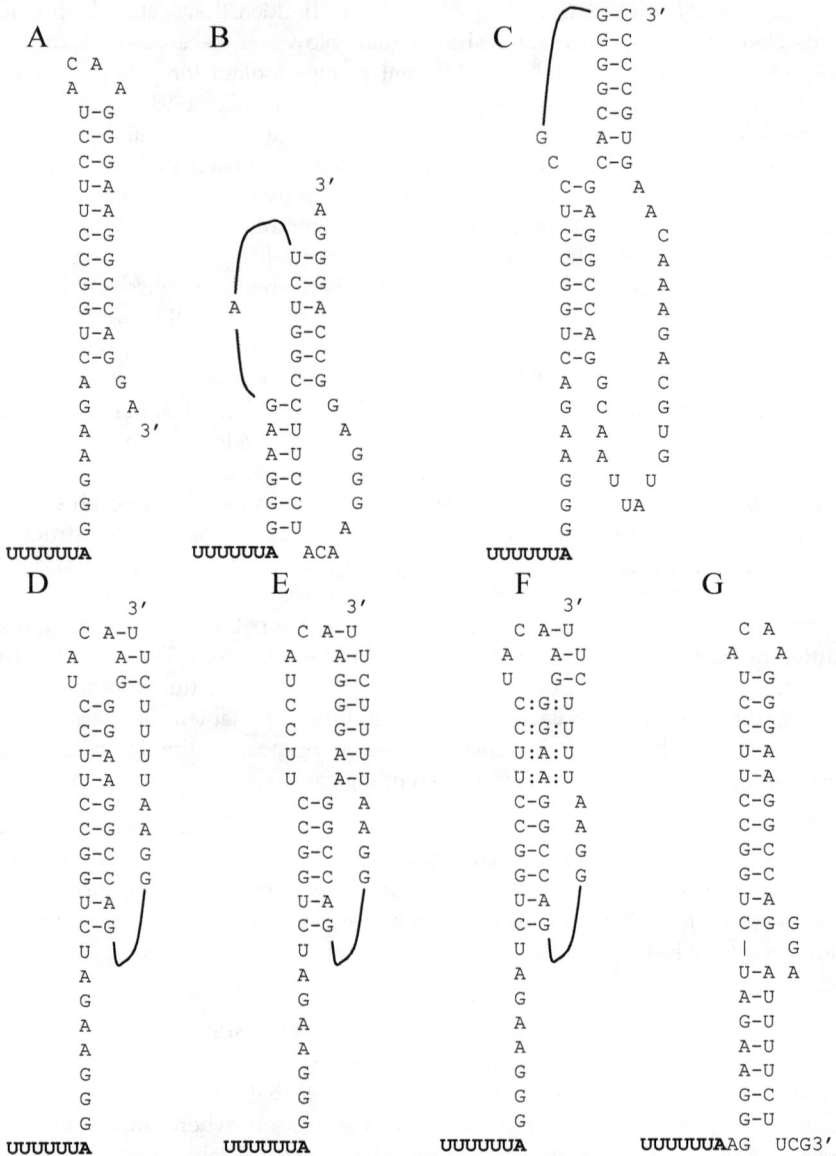

Fig. 3. Different stimulatory elements reported for HIV frameshift signals. There are some minor sequence variations between the subtypes. The stem-loop structure proposed by Jacks et al., (1988a) and pseudoknot proposed by Du et al., (1996) are shown in panels A and B respectivley. Panel C shows the pseudoknots described by Baril et al., (2003) for group O isolates. Panels D and E show two pseudoknots proposed by Taylor et al., (1994) and panel F shows the pseudoknot described by Dinman et al., (2002) using the same sequence. The extended stem-loop structure with a 5'-ACAA-3' tetraloop described by Dulude et al., (2002) is shown in panel G.

4. Elucidation of frameshifting mechanisms by analysis of viral frameshift signals

Ribosomes translate messager RNA with very low error rates to produce proteins. The availability of error-inducing frameshift-stimulating sequences from viruses has facilitated the study of ribosome function. Next I describe how the studies of virally encoded frameshift signals have had a broad impact on our understanding of protein translation. Further, these analyses have enhanced our understanding of viral replication and opened the door to new antivirals (discussed in section 5).

Translation of mRNA sequence into protein is a universal requirement for living organisms from the smallest single cell organisms through to the most complex mammals. The reaction, directed by the ribosome, is very specific with few errors (reviewed in Ogle & Ramakrishnan, 2005). In some instances, small viral genomes for example, the mRNA is polycistronic, encoding more than one protein. Tryptic analysis of a number of retroviral proteins in the late 1970s indicated that the gag and gag-pol proteins of these RNA viruses had the same amino terminus. It was proposed that the identical amino termini could be achieved by a number of mechanisms including recoding (Pawson et al., 1976). Subsequently gag-pol production for a number of retroviruses and retrotransposons was shown to be due to a recoding event, now known as frameshifting (Jacks & Varmus, 1985; Wilson et al., 1986; Farabaugh et al., 1993). The direction of frameshifting and the mRNA sequences on which frameshifting occurs differ amongst retroelements even though some infect the same host. There are several possible ways that frameshifting sequences overcome ribosome fidelity. Elucidation of these mechanisms has been possible through study of various viral frameshift signals.

The diverse sequence characteristics of different frameshift stimulating signals led many groups to postulate different mechanisms for frameshifting; some are specific to a single frameshift signal, others are more general mechanisms. Frameshifting occurs in the decoding center of the ribosome as part of protein translation. Each translation elongation cycle involves the selection of an aminoacylated-tRNA and it's accommodation into the ribosomal A-site. The amino acid is added to the elongating peptide in the peptidyltransferase center and the tRNA, now referred to as the peptidyl-tRNA, is moved into the P-site of the ribosome. After the addition of the next amino acid, or the termination of protein synthesis, the peptidyl-tRNA is moved into the E-site where it can exit the ribosome. During the elongation cycle the anticodon loop of the tRNA is hydrogen bonded to the corresponding triplex codon on the mRNA in the decoding center of the ribosome. Programmed ribosomal frameshifting is the regulated un-pairing of at least one of the tRNAs from the mRNA and re-pairing of the tRNAs to a different position on the mRNA (see Figure 2).

4.1 Simultaneous slippage

Frameshifting occurs when the anticodon loop of a tRNA binds to an out of frame codon on the mRNA. The first frameshift stimulating sequences described all required a heptameric slippery site suggesting that both the aminoacyl- and peptidyl-tRNAs bind out of frame (Jacks & Varmus, 1985; Jacks et al., 1987, 1988a). Other viruses and retroelements suspected of using programmed –1 ribosomal frameshifting (-1 PRF) also showed a preference for certain heptameric sequences and analysis of these sequences revealed that both the A-site and P-site tRNAs could potentially un-pair from the zero frame and re-pair in the –1 frame while maintaining "a two-out-of-three base pair, anticodon-codon configuration" (Jacks et al., 1988b). Thus the basis for this model was founded: both the aminoacyl- and peptidyl-tRNAs 'slipped' together on the mRNA one nucleotide from the zero frame register in a 5' direction to the –1 frame. A 3' stimulatory element was also required and integrated into the model even though it was not known how it contributed frameshifting at that time.

A 3' stimulatory structure is required in the simultaneous slippage model and is present in some examples of the P-site slippage model discussed below. Researchers soon began to investigate how these structures were involved in –1 PRF. An early hypothesis was that it could be "the binding site for a ribosomal protein or RNA or soluble elongation factor; this binding could then affect the fidelity of the ribosome-tRNA interaction at the decoding sites" (Jacks et al., 1988b). This is a hypothesis that has been difficult to prove or disprove and only one experiment stands out: tem Dam et al., (1994) added back RNA corresponding to the pseudoknot so that it might quench the supply of potential pseudoknot binding proteins and thus reduce the frequency of –1 PRF in their reporter in an *in vitro* assay. No change in frameshifting efficiency was observed indicating that there is not a pseudoknot binding protein. However, even if there is no pseudoknot binding protein the stimulatory affect of proteins on frameshifting cannot be totally disregarded. An iron response element (stem-loop structure) was positioned downstream from the HIV slippery site and frameshifting promoted when iron regulatory proteins were present (Kollmus et al., 1996). Because this result is from an artificial construct the results may be suggesting a steric interaction with the ribosome is stimulating frameshifting rather than more specific interactions. A single protein (or RNA molecule) capable of binding the diverse array of pseudoknots to direct frameshifting also seems unlikely.

As more frameshift stimulating pseudoknots were discovered, it became apparent that there were different types of pseudoknots that stimulated frameshifting in both eubacterial and eukaryotic systems. The differing size and structures of pseudoknots may reflect specificities for host ribosomes. This is evident from many experiments demonstrating that frameshifting efficiencies from one particular signal are altered when different lysates or cell lines are used to assay frameshifting efficiency (Barry & Miller, 2002; Cassan et al., 1994; Dulude et al., 2002; Garcia et al., 1993; Kim et al., 1999; Lewis & Matsui, 1996; Napthine et al., 2003; Parkin et al., 1992; Plant & Dinman, 2006; Stahl et al., 1995; Tzeng et al., 1992). But even in one host different viruses use different stimulatory elements. For example, the structures described for coronaviruses infecting humans include two-stem pseudoknots, three-stem pseudoknots and kissing loops (see Figure 1).

Another hypothesis was that the function of the 3' stimulatory structure was simply to cause a translational pause. Adjacent stop codons have also been suggested to promote translational pausing during decoding of the terminator (Rice et al., 1985; Jacks et al., 1988b). The latter suggestion was later supported by the Potato Virus M and Measles virus signals described below (Gramstat et al., 1994; Liston & Briedis, 1995). The ribosome is known to pause at certain positions along an mRNA (Wolin and Walter, 1988) and even pseudoknots (Tu et al., 1992). Time course experiments to demonstrate that ribosomes paused at frameshift stimulating pseudoknots were performed (Somogyi et al., 1993). Translation was stopped by addition of edeine, which hinders mRNA binding and prevents productive peptidyl-tRNA interactions, and methionine incorporation was monitored. Conditions were altered to change the ability of the pseudoknot structure to form (higher magnesium to promote formation and higher temperatures to discourage formation) and all the results correlated with the formation of a pseudoknot (known to stimulate frameshifting) causing the ribosomes to pause. The final variation in experimental conditions was to test the ability of a stem-loop, predicted to be more stable than the pseudoknot, to stimulate pausing and frameshifting. Although the structure could stimulate efficient pausing like the pseudoknot, it did not promote efficient –1 PRF. In a follow up paper (Kontos et al., 2001) heelprinting was performed to show that the ribosomes paused over the A- and P-site codons with a variety of pseudoknots. The conclusion from these experiments was that a pause was necessary for efficient –1 PRF but insufficient, as stem-loops could also promote pausing.

Experiments that replaced the A-site codon of the HIV slippery sequence with a termination codon also elicited a frameshift event in E. coli (Horsfield et al., 1995). This suggests that a switch from translation elongation to termination can also result in a pause sufficient for frameshifting to occur. However, if the A-site codon has been replaced with a termination codon then frameshifting must occur at the P-site alone, or at the P- and E-sites. Mutagenesis of the HIV frameshift signal by Horsfield et al. (1995) indicated that frameshifting required both the A- and E-site codons in this context. Experiments by a different group using the native HIV stimulatory structure also led to a similar conclusion (Leger et al., 2007). The E-site requirement is also supported by the conservation of certain nucleotides in the E-site position upstream of heptameric slippery sites (Bekaert and Rousset, 2005). Sequencing frameshift protein products showed that the 0-frame or the -1 frame A-site could be decoded (Jacks et al., 1988a, Yeverton et al., 1994, Ivanov et al., 1998). Thus, -1 PRF can occur before the 0-frame A-site is decoded, or if it occurs after the 0-frame A-site decoding, then the aminoacyl-tRNA could be removed and a new tRNA that better matches the -1 frame A-site is inserted before peptidyltransfer occurs. These analyses also show that multiple mechanisms of resolving a frameshift signal with a heptameric slippery site are employed in any one cellular system. The timing of frameshifting within the translation elongation cycle has been further delineated by Liao et al. (2010) using the HIV and HTLV-1 frameshift signals. Their work suggests three different pathways are used when heptameric slippery sites are encountered.

4.2 P-site slippage

When any rule is made, the exceptions are usually revealed shortly thereafter and this was the case with simultaneous slippage. The Potato Virus M (PVM) and Measles virus slippery sites did not conform to the heptameric slippery site rules of the simultaneous slippage model, only one tRNA could re-pair in the new frame while maintaining more than two out of three codon:anticodon base pair interactions. Gramstat et al., (1994) were able to demonstrate that a frameshift protein was made by slippage on a four nucleotide sequence from PVM. This sequence was flanked by two termination codons, the first in the –1 frame and the latter in the zero-frame (U AGA AAA UGA; spacing indicates the zero-frame and stop codons are underlined). Frameshifting was abolished if a single point mutation were made in the poly-A stretch, but maintained at wild type levels if the poly-A stretch was substituted with poly-U. These results demonstrate that programmed frameshifting can occur by slippage of just the peptidyl-tRNA.

Based on these results Gramstat et al., (1994) proposed the P-site slippage model. In this model decoding of the AAA codon by the lysine tRNA occurs in a normal manner and is followed by pepdidyl transfer and translocation. While the peptidyl-tRNA is in the P-site waiting for decoding of the UGA stop codon in the A-site, a –1 frameshift event occurs at the P-site and translation elongation resumes in the new frame. The peptidyl-tRNA is able to re-pair in the –1 frame with the same amount of hydrogen bonding as the zero-frame. Unlike the frameshift signals that stimulate simultaneous slippage, the PVM frameshift signal does not require a 3′ secondary structure, instead the 3′ termination signal in the zero-frame seems to perform the same function.

A second example of P-site slippage was described that requires a 3′ stimulatory structure. Liston and Briedis (1995) used protein sequencing to pinpoint the frameshift site in the Measles virus P protein coding region to a CCG proline codon in the C UCC CCG sequence. The preceding zero-frame serine codon (UCC) could allow slippage of the decoding proline tRNA while maintaining two of three base-pairing interactions on the –1 frame CCC sequence. However, the nucleotide 5′ of the serine codon is a cytosine so that if the tRNA^{Ser} slipped to the -1 frame CUC codon it would only retain a Watson-Crick interaction in the third codon:anticodon position, the wobble position. Liston and Briedis used deletion mutants to show that –1 PRF in the measles virus requires a downstream stem-loop structure for efficient frameshifting. Thus, the P-site slippage model posits that the A-site is decoded slowly because of a downstream feature or competition from the termination factors and this allows the peptidyl-tRNA time to un-pair from the zero frame and re-pair in an alternative reading frame.

One feature of the P-site slippage model is that if the A-site is not occupied then it is possible that slippage could occur in either direction (Baranov et al., 2004). There is some evidence that this can happen in wheat germ and rabbit reticulocyte lysates and in human hepatoma cells (Choi et al., 2003). The experiments that demonstrate this were performed using a frameshift signal derived from a viral internal ribosome entry site. A protein from the Hepatitis C Virus (HCV) was identified and labeled "F" (Walewski et al., 2001; Xu et al., 2001). It was initially thought that this viral protein was derived from a frameshifting event but this notion was later disproved (Vassilaki and Mavromara, 2003) and it is now thought that protein F is translated from an internal ribosome entry site (IRES) (Baril and Brakier-Gingras, 2005). However, before the IRES function of the HCV RNA was revealed, Choi et al., (2003) set up an *in vitro* expression system that lacked most of the IRES but from which transframe proteins (both –1 frame and a +1 frame products) were detected. They hypothesized that frameshifting occurred at a poly-A stretch (A AAA AAA AAC) and demonstrated that the frequency of either –1 or +1 frameshifting was approximately 2% in Huh7 cells. When the slippery site was mutated such that slippage could occur at only one position (A AGA AAA ACC) +1 and –1 frameshifting still occurred albeit at a reduced efficiency. This lower level of frameshifting could be increased 2-3 fold by adding the elongation inhibitor puromycin, an antibiotic that allows more time for frameshifting to occur by slowing the completion of the elongation cycle. This demonstrated that frameshifting of the P-site tRNA can occur, in either direction, before an aminoacyl-tRNA is accommodated in the A-site.

Some yeast ribosomal L5 mutant alleles affected the frameshifting efficiency of both L-A and Ty-1 and had decreased affinity for the peptidyl-tRNA suggesting that P-site fidelity affects -1 and +1 PRF (Meskauskas and Dinman, 2001). Thus, the nature of the tRNA:mRNA interaction and the ribosome's affinity for the tRNA at the P-site, along with the length of time this interaction is maintained are important determinants of frameshifting efficiency.

4.3 Out of frame binding and +1 slippage

At the same time that evidence for the simultaneous slippage and P-site slippage models were being unraveled a new class of frameshift signals emerged. These were characterized by their lack of homogeneity in the heptameric slippery sites: the peptidyl-tRNAs could not re-pair in the new (+1) reading frame and establish more than one of the three possible codon:anticodon base pair interactions. It was postulated that some frameshift events are due to the incoming aminoacyl-tRNA binding out of frame rather than slipping. The concepts from the initial discoveries and comparisons of frameshifting in yeast retrotransposons can be extended to help explain the frameshifting mechanisms in genomic, mitochondrial and cilate genes as well as those in viruses.

In 1986 Wilson et al. established that the yeast retrotransposon Ty1 used a +1 frameshift to express the protein encoded by the second open reading frame (ORF) TYB. They showed that a 31 nucleotide stretch of sequence from the overlap of ORFs TYA and TYB was required and sequence comparisons suggested that a conserved 11 nucleotide sequence was important. Belcourt and Farabaugh (1990) later showed that frameshifting required only a heptameric sequence (CCU AGG C) and that disruption of either zero frame codon reduced frameshifting. It was established that the leucyl-tRNA in the P-site slipped forward before the A-site is decoded. The proposal that a pause was required was demonstrated when Xu and Boeke (1990) showed that Ty1 retrotransposition, which requires frameshifting, was reduced when a rare tRNA^{Arg} that decodes the A-site codon was expressed at higher levels. In a complementary experiment, frameshifting was reduced when the *HSX1* gene encoding tRNA^{Arg} was deleted (Kawakami et al., 1993). No other *cis*-acting elements are apparently necessary for +1 frameshifting in these constructs.

While sequence homology between the yeast retrotransposons Ty3 and Ty1 initially suggested that the +1 frameshift mechanism might be similar in both retrotransposons (Hansen et al., 1988) it was later shown that the mechanism is quite different. The Ty3 heptameric frameshift sequence (GCG AGU U) is decoded by a peptidyl-tRNA^{Ala} that is unable to slip forward (Farabaugh et al., 1993) implying that the aminoacyl-tRNA binds out of frame. It was also initially suggested that the +1 frameshift signal (UCC UGA U) in the cellular ornithine decarboxylase transcript was due to out of frame binding at the A-site (Matsufuji et al., 1995). This frameshift signal contains a termination codon in the zero frame A-site. It has subsequently been argued that slippage occurs on the ornithine decarboxylase transcript by detachment of peptidyl tRNA^{Ser} from UCC and re-pairing to CCU (Baranov et al., 2004). The scarcity of the aminoacyl-tRNA for the Ty3 is limiting, indicating that a pause is required. Vimaladithan and Farabaugh (1994) showed that increasing the abundance of the tRNA decreased frameshifting and removal of the tRNA enhanced frameshifting. The nature of the codon:anticodon interaction at the P-site is proposed to leave the first base of the A-site codon unpaired at tRNA accommodation (Sundararajan et al., 1999). Displacement of this nucleotide results in the incoming tRNA recognizing the +1 frame codon. This displacement is facilitated by a stimulatory element.

A 3′ stimulatory element enhances Ty3 +1 frameshifting and has been shown to enhance frameshifting from a variety of heptameric frameshift sites (including the Ty1 frameshift sequence) unless there is a termination codon in the A-site (Li et al., 2001b). The positioning of the stimulator was shown to be crucial, moving it as little as one nucleotide downstream reduced frameshifting. This led to the hypothesis that it may stimulate frameshifting by binding to the 18S rRNA (Li et al., 2001b). A similar stimulatory sequence is present 3' from the antizyme decarboxylase slippery site (Ivanov and Atkins, 2007). There is no apparent homology between the Ty3 stimulatory element and the diverse 3' antizyme sequences, and because the antizyme frameshift sites have a conserved termination signal and specific P-site, requirements which the Ty3 frameshift site lacks, this supports the notion that there are subtle differences in the mechanisms that promote +1 out of frame binding.

4.4 A-site contributions

It has been suggested that the equine arteritis virus slippery site is unusual because it does not conform in sequence to other heptameric slippery sites (G UUA AAC) (den Boon et al., 1991). However, this slippery site has been shown to promote efficient –1 PRF in the context of a coronavirus pseudoknot (Brierley et al., 1992). Because of the existence of this and other proposed slippery sites that did not seem to fit with earlier models of –1 PRF Napthine et al., (2003) investigated the importance of the A-site in driving eukaryotic –1 PRF using a series of constructs based on the IBV frameshift signal. In vitro translations reactions were performed in tRNA depleted wheat germ lysates that were supplemented with either *E. coli* or calf liver tRNA populations. Efficient frameshifting was observed for slippery sites X XXA AAC when the lysate was supplemented with eukayotic tRNAs or the *E. coli* tRNAs. Frameshifting was substantially lower when the calf liver tRNAs and the slippery site X XXA AAG were used but not when the *E. coli* tRNAs were used. The reasons for this are thought to be due to the abundance of tRNAs in the different cell types (Napthine et al., 2003 and references within). *E. coli* contains only one tRNA^{Lys} isoacceptor with the anticodon 3'UUU (which is modified at the wobble position) to decode both AAG and AAA, thus explaining the abundance of slippery sites in bacteria ending with A AAG. Mammalian cells contain an additional two tRNA^{Lys} isoacceptors, one with the anticodon 3'UUC could outcompete the 3'UUU anticodon-containing tRNA for the AAG codon explaining the lack of eukaryote slippery sites utilizing A AAG.

Frameshifting (when either eukaryote or eubacterial tRNAs were added) was reduced when mutations were made to the –1 frame P-site positions (underlined) in the U UUA AAC slippery site. Less of an effect was apparent when the U UUA AAG slippery site and P-site mutations were assayed with the *E. coli* tRNAs. However, frameshifting levels were still quite high prompting the suggestion that slippage may be occurring in the A-site alone as the P-site mutations proposed to minimize post-slippage base-pairing interactions did not actually eliminate frameshifting. This result was most apparent with the X XXA AAG slippery sites and *E. coli* tRNAs (Napthine et al., 2003).

From the above experiments, and in conjunction with the slippery site analysis Brierley et al. (1992) performed in rabbit reticulocyte lysates, it appears that the sequence requirements are not absolute. These results suggest that the A-site codon:anticodon interactions are more important for slippage in the system described and that the P-site interactions or peptidyl-tRNAs do not play a pivotal role in –1 PRF. Another possible explanation of these results could be that –1 PRF requires some interaction between the aminoacy-tRNA and the peptidyl-tRNA in addition to the codon:anticodon interactions. That is, the efficiency of frameshifting is dependent, in part, on the type of tRNA used, particularly in the A-site.

5. Frameshift signals as antiviral targets

Most frameshift signals regulate production of an enzyme involved in viral replication. Altering this regulation can disrupt replication. Thus, frameshift signals are ideal targets for antivirals. The diversity of Programmed Ribosomal Frameshift (PRF) signals among, and within, different viral families means that each signal provides specific, unique features that can be used as drug targets. Specific features of both HIV-1 and coronaviruses have been used as antiviral targets and are described below. Modulation of frameshifting affects the viability of both types of viruses even though frameshifting regulates different proteins for each virus.

The generality of the frameshifting mechanism suggests that there could be opportunities to modulate frameshifting by targeting cellular factors that affect PRF (Dinman et al., 1997). A possible caveat with this approach is the demonstrated presence of frameshift signals in non-viral, host genomes (Baranov et al., 2011; Belew et al., 2010; Manktelow et al., 2005). Programmed frameshifting on non-viral transcripts has also been linked to the mRNA decay pathway in yeast cells (Plant et al., 2004). Until we have a better understanding of the extent of host cell usage of PRF caution should be used when designing antivirals that affect general frameshifting mechanisms. That said, cellular factors have been identified that are able to modulate PRF (Koybayashi et al., 2010; Meskauskas and Dinman, 2001).

A screen of cellular factors identified some factors which, when knocked down by RNA interference (RNAi), reduced HIV replication. Further investigation of a subset of these factors, specifically those likely to affect translation, resulted in the identification of eukaryote release factor 1 (eRF1) as a protein of interest (Kobayashi et al., 2010). Characterization of the mode of action demonstrated that it was through the up-regulation of frameshifting. Translation termination factor eRF1 has a tertiary structure that mimics a tRNA structure allowing it to recognise all three termination codons when they enter the ribosomal A-site. Kobayashi et al. observed an increase in reverse transcriptase activity compare to HIV gag protein when eRF1 was knocked down by RNAi. The increase in frameshifting was not due to a change in ribosome pausing indicating that the reduced amount of eRF1 was not affecting translation elongation efficiency. Given the description of the A-site contribution to frameshifting described above, it is plausible that the mode of action is due to an interaction between eRF1 and the P-site tRNA. However, this has not been demonstrated so it is possible that eRF1 is modulating -1 PRF through other mechanisms.

5.1 Targeting retrovirus frameshift signals

Retroviral systems like HIV regulate the ratios of structural (gag) protein and enzymatic (pol) protein via PRF. It has been proposed that maintaining the ratio of structural and enzymatic proteins is important for viral propagation. Increasing frameshifting efficiency activated the HIV-1 protease and inhibited the budding and assembly of virus-like particles (Karacostas et al., 1993). Hung et al. (1998) showed that there was a direct correlation between frameshifting and inhibition of HIV-1 replication; a two- to three-fold increase in gag-pol production inhibited particle formation. Also, it has been shown that maintaining the optimal gag/gag-pol ratio is important for HIV RNA dimerization (Shehu-Xhilaga et al., 2001). Biswas et al., (2004) demonstrated that a decrease in frameshifting levels below 10% of normal activity abolished virus production. It has been shown for another lentivirus, Equine anemia infectious virus, that an 80% reduction in frameshifting abolishes viral replication (Chen and Montelaro, 2003), strengthening the assertion that optimal ratios of gag protein to gag-pol polyprotein are required for successful retroviral replication.

An RNA feature of the HIV frameshift signal has been used as an antiviral target. It has been shown by NMR that the 3' stimulatory structure for some HIV strains is a stem-loop (Gaudin et al., 2005; Staple & Butcher, 2005). This stem-loop is often refered to as the HIV frameshift stimulatory sequence (FSS) in the literature. The FSS sequence is conserved in most group M subtypes (Baril et al., 2003). It has an uncommon 5'-ACAA-3' tetraloop capping the stem and a purine bulge that creates a bend in the structure (Figure 3G). A bent stem has also been observed in some other viral frameshift stimulating structures (Chen et al., 1996; Chung et al., 2010). A well defined RNA molecule like the HIV-1 FSS makes a suitable target to screen for compounds able to bind. Indeed, it was found that some compounds are able to bind the upper portion of the HIV stem-loop with high affinity (Palde et al., 2010; Staple et al., 2008). Increasing or decreasing the stability of this stem-loop will likely affect frameshifting. The lower stem was not used in screens as it unfolds more readily than the upper stem-loop (Mazauric et al., 2009) and must be single-stranded to fit into the mRNA entry tunnel of the ribosome so that the slippery site is correctly positioned.

Some molecules have been identified that bind to the FSS and alter the stability of the structure. Staple et al., (2008) investigated the binding of a modified synthetic aminoglycoside to the HIV FSS. Using NMR they were able to show that Guanidinoneomycin B bound to the major groove of the upper stem-loop and this resulted in the repositioning of the 5'-ACAA-3' tetraloop. It remains to be determined if the binding of the molecule results in a change in frameshifting efficiency. Oligonucleotides designed to match different portions of the HIV FSS also bound to the target (Vickers & Ecker, 1992). Interestingly, only those binding to the 3' portion of the FSS altered frameshifting. This demonstrates how our lack of understanding of how specific stimulatory elements facilitate frameshifting makes antiviral design difficult.

Marcheschi et al., (2009) replaced the guanosine in the HIV FSS purine bulge with a fluorescent analog, 2-aminopurine. By measuring fluorescence they were able to monitor the stability of the HIV stem-loop as different compounds were applied. This screen led to the identification of Doxorubicin as a frameshift inhibitor. Although it had good affinity with a K_D of 2.8μM Doxorubicin is also unfortunately a general translation inhibitor. However, the identification of a lead compound with some selectivity and specificity demonstrates the utility of this approach.

Another good lead compound was identified by a screen of a resin bound dynamic combinatorial library with the upper stem of the HIV FSS (McNaughton et al., 2007). Subsequent analysis of this compound and derivatives of it demonstrated that it had high affinity and good selectivity (Palde et al., 2010). In solution the K_D of one derivative was 0.18μM. Additionally it was shown that this compound was non-toxic to human cells at concentrations up to 1.0mM. Although it remains to be determined if any of these compounds significantly alter frameshifting and are detrimental to HIV replication, the identification of these compounds, and knowledge of the features that affect selectivity and affinity is encouraging.

5.2 Targeting coronavirus frameshift signals

In contrast to the retroviruses and Totiviruses, which regulate structural and enzymatic protein production by frameshifting, coronaviruses use frameshifting to modulate the ratio of two large polyproteins that both have enzymatic functions. Together two overlapping open reading frames form the replicase gene. The polymerase encoded by coronaviruses is in the second open reading frame, downstream from a frameshift signal near the end of the first open reading frame. Experiments targeting the frameshift regions of the SARS coronavirus and the feline coronavirus have been performed. For both viruses, disrupting the frameshifting mechanism results in a decrease in infectivity. Given the complexity of the coronavirus lifecycle it is not certain how a reduction in frameshifting leads to the loss of infectivity, but clearly the abundance of proteins translated by the frameshifting mechanism is important.

A minimal level of frameshifting has been shown to be essential for SARS coronavirus replication. Using reverse genetics mutant viruses with varying levels of frameshifting efficacy were made. Reduction of frameshifting to 10% of the normal level completely abrogated the production of infectious virus particles (Plant et al., 2010). Suboptimal frameshifting in the SARS coronavirus reduces the amount of genomic RNA more so than the reduction of subgenomic RNA (Plant et al., 2010). However, this effect was not apparent for the feline coronavirus which has fewer subgenomic RNAs (McDonagh et al., 2011) indicating that more work is needed to understand the mechanics of frameshifting in the context as an antiviral target. Frameshifting occurs during the initial translation of the coronavirus genome early in infection and it is not known if the levels of frameshifting remain the same throughout infection. Therefore timing of an anti-frameshifting treatment may be an important consideration.

The SARS frameshift signal has been used as an antiviral target by several groups using different approaches. Frameshifting efficiency has been altered by four different approaches so far; RNA interference, peptide-conjugated antisense morpholino oligomers (P-PMOs), antisense peptide nucleic acids (PNAs), and antiviral compounds.

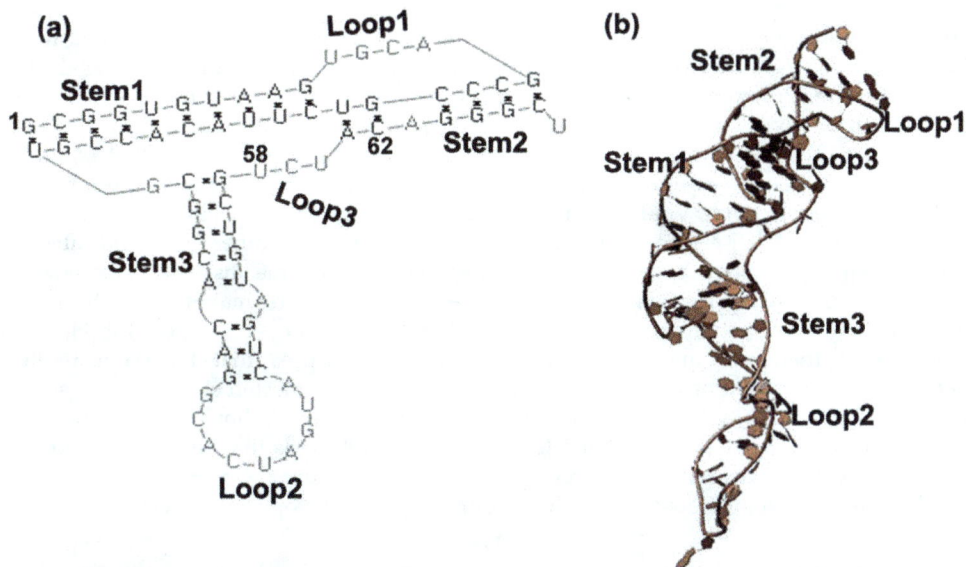

Fig. 4. A) 2-D diagram of the SARS pseudoknot structure. B) 3-D model determined by molecular dynamics simulation. Reprinted with permission from Park et al., 2011, Identification of RNA Pseudoknot-Binding Ligand That Inhibits the -1 Ribosomal Frameshifting of SARS-Coronavirus by Structure-Based Virtual Screening. Copyright 2011 American Chemical Society.

An RNAi approach to inhibit feline coronavirus replication was established by McDonagh et al. (2011). Short interfering RNAs (siRNAs) were designed to target different regions of the genome including the replicase gene. One siRNA targeted the frameshift signal and another targeted the region upstream of the frameshift signal. Both siRNAs reduced the copy number of genomic and subgenomic RNA to a similar extent. However, although both siRNAs also reduced the relative viral titer, the siRNA targeting the frameshift signal caused a significantly greater reduction (McDonagh et al., 2011). P-PMOs directed at the SARS coronavirus frameshift signal and the initiating codon of the replicase gene have been investigated (Neuman et al., 2005). Both P-PMOs resulted in a similar reduction in virus titer. However, in this instance, the reduction in plaque size was more pronounced for P-PMOs targeting the initiating codon of the replicase gene. This suggests that disruption of translation initiation is more effective than an RNAi target within the ORF at limiting viral propagation. Even so, PNAs targeting the pseudoknot caused a significant reduction in the replication of a SARS replicon at concentrations in the low µM range validating the selection of the pseudoknot as an antiviral target (Ahn et al., 2011).

Using molecular dynamics simulation Park et al., (2011) generated a model for the SARS coronavirus frameshift-stimulating pseudoknot. They then used this to screen a database for compounds that might be able to bind to the structure. Several hits were experimentally tested and found to bind to the pseudoknot RNA. The lead compound bound with high affinity and disrupted frameshifting. The effect of this compound on coronavirus replication has not be determined yet.

6. Conclusion

Viruses manipulate host cells in a number of different ways in order to replicate and thrive. The mechanisms used by viruses to manipulate the host cell are varied and evolve as the host defenses evolve. Programmed ribosomal frameshifting is one approach to manipulate host ribosomes that is used by a wide variety of viruses. There are multiple mechanisms that drive frameshifting and a diverse array of viral sequences and structures that stimulate frameshifting. While the position of some frameshift signals upstream of an RNA-dependent RNA polymerase in some viral genomes suggests these signals may have evolved from a common ancestor, the presence of frameshift signals upstream of other genes indicates that frameshifting signals may have developed independently in some instances. The essential nature of frameshifting for many of the viruses described here makes frameshifting an attractive target for antivirals. While some antivirals are being developed for HIV and coronaviruses there are many agriculturally important viruses that have not received attention. Current limitations stem from the diverse nature of frameshift signals amongst virus groups and the different mechanisms driving frameshifting. Although the signals and mechanisms are diverse, an underlying feature of frameshifting is that all these signals alter ribosome fidelity. A clearer understanding of the critical features of frameshift signals and how these alter ribosome fidelity will enhance our ability to develop new antivirals.

7. Acknowledgment

I would like to thank my mentors who introduced me to ribosome recoding and virology. The findings and conclusions in this chapter have not been formally disseminated by the Food and Drug Administration and should not be construed to represent an Agency determination or policy.

8. References

Ahn DG, Lee W, Choi JK, Kim SJ, Plant EP, Almazan F, Taylor DR, Enjuanes L, and Oh JW. 2011. Interference of ribosomal frameshifting by antisense peptide nucleic acids suppresses SARS coronavirus replication. *Antiviral Res.*

Alam SL, Wills NM, Ingram JA, Atkins JF, and Gesteland RF. 1999. Structural studies of the RNA pseudoknot required for readthrough of the gag-termination codon of murine leukemia virus. *J Mol. Biol.* 288 (5): 837-852.

Baranov PV, Gesteland RF, and Atkins JF. 2004. P-site tRNA is a crucial initiator of ribosomal frameshifting. *RNA* 10 (2): 221-230.

Baranov PV, Henderson CM, Anderson CB, Gesteland RF, Atkins JF, and Howard MT. 2005. Programmed ribosomal frameshifting in decoding the SARS-CoV genome. *Virology* 332 (2): 498-510.

Baranov PV, Wills NM, Barriscale KA, Firth AE, Jud MC, Letsou A, Manning G, and Atkins JF. 2011. Programmed ribosomal frameshifting in the expression of the regulator of intestinal stem cell proliferation, adenomatous polyposis coli (APC). *RNA Biol.* 8 (4).

Baril M, Dulude D, Steinberg SV, and Brakier-Gingras L. 2003. The frameshift stimulatory signal of human immunodeficiency virus type 1 group O is a pseudoknot. *J. Mol. Biol.* 331 (3): 571-583.

Baril M, and Brakier-Gingras L. 2005. Translation of the F protein of hepatitis C virus is initiated at a non-AUG codon in a +1 reading frame relative to the polyprotein. *Nucleic Acids Res.* 33 (5): 1474-1486.

Barry JK, and Miller WA. 2002. A -1 ribosomal frameshift element that requires base pairing across four kilobases suggests a mechanism of regulating ribosome and replicase traffic on a viral RNA. *Proc. Natl. Acad. Sci. U. S. A* 99 (17): 11133-11138.

Bekaert M, and Rousset JP. 2005. An extended signal involved in eukaryotic -1 frameshifting operates through modification of the E site tRNA. *Mol. Cell* 17 (1): 61-68.

Bekaert M, Firth AE, Zhang Y, Gladyshev VN, Atkins JF, and Baranov PV. 2010. Recode-2: new design, new search tools, and many more genes. *Nucleic Acids Res.* 38 (Database issue): D69-D74.

Belcourt MF, and Farabaugh PJ. 1990. Ribosomal frameshifting in the yeast retrotransposon Ty: tRNAs induce slippage on a 7 nucleotide minimal site. *Cell* 62 (2): 339-352.

Belew AT, Hepler NL, Jacobs JL, and Dinman JD. 2008. PRFdb: a database of computationally predicted eukaryotic programmed -1 ribosomal frameshift signals. *BMC. Genomics* 9: 339.

Belew AT, Advani VM, and Dinman JD. 2010. Endogenous ribosomal frameshift signals operate as mRNA destabilizing elements through at least two molecular pathways in yeast. *Nucleic Acids Res.*

Biswas P, Jiang X, Pacchia AL, Dougherty JP, and Peltz SW. 2004. The human immunodeficiency virus type 1 ribosomal frameshifting site is an invariant sequence determinant and an important target for antiviral therapy. *J. Virol.* 78 (4): 2082-2087.

Bredenbeek PJ, Pachuk CJ, Noten AF, Charite J, Luytjes W, Weiss SR, and Spaan WJ. 1990. The primary structure and expression of the second open reading frame of the polymerase gene of the coronavirus MHV-A59; a highly conserved polymerase is expressed by an efficient ribosomal frameshifting mechanism. *Nucleic Acids Res.* 18 (7): 1825-1832.

Brierley I, Boursnell ME, Binns MM, Bilimoria B, Blok VC, Brown TD, and Inglis SC. 1987. An efficient ribosomal frame-shifting signal in the polymerase-encoding region of the coronavirus IBV. *EMBO J* 6 (12): 3779-3785.

Brierley I, Digard P, and Inglis SC. 1989. Characterization of an efficient coronavirus ribosomal frameshifting signal: requirement for an RNA pseudoknot. *Cell* 57 (4): 537-547.

Brierley I, Rolley NJ, Jenner AJ, and Inglis SC. 1991. Mutational analysis of the RNA pseudoknot component of a coronavirus ribosomal frameshifting signal. *J. Mol. Biol.* 220 (4): 889-902.

Brierley I, Jenner AJ, and Inglis SC. 1992. Mutational analysis of the "slippery-sequence" component of a coronavirus ribosomal frameshifting signal. *J. Mol. Biol.* 227 (2): 463-479.

Brown CM, Dinesh-Kumar SP, and Miller WA. 1996. Local and distant sequences are required for efficient readthrough of the barley yellow dwarf virus PAV coat protein gene stop codon. *J Virol.* 70 (9): 5884-5892.

Cassan M, Delaunay N, Vaquero C, and Rousset JP. 1994. Translational frameshifting at the gag-pol junction of human immunodeficiency virus type 1 is not increased in infected T-lymphoid cells. *J. Virol.* 68 (3): 1501-1508.

Castano A, and Hernandez C. 2005. Complete nucleotide sequence and genome organization of Pelargonium line pattern virus and its relationship with the family Tombusviridae. *Arch. Virol.* 150 (5): 949-965.

Chen C, and Montelaro RC. 2003. Characterization of RNA elements that regulate gag-pol ribosomal frameshifting in equine infectious anemia virus. *J. Virol.* 77 (19): 10280-10287.

Chen X, Chamorro M, Lee SI, Shen LX, Hines JV, Tinoco I, Jr., and Varmus HE. 1995. Structural and functional studies of retroviral RNA pseudoknots involved in ribosomal frameshifting: nucleotides at the junction of the two stems are important for efficient ribosomal frameshifting. *EMBO J.* 14 (4): 842-852.

Chen X, Kang H, Shen LX, Chamorro M, Varmus HE, and Tinoco I, Jr. 1996. A characteristic bent conformation of RNA pseudoknots promotes -1 frameshifting during translation of retroviral RNA. *J Mol. Biol.* 260 (4): 479-483.

Choi J, Xu Z, and Ou JH. 2003. Triple decoding of hepatitis C virus RNA by programmed translational frameshifting. *Mol. Cell Biol.* 23 (5): 1489-1497.

Chung BY, Firth AE, and Atkins JF. 2010. Frameshifting in alphaviruses: a diversity of 3' stimulatory structures. *J Mol. Biol.* 397 (2): 448-456.

Cornish PV, Hennig M, and Giedroc DP. 2005. A loop 2 cytidine-stem 1 minor groove interaction as a positive determinant for pseudoknot-stimulated -1 ribosomal frameshifting. *Proc. Natl. Acad. Sci. U. S. A* 102 (36): 12694-12699.

Cowley JA, Dimmock CM, Spann KM, and Walker PJ. 2000. Gill-associated virus of Penaeus monodon prawns: an invertebrate virus with ORF1a and ORF1b genes related to arteri- and coronaviruses. *J. Gen. Virol.* 81 (Pt 6): 1473-1484.

Demler SA, and de Zoeten GA. 1991. The nucleotide sequence and luteovirus-like nature of RNA 1 of an aphid non-transmissible strain of pea enation mosaic virus. *J Gen. Virol.* 72 (Pt 8): 1819-1834.

den Boon JA, Snijder EJ, Chirnside ED, de Vries AA, Horzinek MC, and Spaan WJ. 1991. Equine arteritis virus is not a togavirus but belongs to the coronaviruslike superfamily. *J. Virol.* 65 (6): 2910-2920.

Di R, Dinesh-Kumar SP, and Miller WA. 1993. Translational frameshifting by barley yellow dwarf virus RNA (PAV serotype) in Escherichia coli and in eukaryotic cell-free extracts. *Mol. Plant Microbe Interact.* 6 (4): 444-452.

Dinman JD, Icho T, and Wickner RB. 1991. A -1 ribosomal frameshift in a double-stranded RNA virus of yeast forms a gag-pol fusion protein. *Proc. Natl. Acad. Sci. U. S. A* 88 (1): 174-178.

Dinman JD, and Wickner RB. 1992. Ribosomal frameshifting efficiency and gag/gag-pol ratio are critical for yeast M1 double-stranded RNA virus propagation. *J. Virol.* 66 (6): 3669-3676.

Dinman JD, Ruiz-Echevarria MJ, Czaplinski K, and Peltz SW. 1997. Peptidyl-transferase inhibitors have antiviral properties by altering programmed -1 ribosomal frameshifting efficiencies: development of model systems. *Proc. Natl. Acad. Sci. U. S. A.* 94 (13): 6606-6611.

Dinman JD, Richter S, Plant EP, Taylor RC, Hammell AB, and Rana TM. 2002. The frameshift signal of HIV-1 involves a potential intramolecular triplex RNA structure. *Proc. Natl. Acad. Sci. U. S. A* 99 (8): 5331-5336.

Domier LL, McCoppin NK, Larsen RC, and D'Arcy CJ. 2002. Nucleotide sequence shows that Bean leafroll virus has a Luteovirus-like genome organization. *J Gen. Virol.* 83 (Pt 7): 1791-1798.

Dos RF, Carrasco M, Doyle T, and Brierley I. 2004. Programmed -1 ribosomal frameshifting in the SARS coronavirus. *Biochem. Soc. Trans.* 32 (Pt 6): 1081-1083.

Du Z, Giedroc DP, and Hoffman DW. 1996. Structure of the autoregulatory pseudoknot within the gene 32 messenger RNA of bacteriophages T2 and T6: a model for a possible family of structurally related RNA pseudoknots. *Biochemistry* 35 (13): 4187-4198.

Du Z, Holland JA, Hansen MR, Giedroc DP, and Hoffman DW. 1997. Base-pairings within the RNA pseudoknot associated with the simian retrovirus-1 gag-pro frameshift site. *J. Mol. Biol.* 270 (3): 464-470.

Dulude D, Baril M, and Brakier-Gingras L. 2002. Characterization of the frameshift stimulatory signal controlling a programmed -1 ribosomal frameshift in the human immunodeficiency virus type 1. *Nucleic Acids Res.* 30 (23): 5094-5102.

Falk H, Mador N, Udi R, Panet A, and Honigman A. 1993. Two cis-acting signals control ribosomal frameshift between human T-cell leukemia virus type II gag and pro genes. *J. Virol.* 67 (10): 6273-6277.

Farabaugh PJ, Zhao H, and Vimaladithan A. 1993. A novel programed frameshift expresses the POL3 gene of retrotransposon Ty3 of yeast: frameshifting without tRNA slippage. *Cell* 74 (1): 93-103.

Fehrmann F, Welker R, and Krausslich HG. 1997. Intracisternal A-type particles express their proteinase in a separate reading frame by translational frameshifting, similar to D-type retroviruses. *Virology* 235 (2): 352-359.

Firth AE, Chung BY, Fleeton MN, and Atkins JF. 2008. Discovery of frameshifting in Alphavirus 6K resolves a 20-year enigma. *Virol. J* 5: 108.

Firth AE, and Atkins JF. 2009. A conserved predicted pseudoknot in the NS2A-encoding sequence of West Nile and Japanese encephalitis flaviviruses suggests NS1' may derive from ribosomal frameshifting. *Virol. J* 6: 14.

Garcia A, van DJ, and Pleij CW. 1993. Differential response to frameshift signals in eukaryotic and prokaryotic translational systems. *Nucleic Acids Res.* 21 (3): 401-406.

Gaudin C, Mazauric MH, Traikia M, Guittet E, Yoshizawa S, and Fourmy D. 2005. Structure of the RNA signal essential for translational frameshifting in HIV-1. *J Mol. Biol.* 349 (5): 1024-1035.

Ge Z, Hiruki C, and Roy KL. 1993. Nucleotide sequence of sweet clover necrotic mosaic dianthovirus RNA-1. *Virus. Res* 28 (2): 113-124.

Gendron K, Dulude D, Lemay G, Ferbeyre G, and Brakier-Gingras L. 2005. The virion-associated Gag-Pol is decreased in chimeric Moloney murine leukemia viruses in which the readthrough region is replaced by the frameshift region of the human immunodeficiency virus type 1. *Virology.* 334 (2): 342-352.

Giedroc DP, Cornish PV, and Hennig M. 2003. Detection of scalar couplings involving 2'-hydroxyl protons across hydrogen bonds in a frameshifting mRNA pseudoknot. *J. Am. Chem. Soc.* 125 (16): 4676-4677.

Gramstat A, Prüfer D, and Rohde W. 1994. The nucleic acid-binding zinc finger protein of potato virus M is translated by internal initiation as well as by ribosomal frameshifting involving a shifty stop codon and a novel mechanism of P-site slippage. *Nucleic Acids Res.* 22 (19): 3911-3917.

Hansen LJ, Chalker DL, and Sandmeyer SB. 1988. Ty3, a yeast retrotransposon associated with tRNA genes, has homology to animal retroviruses. *Mol. Cell Biol.* 8 (12): 5245-5256.

Harrell L, Melcher U, and Atkins JF. 2002. Predominance of six different hexanucleotide recoding signals 3' of read-through stop codons. *Nucleic Acids Res* 30 (9): 2011-2017.

Herold J, and Siddell SG. 1993. An 'elaborated' pseudoknot is required for high frequency frameshifting during translation of HCV 229E polymerase mRNA. *Nucleic Acids Res.* 21 (25): 5838-5842.

Hizi A, Henderson LE, Copeland TD, Sowder RC, Hixson CV, and Oroszlan S. 1987. Characterization of mouse mammary tumor virus gag-pro gene products and the ribosomal frameshift site by protein sequencing. *Proc. Natl. Acad. Sci. U. S. A* 84 (20): 7041-7045.

Horsfield JA, Wilson DN, Mannering SA, Adamski FM, and Tate WP. 1995. Prokaryotic ribosomes recode the HIV-1 gag-pol-1 frameshift sequence by an E/P site post-translocation simultaneous slippage mechanism. *Nucleic Acids Res.* 23 (9): 1487-1494.

Hung M, Patel P, Davis S, and Green SR. 1998. Importance of ribosomal frameshifting for human immunodeficiency virus type 1 particle assembly and replication. *J. Virol.* 72 (6): 4819-4824.

Isawa H, Kuwata R, Hoshino K, Tsuda Y, Sakai K, Watanabe S, Nishimura M, Satho T, Kataoka M, Nagata N, Hasegawa H, Bando H, Yano K, Sasaki T, Kobayashi M, Mizutani T, and Sawabe K. 2011. Identification and molecular characterization of a new nonsegmented double-stranded RNA virus isolated from Culex mosquitoes in Japan. *Virus. Res.* 155 (1): 147-155.

Ivanov IP, Gesteland RF, Matsufuji S, and Atkins JF. 1998. Programmed frameshifting in the synthesis of mammalian antizyme is +1 in mammals, predominantly +1 in fission yeast, but -2 in budding yeast. *RNA* 4 (10): 1230-1238.

Ivanov IP, and Atkins JF. 2007. Ribosomal frameshifting in decoding antizyme mRNAs from yeast and protists to humans: close to 300 cases reveal remarkable diversity despite underlying conservation. *Nucleic Acids Res.* 35 (6): 1842-1858.

Jacks T, and Varmus HE. 1985. Expression of the Rous sarcoma virus pol gene by ribosomal frameshifting. *Science* 230 (4731): 1237-1242.

Jacks T, Townsley K, Varmus HE, and Majors J. 1987. Two efficient ribosomal frameshifting events are required for synthesis of mouse mammary tumor virus gag-related polyproteins. *Proc. Natl. Acad. Sci. U. S. A* 84 (12): 4298-4302.

Jacks T, Power MD, Masiarz FR, Luciw PA, Barr PJ, and Varmus HE. 1988a. Characterization of ribosomal frameshifting in HIV-1 gag-pol expression. *Nature* 331 (6153): 280-283.

Jacks T, Madhani HD, Masiarz FR, and Varmus HE. 1988b. Signals for ribosomal frameshifting in the Rous sarcoma virus gag-pol region. *Cell* 55 (3): 447-458.

Jiang B, Monroe SS, Koonin EV, Stine SE, and Glass RI. 1993. RNA sequence of astrovirus: distinctive genomic organization and a putative retrovirus-like ribosomal frameshifting signal that directs the viral replicase synthesis. *Proc. Natl. Acad. Sci. U. S A* 90 (22): 10539-10543.

Kang H. 1998. Direct structural evidence for formation of a stem-loop structure involved in ribosomal frameshifting in human immunodeficiency virus type 1. *Biochim. Biophys. Acta* 1397 (1): 73-78.

Karacostas V, Wolffe EJ, Nagashima K, Gonda MA, and Moss B. 1993. Overexpression of the HIV-1 gag-pol polyprotein results in intracellular activation of HIV-1 protease and inhibition of assembly and budding of virus-like particles. *Virology* 193 (2): 661-671.

Kawakami K, Pande S, Faiola B, Moore DP, Boeke JD, Farabaugh PJ, Strathern JN, Nakamura Y, and Garfinkel DJ. 1993. A rare tRNA-Arg(CCU) that regulates Ty1 element ribosomal frameshifting is essential for Ty1 retrotransposition in Saccharomyces cerevisiae. *Genetics* 135 (2): 309-320.

Kim KH, and Lommel SA. 1994. Identification and analysis of the site of -1 ribosomal frameshifting in red clover necrotic mosaic virus. *Virology* 200 (2): 574-582.

Kim SN, Choi JH, Park MW, Jeong SJ, Han KS, and Kim HJ. 2005. Identification of the +1 ribosomal frameshifting site of LRV1-4 by mutational analysis. *Arch. Pharm. Res.* 28 (8): 956-962.

Kim YG, Su L, Maas S, O'Neill A, and Rich A. 1999. Specific mutations in a viral RNA pseudoknot drastically change ribosomal frameshifting efficiency. *Proc. Natl. Acad. Sci. U. S. A* 96 (25): 14234-14239.

Kim YG, Maas S, Wang SC, and Rich A. 2000. Mutational study reveals that tertiary interactions are conserved in ribosomal frameshifting pseudoknots of two luteoviruses. *RNA.* 6 (8): 1157-1165.

Kobayashi Y, Zhuang J, Peltz S, and Dougherty J. 2010. Identification of a cellular factor that modulates HIV-1 programmed ribosomal frameshifting. *J. Biol. Chem.* 285 (26): 19776-19784.

Kolk MH, van der Graaf M, Wijmenga SS, Pleij CW, Heus HA, and Hilbers CW. 1998. NMR structure of a classical pseudoknot: interplay of single- and double-stranded RNA. *Science* 280 (5362): 434-438.

Kollmus H, Honigman A, Panet A, and Hauser H. 1994. The sequences of and distance between two cis-acting signals determine the efficiency of ribosomal frameshifting in human immunodeficiency virus type 1 and human T-cell leukemia virus type II in vivo. *J Virol.* 68 (9): 6087-6091.

Kollmus H, Hentze MW, and Hauser H. 1996. Regulated ribosomal frameshifting by an RNA-protein interaction. *RNA.* 2 (4): 316-323.

Kontos H, Napthine S, and Brierley I. 2001. Ribosomal pausing at a frameshifter RNA pseudoknot is sensitive to reading phase but shows little correlation with frameshift efficiency. *Mol. Cell Biol.* 21 (24): 8657-8670.

Kujawa AB, Drugeon G, Hulanicka D, and Haenni AL. 1993. Structural requirements for efficient translational frameshifting in the synthesis of the putative viral RNA-dependent RNA polymerase of potato leafroll virus. *Nucleic Acids Res.* 21 (9): 2165-2171.

Leger M, Dulude D, Steinberg SV, and Brakier-Gingras L. 2007. The three transfer RNAs occupying the A, P and E sites on the ribosome are involved in viral programmed -1 ribosomal frameshift. *Nucleic Acids Res.* 35 (16): 5581-5592.

Lewis TL, and Matsui SM. 1996. Astrovirus ribosomal frameshifting in an infection-transfection transient expression system. *J. Virol.* 70 (5): 2869-2875.

Li L, Wang AL, and Wang CC. 2001a. Structural analysis of the -1 ribosomal frameshift elements in giardiavirus mRNA. *J. Virol.* 75 (22): 10612-10622.

Li Z, Stahl G, and Farabaugh PJ. 2001b. Programmed +1 frameshifting stimulated by complementarity between a downstream mRNA sequence and an error-correcting region of rRNA. *RNA* 7 (2): 275-284.

Liao PY, Choi YS, Dinman JD, and Lee KH. 2011. The many paths to frameshifting: kinetic modelling and analysis of the effects of different elongation steps on programmed -1 ribosomal frameshifting. *Nucleic Acids Res.* 39 (1): 300-312.

Liston P, and Briedis DJ. 1995. Ribosomal frameshifting during translation of measles virus P protein mRNA is capable of directing synthesis of a unique protein. *J. Virol.* 69 (11): 6742-6750.

Manktelow E, Shigemoto K, and Brierley I. 2005. Characterization of the frameshift signal of Edr, a mammalian example of programmed -1 ribosomal frameshifting. *Nucleic Acids Res.* 33 (5): 1553-1563.

Marcheschi RJ, Mouzakis KD, and Butcher SE. 2009. Selection and characterization of small molecules that bind the HIV-1 frameshift site RNA. *ACS Chem. Biol.* 4 (10): 844-854.

Marczinke B, Bloys AJ, Brown TD, Willcocks MM, Carter MJ, and Brierley I. 1994. The human astrovirus RNA-dependent RNA polymerase coding region is expressed by ribosomal frameshifting. *J. Virol.* 68 (9): 5588-5595.

Marczinke B, Fisher R, Vidakovic M, Bloys AJ, and Brierley I. 1998. Secondary structure and mutational analysis of the ribosomal frameshift signal of rous sarcoma virus. *J. Mol. Biol.* 284 (2): 205-225.

Matsufuji S, Matsufuji T, Miyazaki Y, Murakami Y, Atkins JF, Gesteland RF, and Hayashi S. 1995. Autoregulatory frameshifting in decoding mammalian ornithine decarboxylase antizyme. *Cell* 80 (1): 51-60.

Mazauric MH, Seol Y, Yoshizawa S, Visscher K, and Fourmy D. 2009. Interaction of the HIV-1 frameshift signal with the ribosome. *Nucleic Acids Res.* 37 (22): 7654-7664.

McDonagh P, Sheehy PA, and Norris JM. 2011. In vitro inhibition of feline coronavirus replication by small interfering RNAs. *Vet. Microbiol.*

McNaughton BR, Gareiss PC, and Miller BL. 2007. Identification of a selective small-molecule ligand for HIV-1 frameshift-inducing stem-loop RNA from an 11,325 member resin bound dynamic combinatorial library. *J Am. Chem. Soc.* 129 (37): 11306-11307.

Melian EB, Hinzman E, Nagasaki T, Firth AE, Wills NM, Nouwens AS, Blitvich BJ, Leung J, Funk A, Atkins JF, Hall R, and Khromykh AA. 2010. NS1' of flaviviruses in the Japanese encephalitis virus serogroup is a product of ribosomal frameshifting and plays a role in viral neuroinvasiveness. *J. Virol.* 84 (3): 1641-1647.

Meskauskas A, and Dinman JD. 2001. Ribosomal protein L5 helps anchor peptidyl-tRNA to the P-site in Saccharomyces cerevisiae. *RNA* 7 (8): 1084-1096.

Michiels PJ, Versleijen AA, Verlaan PW, Pleij CW, Hilbers CW, and Heus HA. 2001. Solution structure of the pseudoknot of SRV-1 RNA, involved in ribosomal frameshifting. *J. Mol. Biol.* 310 (5): 1109-1123.

Moosmayer D, Reil H, Ausmeier M, Scharf JG, Hauser H, Jentsch KD, and Hunsmann G. 1991. Expression and frameshifting but extremely inefficient proteolytic processing of the HIV-1 gag and pol gene products in stably transfected rodent cell lines. *Virology* 183 (1): 215-224.

Morikawa S, and Bishop DH. 1992. Identification and analysis of the gag-pol ribosomal frameshift site of feline immunodeficiency virus. *Virology* 186 (2): 389-397.

Nam SH, Copeland TD, Hatanaka M, and Oroszlan S. 1993. Characterization of ribosomal frameshifting for expression of pol gene products of human T-cell leukemia virus type I. *J. Virol.* 67 (1): 196-203.

Napthine S, Liphardt J, Bloys A, Routledge S, and Brierley I. 1999. The role of RNA pseudoknot stem 1 length in the promotion of efficient -1 ribosomal frameshifting. *J. Mol. Biol.* 288 (3): 305-320.

Napthine S, Vidakovic M, Girnary R, Namy O, and Brierley I. 2003. Prokaryotic-style frameshifting in a plant translation system: conservation of an unusual single-tRNA slippage event. *EMBO J.* 22 (15): 3941-3950.

Neuman BW, Stein DA, Kroeker AD, Churchill MJ, Kim AM, Kuhn P, Dawson P, Moulton HM, Bestwick RK, Iversen PL, and Buchmeier MJ. 2005. Inhibition, escape, and attenuated growth of severe acute respiratory syndrome coronavirus treated with antisense morpholino oligomers. *J. Virol.* 79 (15): 9665-9676.

Nibert ML. 2007. '2A-like' and 'shifty heptamer' motifs in penaeid shrimp infectious myonecrosis virus, a monosegmented double-stranded RNA virus. *J. Gen. Virol.* 88 (Pt 4): 1315-1318.

Nixon PL, Rangan A, Kim YG, Rich A, Hoffman DW, Hennig M, and Giedroc DP. 2002. Solution structure of a luteoviral P1-P2 frameshifting mRNA pseudoknot. *J. Mol. Biol.* 322 (3): 621-633.

Ogle JM, and Ramakrishnan V. 2005. Structural insights into translational fidelity. *Annu. Rev. Biochem.* 74: 129-177.

Okamoto K, Nagano H, Iwakawa H, Mizumoto H, Takeda A, Kaido M, Mise K, and Okuno T. 2008. cis-Preferential requirement of a -1 frameshift product p88 for the replication of Red clover necrotic mosaic virus RNA1. *Virology* 375 (1): 205-212.

Palde PB, Ofori LO, Gareiss PC, Lerea J, and Miller BL. 2010. Strategies for recognition of stem-loop RNA structures by synthetic ligands: application to the HIV-1 frameshift stimulatory sequence. *J. Med. Chem.* 53 (16): 6018-6027.

Pallan PS, Marshall WS, Harp J, Jewett FC, III, Wawrzak Z, Brown BA, Rich A, and Egli M. 2005. Crystal structure of a luteoviral RNA pseudoknot and model for a minimal ribosomal frameshifting motif. *Biochemistry* 44 (34): 11315-11322.

Park SJ, Kim YG, and Park HJ. 2011. Identification of RNA Pseudoknot-Binding Ligand That Inhibits the -1 Ribosomal Frameshifting of SARS-Coronavirus by Structure-Based Virtual Screening. *J Am. Chem. Soc.*

Parkin NT, Chamorro M, and Varmus HE. 1992. Human immunodeficiency virus type 1 gag-pol frameshifting is dependent on downstream mRNA secondary structure: demonstration by expression in vivo. *J. Virol.* 66 (8): 5147-5151.

Paul CP, Barry JK, Dinesh-Kumar SP, Brault V, and Miller WA. 2001. A sequence required for -1 ribosomal frameshifting located four kilobases downstream of the frameshift site. *J. Mol. Biol.* 310 (5): 987-999.

Pawson T, Martin GS, and Smith AE. 1976. Cell-free translation of virion RNA from nondefective and transformation-defective Rous sarcoma viruses. *J. Virol.* 19 (3): 950-967.

Pennell S, Manktelow E, Flatt A, Kelly G, Smerdon SJ, and Brierley I. 2008. The stimulatory RNA of the Visna-Maedi retrovirus ribosomal frameshifting signal is an unusual pseudoknot with an interstem element. *RNA.* 14 (7): 1366-1377.

Plant EP, Wang P, Jacobs JL, and Dinman JD. 2004. A programmed -1 ribosomal frameshift signal can function as a cis-acting mRNA destabilizing element. *Nucleic Acids Res.* 32 (2): 784-790.

Plant EP, Perez-Alvarado GC, Jacobs JL, Mukhopadhyay B, Hennig M, and Dinman JD. 2005. A three-stemmed mRNA pseudoknot in the SARS coronavirus frameshift signal. *PLoS. Biol.* 3 (6): e172.

Plant EP, and Dinman JD. 2006. Comparative study of the effects of heptameric slippery site composition on -1 frameshifting among different eukaryotic systems. *RNA* 12 (4): 666-673.

Plant EP, Rakauskaite R, Taylor DR, and Dinman JD. 2010. Achieving a golden mean: mechanisms by which coronaviruses ensure synthesis of the correct stoichiometric ratios of viral proteins. *J. Virol.* 84 (9): 4330-4340.

Prüfer D, Tacke E, Schmitz J, Kull B, Kaufmann A, and Rohde W. 1992. Ribosomal frameshifting in plants: a novel signal directs the -1 frameshift in the synthesis of the putative viral replicase of potato leafroll luteovirus. *EMBO J.* 11 (3): 1111-1117.

Rakauskaite R, Liao PY, Rhodin MH, Lee K, and Dinman JD. 2011. A rapid, inexpensive yeast-based dual-fluorescence assay of programmed--1 ribosomal frameshifting for high-throughput screening. *Nucleic Acids Res.*

Reil H, Kollmus H, Weidle UH, and Hauser H. 1993. A heptanucleotide sequence mediates ribosomal frameshifting in mammalian cells. *J. Virol.* 67 (9): 5579-5584.

Rice NR, Stephens RM, Burny A, and Gilden RV. 1985. The gag and pol genes of bovine leukemia virus: nucleotide sequence and analysis. *Virology* 142 (2): 357-377.

Salem NM, Miller WA, Rowhani A, Golino DA, Moyne AL, and Falk BW. 2008. Rose spring dwarf-associated virus has RNA structural and gene-expression features like those of Barley yellow dwarf virus. *Virology*. 375 (2): 354-360.

Shehu-Xhilaga M, Crowe SM, and Mak J. 2001. Maintenance of the Gag/Gag-Pol ratio is important for human immunodeficiency virus type 1 RNA dimerization and viral infectivity. *J. Virol*. 75 (4): 1834-1841.

Shen LX, and Tinoco I, Jr. 1995. The structure of an RNA pseudoknot that causes efficient frameshifting in mouse mammary tumor virus. *J. Mol. Biol*. 247 (5): 963-978.

Sivakumaran K, Fowler BC, and Hacker DL. 1998. Identification of viral genes required for cell-to-cell movement of southern bean mosaic virus. *Virology* 252 (2): 376-386.

Snijder EJ, den Boon JA, Bredenbeek PJ, Horzinek MC, Rijnbrand R, and Spaan WJ. 1990. The carboxyl-terminal part of the putative Berne virus polymerase is expressed by ribosomal frameshifting and contains sequence motifs which indicate that toro- and coronaviruses are evolutionarily related. *Nucleic. Acids Res*. 18 (15): 4535-4542.

Somogyi P, Jenner AJ, Brierley I, and Inglis SC. 1993. Ribosomal pausing during translation of an RNA pseudoknot. *Mol. Cell Biol*. 13 (11): 6931-6940.

Stahl G, Bidou L, Rousset JP, and Cassan M. 1995. Versatile vectors to study recoding: conservation of rules between yeast and mammalian cells. *Nucleic Acids Res*. 23 (9): 1557-1560.

Staple DW, and Butcher SE. 2005. Solution structure and thermodynamic investigation of the HIV-1 frameshift inducing element. *J. Mol. Biol*. 349 (5): 1011-1023.

Staple DW, Venditti V, Niccolai N, Elson-Schwab L, Tor Y, and Butcher SE. 2008. Guanidinoneomycin B recognition of an HIV-1 RNA helix. *Chembiochem*. 9 (1): 93-102.

Su L, Chen L, Egli M, Berger JM, and Rich A. 1999. Minor groove RNA triplex in the crystal structure of a ribosomal frameshifting viral pseudoknot. *Nat. Struct. Biol*. 6 (3): 285-292.

Su MC, Chang CT, Chu CH, Tsai CH, and Chang KY. 2005. An atypical RNA pseudoknot stimulator and an upstream attenuation signal for -1 ribosomal frameshifting of SARS coronavirus. *Nucleic Acids Res*. 33 (13): 4265-4275.

Sundararajan A, Michaud WA, Qian Q, Stahl G, and Farabaugh PJ. 1999. Near-cognate peptidyl-tRNAs promote +1 programmed translational frameshifting in yeast. *Mol. Cell* 4 (6): 1005-1015.

Sung D, and Kang H. 1998. Mutational analysis of the RNA pseudoknot involved in efficient ribosomal frameshifting in simian retrovirus-1. *Nucleic Acids Res*. 26 (6): 1369-1372.

Tamm T, Suurvali J, Lucchesi J, Olspert A, and Truve E. 2009. Stem-loop structure of Cocksfoot mottle virus RNA is indispensable for programmed -1 ribosomal frameshifting. *Virus Res*. 146 (1-2): 73-80.

Taylor EW, Ramanathan CS, Jalluri RK, and Nadimpalli RG. 1994. A basis for new approaches to the chemotherapy of AIDS: novel genes in HIV-1 potentially encode selenoproteins expressed by ribosomal frameshifting and termination suppression. *J Med. Chem*. 37 (17): 2637-2654.

Telenti A, Martinez R, Munoz M, Bleiber G, Greub G, Sanglard D, and Peters S. 2002. Analysis of natural variants of the human immunodeficiency virus type 1 gag-pol frameshift stem-loop structure. *J. Virol*. 76 (15): 7868-7873.

ten Dam E, Brierley I, Inglis S, and Pleij C. 1994. Identification and analysis of the pseudoknot-containing gag-pro ribosomal frameshift signal of simian retrovirus-1. *Nucleic Acids Res.* 22 (12): 2304-2310.

Tu C, Tzeng TH, and Bruenn JA. 1992. Ribosomal movement impeded at a pseudoknot required for frameshifting. *Proc. Natl. Acad. Sci. U. S. A* 89 (18): 8636-8640.

Tzeng TH, Tu CL, and Bruenn JA. 1992. Ribosomal frameshifting requires a pseudoknot in the Saccharomyces cerevisiae double-stranded RNA virus. *J. Virol.* 66 (2): 999-1006.

Vassilaki N, and Mavromara P. 2003. Two alternative translation mechanisms are responsible for the expression of the HCV ARFP/F/core+1 coding open reading frame. *J. Biol. Chem.* 278 (42): 40503-40513.

Vickers TA, and Ecker DJ. 1992. Enhancement of ribosomal frameshifting by oligonucleotides targeted to the HIV gag-pol region. *Nucleic Acids Res.* 20 (15): 3945-3953.

Vimaladithan A, and Farabaugh PJ. 1994. Special peptidyl-tRNA molecules can promote translational frameshifting without slippage. *Mol. Cell Biol.* 14 (12): 8107-8116.

Walewski JL, Keller TR, Stump DD, and Branch AD. 2001. Evidence for a new hepatitis C virus antigen encoded in an overlapping reading frame. *RNA* 7 (5): 710-721.

Wills NM, Gesteland RF, and Atkins JF. 1994. Pseudoknot-dependent read-through of retroviral gag termination codons: importance of sequences in the spacer and loop 2. *EMBO J* 13 (17): 4137-4144.

Wilson W, Malim MH, Mellor J, Kingsman AJ, and Kingsman SM. 1986. Expression strategies of the yeast retrotransposon Ty: a short sequence directs ribosomal frameshifting. *Nucleic Acids Res.* 14 (17): 7001-7016.

Wilson W, Braddock M, Adams SE, Rathjen PD, Kingsman SM, and Kingsman AJ. 1988. HIV expression strategies: ribosomal frameshifting is directed by a short sequence in both mammalian and yeast systems. *Cell* 55 (6): 1159-1169.

Wolin SL, and Walter P. 1988. Ribosome pausing and stacking during translation of a eukaryotic mRNA. *EMBO J.* 7 (11): 3559-3569.

Xiong Z, Kim KH, Kendall TL, and Lommel SA. 1993. Synthesis of the putative red clover necrotic mosaic virus RNA polymerase by ribosomal frameshifting in vitro. *Virology* 193 (1): 213-221.

Xu H, and Boeke JD. 1990. Host genes that influence transposition in yeast: the abundance of a rare tRNA regulates Ty1 transposition frequency. *Proc. Natl. Acad. Sci. U. S. A* 87 (21): 8360-8364.

Xu Z, Choi J, Yen TS, Lu W, Strohecker A, Govindarajan S, Chien D, Selby MJ, and Ou J. 2001. Synthesis of a novel hepatitis C virus protein by ribosomal frameshift. *EMBO J.* 20 (14): 3840-3848.

Yu ET, Zhang Q, and Fabris D. 2005. Untying the FIV frameshifting pseudoknot structure by MS3D. *J. Mol. Biol.* 345 (1): 69-80.

Zhai Y, Attoui H, Mohd JF, Wang HQ, Cao YX, Fan SP, Sun YX, Liu LD, Mertens PP, Meng WS, Wang D, and Liang G. 2010. Isolation and full-length sequence analysis of Armigeres subalbatus totivirus, the first totivirus isolate from mosquitoes representing a proposed novel genus (Artivirus) of the family Totiviridae. *J. Gen. Virol.* 91 (Pt 11): 2836-2845.

Ziebuhr J. 2005. The coronavirus replicase. *Curr. Top. Microbiol. Immunol.* 287: 57-94.

7

Hepatitis B Virus X Protein:
A Key Regulator of the Virus Life Cycle

Julie Lucifora and Ulrike Protzer
*Institute of Virology, Technische Universität
München / Helmholtz Zentrum München*
Germany

1. Introduction

Hepatitis B virus (HBV) is one of the most important human pathogens. The outcome of HBV infection as well as the severity of HBV-induced liver disease varies widely from one patient to another. In around 90-95% of adults, exposure to HBV leads to an acute infection which is rapidly cleared without long-term consequences. The remaining 5-10% fail to control viral infection that consequently evolves to chronicity. The rate of chronicity of viral infection is dramatically higher (up to 90%) in neonates born from infected mothers, suggesting that infection around birth successfully induces peripheral tolerance to viral antigens which prevents clearance. About 2 billion humans have been infected by HBV worldwide and more than 350 million are chronic carriers. The latter have high risk to develop severe liver disease, including liver cirrhosis and hepatocellular carcinoma. Around 600,000 persons die each year due to consequences of hepatitis B infection. As HBV is a non-cytopatic virus, HBV-related liver damage very likely results from the immune response against infected hepatocytes which is activated but not strong enough to clear infection.

Our knowledge of the molecular biology of HBV has increased considerably over the past decades, leading to the development of very effective prophylactic vaccines and to the development of direct antivirals active against HBV. Five nucleos(t)ide analogs are currently approved to treat chronic hepatitis B. Belonging to the same class of nucleosidic reverse transcriptase inhibitors, they specifically inhibit viral polymerase activity and thus suppress HBV replication, significantly improving liver histology and the clinical outcomes of the disease after one year of treatment (Liaw *et al.*, 2004). Unfortunately, nucleos(t)ide analogs act at a late stage in the HBV life cycle (i.e. maturation of newly formed viral capsids by reverse transcription of pregenomic RNA) and do neither prevent formation and nuclear establishment nor activity of the HBV transcription template, the so called HBV covalently closed circular (ccc) DNA.

Long-term treatments with nucleos(t)ide analogs are thus necessary to cure HBV infected cells and unfortunately lead to the selection of HBV drug-resistant strains (Zoulim, 2006). Even very effective antivirals such as Tenofovir lead to HBsAg seroconversion in only 3 to 8% of patients over three years (Heathcote *et al.*, 2011; van Bommel *et al.*, 2010). Pegylated (PEG)-IFN-α is an established treatment alternative and acts as an antiviral but also enhances the host's immune defense. However, only 30% of PEG-IFN-α -treated patients achieve a sustained antiviral response (Karayiannis, 2003), and only about 8-10 % of patients clear the virus (Marcellin *et al.*, 2009) with slightly increasing rates during long-term follow-up (Moucari *et al.*, 2009). New therapeutic approaches that target other viral proteins, besides viral polymerase, are needed to decrease viral drug resistance and improve treatments against HBV.

This chapter will particularly focus on the hepatitis B virus X protein (HBx) that is essential to initiate and maintain transcription of HBV RNA from nuclear cccDNA and thus is a key regulator of the virus life cycle. Due to its central role, HBx represents a very promising new target for antiviral strategies against HBV.

1.1 Hepatitis B virus structure and proteins

HBV belongs to the family *hepadnaviridae*. It is a small, enveloped DNA virus that replicates via reverse transcription of an RNA intermediate. HBV virions, also called Dane particles, are spherical lipid-containing structures with a diameter of ~42 nm (Fig. 1). The inner shell

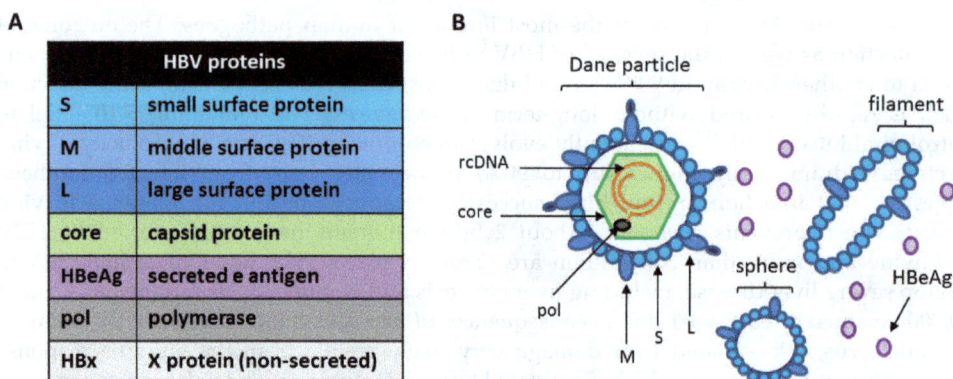

A

HBV proteins	
S	small surface protein
M	middle surface protein
L	large surface protein
core	capsid protein
HBeAg	secreted e antigen
pol	polymerase
HBx	X protein (non-secreted)

B

Fig. 1. HBV proteins and virion structure. (A) List of all HBV proteins. (B) Viral particles present in the serum of HBV-infected patients are schematically represented. The so-called "Dane particles" are fully infectious viral particles containing the HBV capsid and one rcDNA genome copy with the viral polymerase attached. Subviral particles of spherical or filamentous shapes consist of empty viral envelopes. Together, Dane particles, spheres, and filaments are recognized as HBsAg. The precore protein is secreted as HBeAg.

of the virus consists of an icosahedral capsid, which is assembled from 180 or 240 subunits of the core protein. The capsid is covered by a lipid bilayer membrane densely packed with the three envelope proteins, large (L), middle (M), and predominantly small (S) protein, and is acquired by budding into the endoplasmic reticulum. They are translated from individual start codons but share the open reading frame and the same C-terminal amino acids, called the S domain. As a consequence, the M protein shares the S and has an extra N-terminal domain called preS2, and the L protein encompasses the S and two extra domains: preS2 and preS1. Capsids contain a single copy of the HBV genome consisting of a 3.2-kb partially double-stranded relaxed circular (rc) DNA molecule. The viral polymerase serves as a protein primer and remains covalently linked to the 5' end of the complete strand, also called viral (–) strand DNA of the rcDNA after reverse transcription. Besides virions, HBV infection leads to secretion of huge amounts of subviral particles, which consist of empty viral envelopes with filamentous or spherical shapes (Fig. 1) containing mainly S and little L protein. Subviral particles are the most abundant HBV structures released into the bloodstream, are commonly defined as hepatitis B surface (HBs) antigen and are thought to facilitate virus spread and persistence in the host by adsorbing virus-neutralizing antibodies and tolerizing T cell responses.

In addition to polymerase and the structural proteins, the HBV genome also encodes for two non-structural proteins, which have less well-defined functions. Secreted HBeAg may have immunoregulatory functions (Bertoletti & Gehring, 2006; Chen *et al.*, 2005; Chen *et al.*, 2004; Visvanathan *et al.*, 2007), whereas HBx seems to have multiple key functions as it will be detailed later.

1.2 Overview of the hepatitis B virus life cycle

A schematic overview of the HBV life cycle is depicted below in Fig. 2. HBV infection is restricted to hepatocytes. HBV entry into these cells is thought to be a multistep process. Virions are first trapped at the surface of the cell by heparan sulfate proteoglycans (Schulze *et al.*, 2007) and then bind to a receptor allowing uptake into the cells via an endocytosis process (Kott, 2010; Leistner *et al.*, 2008). So far, this cellular receptor as not been identified. Proteolytic cleavage of the surface protein occurs within the endosomal compartment, probably resulting in a conformational change that exposes some translocation motifs at the surface of the viral particle allowing fusion of viral and cellular membranes and release of the capsid into the cytosol (Stoeckl *et al.*, 2006).

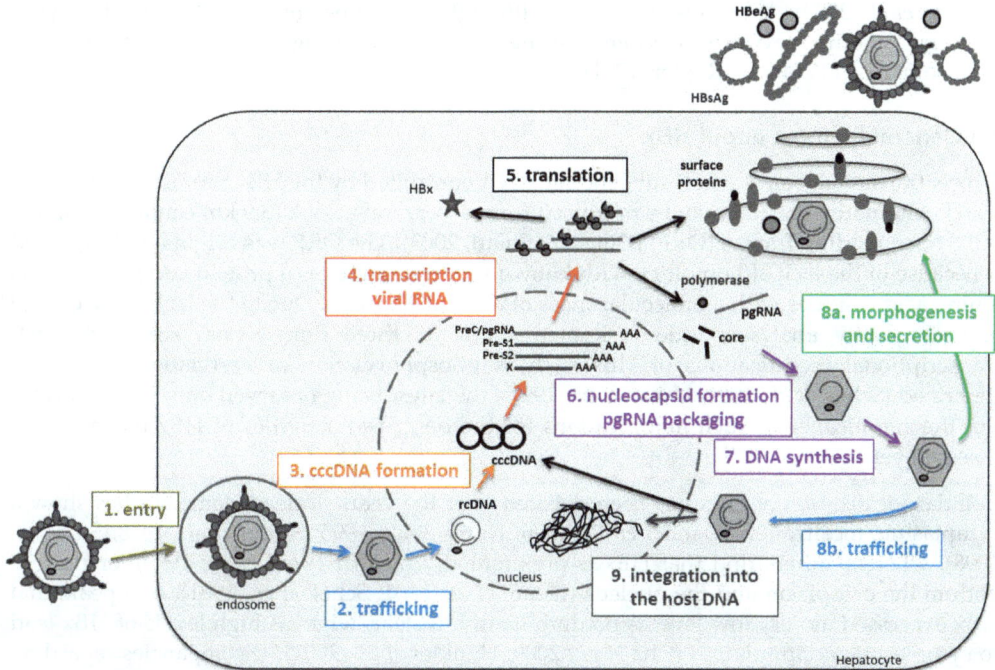

Fig. 2. Schematic overview of the HBV life cycle.

The naked capsid is then directed towards the nucleus, and the HBV genome is translocated to the nucleus (Rabe *et al.*, 2006). In the nucleus, the rcDNA genome is converted by cellular enzymes into a covalently closed circular DNA (cccDNA), the episomal persistance form of the virus serving as transcription template. The 3.5 kb RNA species serves as pregenomic RNA (pgRNA) and as messenger RNAs for the synthesis of polymerase and core proteins as well as HBeAg. The 2.1 and 2.4 kb subgenomic RNAs encode for the three viral envelope proteins, a small 0.7 kbRNA for the HBx. The pgRNA is exported in an unspliced form, encapisidated together with the viral polymerase and used as a template for reverse transcription. The capsid spontaneously self-assembles from core dimers present in the cytoplasm (Zlotnick *et al.*, 1999) due to the nucleic acid-binding domain of the core protein. Specific packaging of pgRNA into the capsid is mediated by binding of the primer region of the viral polymerase to the ε stem-loop in the 5′ region of pgRNA (Hirsch *et al.*, 1990; Junker-Niepmann *et al.*, 1990; Knaus & Nassal, 1993; Nassal, 1992; Porterfield *et al.*, 2010). The pgRNA is then reverse transcribed by the reverse transcriptase domain of the polymerase within the capsid in the cytoplasm of the infected cell. Upon minus and then plus strand DNA synthesis the capsid matures and can be enveloped or reimported into the nucleus to fill up a cccDNA pool.

HBV budding has been shown to be strictly dependent on the L protein (Bruss & Vieluf, 1995): when the ratio between L proteins and nucleocapsids is not optimal, the latter are preferentially targeted to the nucleus to amplify the cccDNA pool (Summers *et al.*, 1990). Whether HBV virions bud into the endoplasmic reticulum or late endosomes or multivesicular bodies, before they exit the cell via the exosome pathway, is not entirely clear (Patient *et al.*, 2009). As an alternative and although it is not essential for the HBV life cycle, the viral genome may also integrate into the host genome using cellular enzymes such as topoisomerase I (Wang & Rogler, 1991).

1.3 General features about HBx

HBx is translated from a small subgenomic RNA controlled by the HBx promoter (Guo *et al.*, 1991). Alternatively, HBx may be produced form a very long RNA (3.9 kb) containing all the HBV open reading frames (ORF) (Doitsh & Shaul, 2003). The ORF was originally designated X because of the lack of homology with known sequences. HBx is a protein composed of 154 amino acid residues with a molecular mass of around 17.5 kDa. Due to the lack of successful crystallography analyses, little is known about its three dimensional structure. Post-transcriptional modifications of HBx such as phosphorylation or acetylation have been described (Schek *et al.*, 1991; Urban *et al.*, 1997), the latest being observed only in insect cells. But the significance of such modifications for the described activities of HBx has not been assessed yet.

Cellular localization of HBx has been debated over the years. Indeed, some studies show a cytoplasmic localization (Dandri *et al.*, 1996; Doria *et al.*, 1995; Sirma *et al.*, 1998; Su *et al.*, 1998), whereas others find that HBx is preferentially nuclear (Weil *et al.*, 1999), or present both in the cytoplasm and the nucleus (Hoare *et al.*, 2001; Schek *et al.*, 1991). It appears that HBx expressed at very low level is predominantly nuclear, whereas high levels of HBx lead to cytoplasmic accumulation (Cha *et al.*, 2009; Henkler *et al.*, 2001). Discrepancies regarding HBx localization could thus be attributed to variations of HBx expression levels according to the models used for the experiments in the different studies.

Cellular localization of HBx was shown to influence the half-life of the protein. Indeed, the pool of HBx associated with the cytoskeleton and nuclear framework has a longer half-life (around 3 h) than the one associated with the cytosolic fraction (15 to 20 min) (Dandri *et al.*, 1998; Schek *et al.*, 1991). Both ubiquitin-dependent and ubiquitin-independent mechanisms have been involved in HBx turnover (Hu *et al.*, 1999; Kim *et al.*, 2008).

2. Importance of HBx for HBV infection

In the woodchuck model of HBV infection, it was shown that the woodchuck hepatitis virus (WHV) X protein (WHx) is essential for the establishment of viral infection *in vivo* (Chen *et al.*, 1993; Zoulim *et al.*, 1994). Indeed, injection of WHV wild type genomes into the liver of woodchuck lead to WHV infection of all the tested animals whereas no replication was observed when genomes deficient for WHx expression were injected (Chen *et al.*, 1993; Zoulim *et al.*, 1994). Few years later, it was observed that animals injected with WHx-defective mutants eventually developed a low viremia after an extended period of time (Zhang *et al.*, 2001), suggesting that this WHx-defective mutant were not completely defective but largely attenuated for HBV replication *in vivo*. Accordingly, genotypic reversions to wild type WHV were observed in all animals inoculated with WHx-deficient mutants (Zhang *et al.*, 2001). Taken together, these results point out the importance of WHx for a productive and long lasting WHV infection.

In addition, it was shown that HBx-deficient HBV genomes are somewhat compromised for HBV replication using HBV hydrodynamically-injected mice (Keasler *et al.*, 2007; 2009) or cell culture models (Belloni *et al.*, 2009; Blum *et al.*, 1992; Keasler *et al.*, 2007; Leupin *et al.*, 2005). Surprisingly, the absence of HBx had no effect on HBV replication in human hepatoma Huh7 cell lines, but impaired replication in HepG2 cells (Blum *et al.*, 1992; Keasler *et al.*, 2007; 2009; Leupin *et al.*, 2005). Accordingly, data in HBV transgenic mice are contradictory with some mouse lines showing reduced replication (Xu *et al.*, 2002), whereas others replicate HBV to high levels (Dumortier *et al.*, 2005).

The importance of HBx in the context of human HBV infection was demonstrated very recently using human hepatocyte chimeric mice and relevant cellular models of HBV infection. Indeed, it was observed that mice injected with HBx deficient HBV virus developed measurable viremia only in HBx-expressing livers (Tsuge *et al.*, 2010). Moreover, using primary human hepatocyte (Schulze-Bergkamen *et al.*, 2003) and differentiated HepaRG cells (Gripon *et al.*, 2002), that are the only two models of HBV infection *in vitro*, we recently demonstrated that HBx is essential to initiate and constantly required to maintain productive HBV infection (Lucifora *et al.*, 2011).

This latter study highlighted the importance of performing experiments in relevant *in vitro* and *in vivo* models. Indeed, results obtained with *in vitro* HBV infection models (i.e. primary human hepatocyte and differentiated HepaRG cells) (Lucifora *et al.*, 2011) support and explain the above mentioned observations obtained in mouse livers (Keasler *et al.*, 2007; Tsuge *et al.*). However, they differ from results obtained by transfection of linearized HBV genomes into transformed cells (Blum *et al.*, 1992; Leupin *et al.*, 2005) especially when HBx is overexpressed to non-physiological levels. Solving this apparent discrepancy, we were able to demonstrate that HBx is essential when HBV transcription is initiated from its natural transcription circular template (cccDNA) but not from a linearized 1.3-fold genome length

HBV genome (Lucifora *et al.*, 2011) containing a duplicate copy of the HBx open reading frame 5′ of the HBV genome (Reifenberg *et al.*, 2002; Sprinzl *et al.*, 2001; Zhang *et al.*, 2004) – irrespective of whether the linearized HBV genome is integrated or episomal.

3. Functions of HBx in the HBV life cycle

Different functions have been attributed to HBx regarding HBV life cycle (Fig. 3).

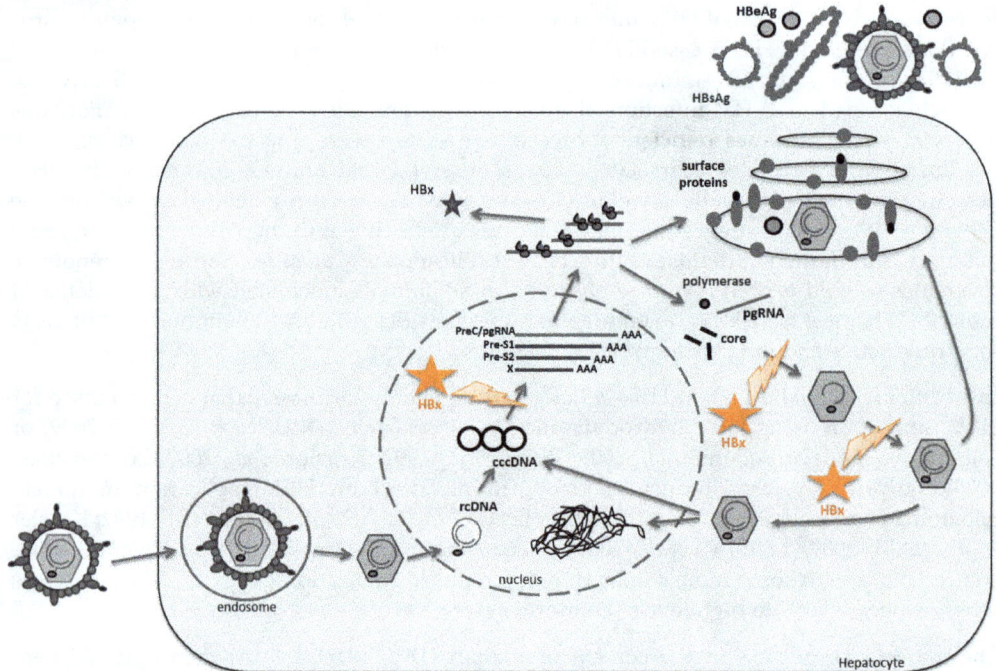

Fig. 3. Functions attributed to HBx in the HBV life cycle. HBx is an important regulator of HBV transcription. Moreover, it might also enhance pgRNA encapsidation and viral polymerase activity.

Several studies have shown that HBx can stimulate HBV replication by activating viral transcription (Cha *et al.*, 2009; Leupin *et al.*, 2005; Tang *et al.*, 2005; Zhang *et al.*, 2004; Zhang *et al.*, 2001) or enhancing viral polymerase activity via calcium signalling pathways (Bouchard *et al.*, 2003; Bouchard *et al.*, 2001; Klein *et al.*, 1999). HBx was also proposed to enhance pgRNA encapsidation by increasing phosphorylation of the viral core protein (Melegari *et al.*, 2005) although these results were recently challenged (Cha *et al.*, 2009).

We recently showed that HBx does not determine the ability of HBV to enter the host cell or to deposit functional nuclear cccDNA but is essential for viral transcription from its natural transcription template, the nuclear HBV cccDNA (Lucifora *et al.*, 2011). Indeed primary human hepatocytes or differentiated HepaRG cells inoculated with different HBV virions, HBV(wt) and HBV(x-) established comparable amounts of nuclear transcription templates but in contrast to HBV(wt), transcription of HBV RNAs and expression of HBV proteins was dramatically impaired in cells inoculated with HBV(x-) (Lucifora *et al.*, 2011). Trans-complementation of HBx in HBV(x-)-infected cells was able to rescue HBV transcription, antigen secretion and replication even weeks after infection. This demonstrated that HBx-deficient cccDNA is fully functional and very stable, but also that HBx is necessary to initiate and maintain HBV replication after infection of human hepatocytes (Lucifora *et al.*, 2011).

Our results complement a series of data indicating that HBx has an important role in epigenetic regulation of HBV transcription from cccDNA. Indeed, cccDNA can persist in the cell nucleus as a stable chromatin-like episome (Bock *et al.*, 2001) and was shown to be submitted to epigenetic modifications such H3 and H4 histone acetylations when HBV was actively replicating (Pollicino *et al.*, 2006). Besides cellular proteins such as histone acetyltransferases and histone deacetylases, HBx is also recruited onto the cccDNA with a kinetic paralleling HBV replication (Belloni *et al.*, 2009). Moreover, in the absence of HBx, the acetylation of cccDNA-bound histones H4 was significantly reduced (Belloni *et al.*, 2009; Lucifora *et al.*, 2011), the recruitment of the histone acetyltransferase p300 was severely impaired whereas the recruitment of the histone deacetylases hSirt1 and HDAC1 was increased and occured at earlier times (Belloni *et al.*, 2009).

The differences mentioned above in the regulation of viral transcription from cccDNA and from linearized HBV genomes (which are present in all plasmid constructs and stable cell lines) may help to explain, why the function of HBx was evaluated differently when different HBV constructs were used (Blum *et al.*, 1992; Bouchard *et al.*, 2001; Melegari *et al.*, 2005; Reifenberg *et al.*, 2002; Sprinzl *et al.*, 2001). However, transcriptional regulation by HBx may also depend on the cell type used, since transformed cells may lack or antagonize cellular proteins with a positive or negative influence on viral transcription.

Although HBx is essential for the expression of the other viral proteins, no evidence for packaging of HBx into the HBV particle has been provided (Lucifora *et al.*, 2011). Therefore, the question of how HBx expression itself is induced and regulated remains open. Different hypotheses may apply. First, HBx mRNA transcription may be specifically regulated and may occur before transcription of the other HBV RNAs. This implies the question whether an early-late shift exists for HBV such as for most other viruses – with HBx as an early protein essential for expression of the remaining (late) proteins. Some studies performed in transfection models support this assumption (Doitsh & Shaul, 2004; Wu *et al.*, 1991) suggesting that HBV may express its gene products in a defined order.

A second hypothesis does not require the presence of HBx in the early phase of HBV infection. If HBV transcription from cccDNA starts shortly after infection independent from HBx, this would lead to the production of all the HBV proteins including HBx. Subsequent activation of a cellular response controlling HBV replication and/or binding of cellular restriction factor(s) could - in the absence of HBx - inhibit HBV transcription from cccDNA. HBx would here be essential to prevent inhibition of HBV transcription by cell-intrinsic mechanisms. Since HBx would have to up-regulate its own expression in a "positive feed-back loop", this would explain why a lag phase is observed before HBV replication starts after infection in all the HBV infection models (Dandri *et al.*, 2005; Gripon *et al.*, 1988; Gripon *et al.*, 2002; Walter *et al.*, 1996; Wieland *et al.*, 2004). Whether one of these hypotheses or a third one explains dependency of HBV replication on HBx is currently investigated.

4. HBx influences many cellular processes

Besides its role in HBV replication, thousands of publications showed that HBx interacts with various cellular partners and modifies many cellular processes including transcription, cell cycle progression, DNA damage repair, apoptosis and carcinogenesis (for review, see Benhenda *et al.*, 2009; Bouchard & Schneider, 2004; Wei *et al.*, 2010). As we will show with the following examples, interactions of HBx with cellular components may represent an attempt of the virus to manipulate the cellular context in order to stimulate virus replication and spread.

HBx has been described to be a weak transactivator able to activate HBV promoters and enhancers as well as many different cellular promoters (Yen, 1996). Whereas, HBx does not seem to directly bind to DNA, its transactivation activity was reported to occur via several DNA binding sites such as NF-KB, AP-1, c-EBP, ATF/CREB, NF-AT, SP1 etc. (for review, see Quasdorff & Protzer, 2010; Yen, 1996).

Different studies have shown an interplay between HBx and apoptosis pathways. Indeed, HBx could sensitize the cells to apoptotic signals such as treatments with TNF or doxorubicin, oxidative stress or growth factor deprivation (for review, see Benhenda *et al.*, 2009; Bouchard & Schneider, 2004; Wei *et al.*, 2010). This may promote hepatocyte regeneration, thus providing a larger reservoir of cells for infection. However HBx may also prevent apoptosis induction since it rapidly blocks spread of HBV progeny (Arzberger *et al.*, 2010).

HBx may also be involved in cell cycle regulation but its relative influence seems to differ according to the models used (for review, see Benhenda *et al.*, 2009; Bouchard & Schneider, 2004; Wei *et al.*, 2010). For example, using primary rat hepatocytes, it was recently demonstrated that HBx induces normally quiescent hepatocytes to enter the G_1 phase of the cell cycle and that this calcium-dependent HBx activity is required for HBV replication (Gearhart & Bouchard, 2010). While this effect of HBx on cell cycle progression can probably lead to carcinogenesis and thus become deleterious for the host, it is believed that it might be important for the virus to induce expansion of available deoxynucleoside triphosphate pools within the cells which it needs for replication (Bouchard *et al.*, 2003). Indeed, using HepG2 cells, it was reported that HBx is sufficient for the induction of the R2 subunit of the ribonucleotide reductase (RNR) (Cohen *et al.*, 2010). RNR is the key enzyme responsible for *de novo* dNTP synthesis and is composed of R1 and R2 subunits (Nordlund & Reichard, 2006). While the R1 subunit is expressed in quiescent cells, the R2 subunit expression is silenced (Chabes *et al.*, 2003). As a consequence of induction of R2 by HBx, the dNTP pool for effective viral production was increased without affecting cell cycle progression (Cohen *et al.*, 2010).

Different groups using different models showed that HBx may localize and interact with the proteasome components thereby influencing proteasome subunit composition (Chen *et al.*, 2001; Fischer *et al.*, 1995; Hu *et al.*, 1999; Zhang *et al.*, 2000). Moreover proteasome inhibition was shown to enhance HBV replication in cell culture and in mice models (Zhang *et al.*, 2004; Zhang *et al.*, 2010). Indeed, in the presence of proteasome inhibitors, the replication of the wild-type virus was not affected, while the replication of the HBx-negative virus was enhanced and restored to the wild-type level (Zhang *et al.*, 2004; Zhang *et al.*, 2010). Thus HBx may functions through the inhibition of proteasome activities to enhance HBV replication.

Finally, several studies have pointed out an interaction between HBx and the DNA repair protein DDB1 that would be essential for HBV infection (Leupin *et al.*, 2005; Sitterlin *et al.*, 2000). However, the exact mechanism by which this interaction may help the virus is still debated.

Of note, most of the interactions of HBx with cellular processes have been studied in many different models often leading to significant overexpression of HBx and outside the context of HBV infection. Thus, it remains important to determine whether similar manipulations of the cellular machinery by HBx would also occur in the context of an authentic HBV infection.

5. Conclusion

Numerous and significant studies have been performed over the past decades to analyze the role of HBx in the HBV life cycle. Many data were generated by using different *in vivo* and *in vitro* models, but contradictory results describing HBx function were obtained. The importance and the precise role of HBx on HBV life cycle thus remained unclear until recently models allowing an authentic HBV infection were used (Lucifora *et al.*, 2011; Tsuge *et al.*, 2010). Most studies, including the most recent, agree that HBx is essential for HBV infection. Besides its importance for HBV transcription from nuclear HBV cccDNA, it may also influence downstream steps of the HBV life cycle possibly by manipulating different cellular machineries. Unfortunately, in the long-term, these manipulations are probably leading to hepatocellular de-differentiation and progression towards liver cancer. As HBx plays a central role in HBV infection and cannot avoid influencing many cellular processes related to disease progression, it may be a very interesting target for new therapies against chronic hepatitis B. Targeting HBx may prevent both: viral replication as well as liver tissue damage and carcinogenesis.

6. Acknowledgement

Julie Lucifora holds a stipend from the European Association for the Study of Liver disease (EASL): "Sheila Sherlock EASL Post-Doc Fellowship".

7. References

Arzberger, S., Hosel, M. & Protzer, U. (2010). Apoptosis of hepatitis B virus-infected hepatocytes prevents release of infectious virus. J Virol 84, 11994-12001.

Belloni, L., Pollicino, T., De Nicola, F., Guerrieri, F., Raffa, G., Fanciulli, M., Raimondo, G. & Levrero, M. (2009). Nuclear HBx binds the HBV minichromosome and modifies the epigenetic regulation of cccDNA function. Proc Natl Acad Sci U S A 106, 19975-19979.

Benhenda, S., Cougot, D., Buendia, M. A. & Neuveut, C. (2009). Hepatitis B virus X protein molecular functions and its role in virus life cycle and pathogenesis. Adv Cancer Res 103, 75-109.

Bertoletti, A. & Gehring, A. J. (2006). The immune response during hepatitis B virus infection. J Gen Virol 87, 1439-1449.

Blum, H. E., Zhang, Z. S., Galun, E., von Weizsacker, F., Garner, B., Liang, T. J. & Wands, J. R. (1992). Hepatitis B virus X protein is not central to the viral life cycle in vitro. J Virol 66, 1223-1227.

Bock, C. T., Schwinn, S., Locarnini, S., Fyfe, J., Manns, M. P., Trautwein, C. & Zentgraf, H. (2001). Structural organization of the hepatitis B virus minichromosome. J Mol Biol 307, 183-196.

Bouchard, M. J., Puro, R. J., Wang, L. & Schneider, R. J. (2003). Activation and inhibition of cellular calcium and tyrosine kinase signaling pathways identify targets of the HBx protein involved in hepatitis B virus replication. J Virol 77, 7713-7719.

Bouchard, M. J. & Schneider, R. J. (2004). The enigmatic X gene of hepatitis B virus. J Virol 78, 12725-12734.

Bouchard, M. J., Wang, L. H. & Schneider, R. J. (2001). Calcium signaling by HBx protein in hepatitis B virus DNA replication. Science 294, 2376-2378.

Bruss, V. & Vieluf, K. (1995). Functions of the internal pre-S domain of the large surface protein in hepatitis B virus particle morphogenesis. J Virol 69, 6652-6657.

Cha, M. Y., Ryu, D. K., Jung, H. S., Chang, H. E. & Ryu, W. S. (2009). Stimulation of hepatitis B virus genome replication by HBx is linked to both nuclear and cytoplasmic HBx expression. J Gen Virol 90, 978-986.

Chabes, A. L., Pfleger, C. M., Kirschner, M. W. & Thelander, L. (2003). Mouse ribonucleotide reductase R2 protein: a new target for anaphase-promoting complex-Cdh1-mediated proteolysis. Proc Natl Acad Sci U S A 100, 3925-3929.

Chen, H. S., Kaneko, S., Girones, R., Anderson, R. W., Hornbuckle, W. E., Tennant, B. C., Cote, P. J., Gerin, J. L., Purcell, R. H. & Miller, R. H. (1993). The woodchuck hepatitis virus X gene is important for establishment of virus infection in woodchucks. J Virol 67, 1218-1226.

Chen, M., Sallberg, M., Hughes, J., Jones, J., Guidotti, L. G., Chisari, F. V., Billaud, J. N. & Milich, D. R. (2005). Immune tolerance split between hepatitis B virus precore and core proteins. J Virol 79, 3016-3027.

Chen, M. T., Billaud, J. N., Sallberg, M., Guidotti, L. G., Chisari, F. V., Jones, J., Hughes, J. & Milich, D. R. (2004). A function of the hepatitis B virus precore protein is to regulate the immune response to the core antigen. Proc Natl Acad Sci U S A 101, 14913-14918.

Chen, W. N., Oon, C. J. & Goo, K. S. (2001). Hepatitis B virus X protein in the proteasome of mammalian cells: defining the targeting domain. Mol Biol Rep 28, 31-34.

Cohen, D., Adamovich, Y., Reuven, N. & Shaul, Y. (2010). Hepatitis B virus activates deoxynucleotide synthesis in nondividing hepatocytes by targeting the R2 gene. Hepatology 51, 1538-1546.

Dandri, M., Burda, M. R., Zuckerman, D. M., Wursthorn, K., Matschl, U., Pollok, J. M., Rogiers, X., Gocht, A., Kock, J., Blum, H. E., von Weizsacker, F. & Petersen, J. (2005). Chronic infection with hepatitis B viruses and antiviral drug evaluation in uPA mice after liver repopulation with tupaia hepatocytes. J Hepatol 42, 54-60.

Dandri, M., Petersen, J., Stockert, R. J., Harris, T. M. & Rogler, C. E. (1998). Metabolic labeling of woodchuck hepatitis B virus X protein in naturally infected hepatocytes reveals a bimodal half-life and association with the nuclear framework. J Virol 72, 9359-9364.

Dandri, M., Schirmacher, P. & Rogler, C. E. (1996). Woodchuck hepatitis virus X protein is present in chronically infected woodchuck liver and woodchuck hepatocellular carcinomas which are permissive for viral replication. J Virol 70, 5246-5254.

Doitsh, G. & Shaul, Y. (2003). A long HBV transcript encoding pX is inefficiently exported from the nucleus. Virology 309, 339-349.

Doitsh, G. & Shaul, Y. (2004). Enhancer I predominance in hepatitis B virus gene expression. Mol Cell Biol 24, 1799-1808.

Doria, M., Klein, N., Lucito, R. & Schneider, R. J. (1995). The hepatitis B virus HBx protein is a dual specificity cytoplasmic activator of Ras and nuclear activator of transcription factors. Embo J 14, 4747-4757.

Dumortier, J., Schonig, K., Oberwinkler, H., Low, R., Giese, T., Bujard, H., Schirmacher, P. & Protzer, U. (2005). Liver-specific expression of interferon gamma following adenoviral gene transfer controls hepatitis B virus replication in mice. Gene Ther 12, 668-677.

Fischer, M., Runkel, L. & Schaller, H. (1995). HBx protein of hepatitis B virus interacts with the C-terminal portion of a novel human proteasome alpha-subunit. Virus Genes 10, 99-102.

Gearhart, T. L. & Bouchard, M. J. (2010). The hepatitis B virus X protein modulates hepatocyte proliferation pathways to stimulate viral replication. J Virol 84, 2675-2686.

Gripon, P., Diot, C., Theze, N., Fourel, I., Loreal, O., Brechot, C. & Guguen-Guillouzo, C. (1988). Hepatitis B virus infection of adult human hepatocytes cultured in the presence of dimethyl sulfoxide. J Virol 62, 4136-4143.

Gripon, P., Rumin, S., Urban, S., Le Seyec, J., Glaise, D., Cannie, I., Guyomard, C., Lucas, J., Trepo, C. & Guguen-Guillouzo, C. (2002). Infection of a human hepatoma cell line by hepatitis B virus. Proc Natl Acad Sci U S A 99, 15655-15660.

Guo, W. T., Wang, J., Tam, G., Yen, T. S. & Ou, J. S. (1991). Leaky transcription termination produces larger and smaller than genome size hepatitis B virus X gene transcripts. Virology 181, 630-636.

Heathcote, E. J., Marcellin, P., Buti, M., Gane, E., De Man, R. A., Krastev, Z., Germanidis, G., Lee, S. S., Flisiak, R., Kaita, K., Manns, M., Kotzev, I., Tchernev, K., Buggisch, P., Weilert, F., Kurdas, O. O., Shiffman, M. L., Trinh, H., Gurel, S., Snow-Lampart, A., Borroto-Esoda, K., Mondou, E., Anderson, J., Sorbel, J. & Rousseau, F. (2011). Three-year efficacy and safety of tenofovir disoproxil fumarate treatment for chronic hepatitis B. Gastroenterology 140, 132-143.

Henkler, F., Hoare, J., Waseem, N., Goldin, R. D., McGarvey, M. J., Koshy, R. & King, I. A. (2001). Intracellular localization of the hepatitis B virus HBx protein. J Gen Virol 82, 871-882.

Hirsch, R. C., Lavine, J. E., Chang, L. J., Varmus, H. E. & Ganem, D. (1990). Polymerase gene products of hepatitis B viruses are required for genomic RNA packaging as wel as for reverse transcription. Nature 344, 552-555.

Hoare, J., Henkler, F., Dowling, J. J., Errington, W., Goldin, R. D., Fish, D. & McGarvey, M. J. (2001). Subcellular localisation of the X protein in HBV infected hepatocytes. J Med Virol 64, 419-426.

Hu, Z., Zhang, Z., Doo, E., Coux, O., Goldberg, A. L. & Liang, T. J. (1999). Hepatitis B virus X protein is both a substrate and a potential inhibitor of the proteasome complex. J Virol 73, 7231-7240.

Junker-Niepmann, M., Bartenschlager, R. & Schaller, H. (1990). A short cis-acting sequence is required for hepatitis B virus pregenome encapsidation and sufficient for packaging of foreign RNA. EMBO J 9, 3389-3396.

Karayiannis, P. (2003). Hepatitis B virus: old, new and future approaches to antiviral treatment. J Antimicrob Chemother 51, 761-785.

Keasler, V. V., Hodgson, A. J., Madden, C. R. & Slagle, B. L. (2007). Enhancement of hepatitis B virus replication by the regulatory X protein in vitro and in vivo. J Virol 81, 2656-2662.

Keasler, V. V., Hodgson, A. J., Madden, C. R. & Slagle, B. L. (2009). Hepatitis B virus HBx protein localized to the nucleus restores HBx-deficient virus replication in HepG2 cells and in vivo in hydrodynamically-injected mice. Virology 390, 122-129.

Kim, J. H., Sohn, S. Y., Benedict Yen, T. S. & Ahn, B. Y. (2008). Ubiquitin-dependent and - independent proteasomal degradation of hepatitis B virus X protein. Biochem Biophys Res Commun 366, 1036-1042.

Klein, N. P., Bouchard, M. J., Wang, L. H., Kobarg, C. & Schneider, R. J. (1999). Src kinases involved in hepatitis B virus replication. EMBO J 18, 5019-5027.

Knaus, T. & Nassal, M. (1993). The encapsidation signal on the hepatitis B virus RNA pregenome forms a stem-loop structure that is critical for its function. Nucleic Acids Res 21, 3967-3975.

Kott, N., König, A., Glebe, D. (2010). Hepatitis B virus (HBV) bypasses classical endocytic pathways to infect primary hepatocytes in vitro. Journal of Hepatology 52, S48-S49.

Leistner, C. M., Gruen-Bernhard, S. & Glebe, D. (2008). Role of glycosaminoglycans for binding and infection of hepatitis B virus. Cell Microbiol 10, 122-133.

Leupin, O., Bontron, S., Schaeffer, C. & Strubin, M. (2005). Hepatitis B virus X protein stimulates viral genome replication via a DDB1-dependent pathway distinct from that leading to cell death. J Virol 79, 4238-4245.

Liaw, Y. F., Sung, J. J., Chow, W. C., Farrell, G., Lee, C. Z., Yuen, H., Tanwandee, T., Tao, Q. M., Shue, K., Keene, O. N., Dixon, J. S., Gray, D. F. & Sabbat, J. (2004). Lamivudine for patients with chronic hepatitis B and advanced liver disease. N Engl J Med 351, 1521-1531.

Lucifora, J., Arzberger, S., Durantel, D., Belloni, L., Strubin, M., Levrero, M., Zoulim, F., Hantz, O. & Protzer, U. (2011). Hepatitis B Virus X protein is essential to initiate and maintain virus replication after infection. J Hepatol. 2011 55, 996-1003.

Marcellin, P., Bonino, F., Lau, G. K., Farci, P., Yurdaydin, C., Piratvisuth, T., Jin, R., Gurel, S., Lu, Z. M., Wu, J., Popescu, M. & Hadziyannis, S. (2009). Sustained response of hepatitis B e antigen-negative patients 3 years after treatment with peginterferon alpha-2a. Gastroenterology 136, 2169-2179 e2161-2164.

Melegari, M., Wolf, S. K. & Schneider, R. J. (2005). Hepatitis B virus DNA replication is coordinated by core protein serine phosphorylation and HBx expression. J Virol 79, 9810-9820.

Moucari, R., Korevaar, A., Lada, O., Martinot-Peignoux, M., Boyer, N., Mackiewicz, V., Dauvergne, A., Cardoso, A. C., Asselah, T., Nicolas-Chanoine, M. H., Vidaud, M., Valla, D., Bedossa, P. & Marcellin, P. (2009). High rates of HBsAg seroconversion in HBeAg-positive chronic hepatitis B patients responding to interferon: a long-term follow-up study. J Hepatol 50, 1084-1092.

Nassal, M. (1992). The arginine-rich domain of the hepatitis B virus core protein is required for pregenome encapsidation and productive viral positive-strand DNA synthesis but not for virus assembly. J Virol 66, 4107-4116.

Nordlund, P. & Reichard, P. (2006). Ribonucleotide reductases. Annu Rev Biochem 75, 681-706.

Patient, R., Hourioux, C. & Roingeard, P. (2009). Morphogenesis of hepatitis B virus and its subviral envelope particles. Cell Microbiol 11, 1561-1570.

Pollicino, T., Belloni, L., Raffa, G., Pediconi, N., Squadrito, G., Raimondo, G. & Levrero, M. (2006). Hepatitis B virus replication is regulated by the acetylation status of hepatitis B virus cccDNA-bound H3 and H4 histones. Gastroenterology 130, 823-837.

Porterfield, J. Z., Dhason, M. S., Loeb, D. D., Nassal, M., Stray, S. J. & Zlotnick, A. (2010). Full-length hepatitis B virus core protein packages viral and heterologous RNA with similarly high levels of cooperativity. J Virol 84, 7174-7184.

Quasdorff, M. & Protzer, U. (2010). Control of hepatitis B virus at the level of transcription. J Viral Hepat 17, 527-536.

Rabe, B., Glebe, D. & Kann, M. (2006). Lipid-mediated introduction of hepatitis B virus capsids into nonsusceptible cells allows highly efficient replication and facilitates the study of early infection events. J Virol 80, 5465-5473.

Reifenberg, K., Nusser, P., Lohler, J., Spindler, G., Kuhn, C., von Weizsacker, F. & Kock, J. (2002). Virus replication and virion export in X-deficient hepatitis B virus transgenic mice. J Gen Virol 83, 991-996.

Schek, N., Bartenschlager, R., Kuhn, C. & Schaller, H. (1991). Phosphorylation and rapid turnover of hepatitis B virus X-protein expressed in HepG2 cells from a recombinant vaccinia virus. Oncogene 6, 1735-1744.

Schulze-Bergkamen, H., Untergasser, A., Dax, A., Vogel, H., Buchler, P., Klar, E., Lehnert, T., Friess, H., Buchler, M. W., Kirschfink, M., Stremmel, W., Krammer, P. H., Muller, M. & Protzer, U. (2003). Primary human hepatocytes--a valuable tool for investigation of apoptosis and hepatitis B virus infection. J Hepatol 38, 736-744.

Schulze, A., Gripon, P. & Urban, S. (2007). Hepatitis B virus infection initiates with a large surface protein-dependent binding to heparan sulfate proteoglycans. Hepatology 46, 1759-1768.

Sirma, H., Weil, R., Rosmorduc, O., Urban, S., Israel, A., Kremsdorf, D. & Brechot, C. (1998). Cytosol is the prime compartment of hepatitis B virus X protein where it colocalizes with the proteasome. Oncogene 16, 2051-2063.

Sitterlin, D., Bergametti, F., Tiollais, P., Tennant, B. C. & Transy, C. (2000). Correct binding of viral X protein to UVDDB-p127 cellular protein is critical for efficient infection by hepatitis B viruses. Oncogene 19, 4427-4431.

Sprinzl, M. F., Oberwinkler, H., Schaller, H. & Protzer, U. (2001). Transfer of hepatitis B virus genome by adenovirus vectors into cultured cells and mice: crossing the species barrier. J Virol 75, 5108-5118.

Stoeckl, L., Funk, A., Kopitzki, A., Brandenburg, B., Oess, S., Will, H., Sirma, H. & Hildt, E. (2006). Identification of a structural motif crucial for infectivity of hepatitis B viruses. Proc Natl Acad Sci U S A 103, 6730-6734.

Su, Q., Schroder, C. H., Hofmann, W. J., Otto, G., Pichlmayr, R. & Bannasch, P. (1998). Expression of hepatitis B virus X protein in HBV-infected human livers and hepatocellular carcinomas. Hepatology 27, 1109-1120.

Summers, J., Smith, P. M. & Horwich, A. L. (1990). Hepadnavirus envelope proteins regulate covalently closed circular DNA amplification. J Virol 64, 2819-2824.

Tang, H., Delgermaa, L., Huang, F., Oishi, N., Liu, L., He, F., Zhao, L. & Murakami, S. (2005). The transcriptional transactivation function of HBx protein is important for its augmentation role in hepatitis B virus replication. J Virol 79, 5548-5556.

Tsuge, M., Hiraga, N., Akiyama, R., Tanaka, S., Matsushita, M., Mitsui, F., Abe, H., Kitamura, S., Hatakeyama, T., Kimura, T., Miki, D., Mori, N., Imamura, M., Takahashi, S., Hayes, C. N. & Chayama, K. (2010). HBx protein is indispensable for development of viraemia in human hepatocyte chimeric mice. J Gen Virol 91, 1854-1864.

Urban, S., Hildt, E., Eckerskorn, C., Sirma, H., Kekule, A. & Hofschneider, P. H. (1997). Isolation and molecular characterization of hepatitis B virus X-protein from a baculovirus expression system. Hepatology 26, 1045-1053.

van Bommel, F., de Man, R. A., Wedemeyer, H., Deterding, K., Petersen, J., Buggisch, P., Erhardt, A., Huppe, D., Stein, K., Trojan, J., Sarrazin, C., Bocher, W. O., Spengler,

U., Wasmuth, H. E., Reinders, J. G., Moller, B., Rhode, P., Feucht, H. H., Wiedenmann, B. & Berg, T. (2010). Long-term efficacy of tenofovir monotherapy for hepatitis B virus-monoinfected patients after failure of nucleoside/nucleotide analogues. Hepatology 51, 73-80.

Visvanathan, K., Skinner, N. A., Thompson, A. J., Riordan, S. M., Sozzi, V., Edwards, R., Rodgers, S., Kurtovic, J., Chang, J., Lewin, S., Desmond, P. & Locarnini, S. (2007). Regulation of Toll-like receptor-2 expression in chronic hepatitis B by the precore protein. Hepatology 45, 102-110.

Walter, E., Keist, R., Niederost, B., Pult, I. & Blum, H. E. (1996). Hepatitis B virus infection of tupaia hepatocytes in vitro and in vivo. Hepatology 24, 1-5.

Wang, H. P. & Rogler, C. E. (1991). Topoisomerase I-mediated integration of hepadnavirus DNA in vitro. J Virol 65, 2381-2392.

Wei, Y., Neuveut, C., Tiollais, P. & Buendia, M. A. (2010). Molecular biology of the hepatitis B virus and role of the X gene. Pathol Biol (Paris) 58, 267-272.

Weil, R., Sirma, H., Giannini, C., Kremsdorf, D., Bessia, C., Dargemont, C., Brechot, C. & Israel, A. (1999). Direct association and nuclear import of the hepatitis B virus X protein with the NF-kappaB inhibitor IkappaBalpha. Mol Cell Biol 19, 6345-6354.

Wieland, S., Thimme, R., Purcell, R. H. & Chisari, F. V. (2004). Genomic analysis of the host response to hepatitis B virus infection. Proc Natl Acad Sci U S A 101, 6669-6674.

Wu, H. L., Chen, P. J., Lin, M. H. & Chen, D. S. (1991). Temporal aspects of major viral transcript expression in Hep G2 cells transfected with cloned hepatitis B virus DNA: with emphasis on the X transcript. Virology 185, 644-651.

Xu, Z., Yen, T. S., Wu, L., Madden, C. R., Tan, W., Slagle, B. L. & Ou, J. H. (2002). Enhancement of hepatitis B virus replication by its X protein in transgenic mice. J Virol 76, 2579-2584.

Yen, T. S. (1996). Hepadnaviral X Protein:Review of Recent Progress. J Biomed Sci 3, 20-30.

Zhang, Z., Protzer, U., Hu, Z., Jacob, J. & Liang, T. J. (2004). Inhibition of cellular proteasome activities enhances hepadnavirus replication in an HBX-dependent manner. J Virol 78, 4566-4572.

Zhang, Z., Sun, E., Ou, J. H. & Liang, T. J. (2010). Inhibition of cellular proteasome activities mediates HBX-independent hepatitis B virus replication in vivo. J Virol 84, 9326-9331.

Zhang, Z., Torii, N., Furusaka, A., Malayaman, N., Hu, Z. & Liang, T. J. (2000). Structural and functional characterization of interaction between hepatitis B virus X protein and the proteasome complex. J Biol Chem 275, 15157-15165.

Zhang, Z., Torii, N., Hu, Z., Jacob, J. & Liang, T. J. (2001). X-deficient woodchuck hepatitis virus mutants behave like attenuated viruses and induce protective immunity in vivo. J Clin Invest 108, 1523-1531.

Zlotnick, A., Johnson, J. M., Wingfield, P. W., Stahl, S. J. & Endres, D. (1999). A theoretical model successfully identifies features of hepatitis B virus capsid assembly. Biochemistry 38, 14644-14652.

Zoulim, F., Lucifora, J. (2006). Hepatitis B virus and drug resistance: implications for treatment. Future Virology 1, 361-376.

Zoulim, F., Saputelli, J. & Seeger, C. (1994). Woodchuck hepatitis virus X protein is required for viral infection in vivo. J Virol 68, 2026-2030.

Part 3

Genomic Sequence Diversity and Evolution

8

Microarray Techniques for Evaluation of Genetic Stability of Live Viral Vaccines

Majid Laassri[1], Elena Cherkasova[2],
Mones S. Abu-Asab[3] and Konstantin Chumakov[1]
[1]*Center for Biologics Evaluation and Research,*
U.S. Food and Drug Administration, Rockville,
[2]*National Heart, Lung, and Blood Institute,*
National Institutes of Health, Bethesda,
[3]*National Cancer Institute, National Institutes of Health, Bethesda*
USA

1. Introduction

Recent advances in biotechnology gave rise to a set of microarray technologies that became ubiquitous in research, medicine, and industry for tens of applications. Microarray technology has centered on providing platforms for analyzing, in a single experiment, tens or even hundreds of samples from different biologic sources. Its rapid and global adoption has been predicated on its simplicity and efficiency in quickly providing relevant data generated by simultaneous testing of biological samples with a large number of probes. In this chapter, we describe new microarray approaches that have considerably simplified the characterization of viral genes and genomes with specific emphasis on analysis of the genetic stability of live viral vaccines.

There are different types of vaccines that immunize against viruses: whole viral vaccines (either live attenuated or inactivated), subunit vaccines; purified or recombinant viral antigen vaccines, and DNA vaccines.

Genetic instability and plasticity of genomes are inherent properties of viruses, especially RNA viruses, with many profound implications for their replication, evolution, and pathogenesis. Because of the presence of a large number of mutants, populations of viruses are often described as quasispecies (Domingo et al., 1985; Hansen et al., 2004). Most mutants are present at a relatively low level, making them difficult to detect using conventional sequencing methods.

Genetic stability of live viral vaccines, including recombinant virus vaccines, is a key element of their safety and protective efficacy. Assessment of genetic stability is an important part of pre-licensure evaluation and quality control of a live viral vaccine, both during its manufacturing and after its administration. Spontaneous mutations easily emerge during viral replication and accumulation of mutants must be identified to ensure the safety of live vaccines.

To ensure maximum genetic stability and to optimize genetic structure of prospective live vaccine strains, it is important to identify the mutations that accumulate both during manufacturing and replication in vaccine recipients. Incorporation of mutations that increase virus fitness and do not affect its attenuation into the genetic makeup of the new vaccine strain may increase its potency and contribute to genetic stability

Most new viral vaccines are produced by propagation in cell cultures that do not necessarily represent the natural substrate for the virus, raising the possibility of introducing undesirable mutations in the course of virus adaptation. RNA viral vaccines mutate easily upon passage in cell cultures, which can change the phenotype (Amexis et al., 2001), leading to increased pathogenicity. That occurred with pSPBNGA-GA, a live rabies virus recombinant vaccine candidate, which was obtained via reverse genetics (Dietzschold et al., 2004). Additionally, it was demonstrated that some deletions in HIV-1 vaccine strains can evolve into fast-replicating variants by multiplication of remaining sequence motifs, and their safety is therefore not guaranteed (Berkhout et al., 1999), and the presence of even a small fraction of viral mutants in an oral poliovirus vaccine can have negative effect on its safety (Chumakov et al., 1991), suggesting that genetic consistency must be carefully monitored to ensure that accumulated mutants do not adversely impair the safety and efficacy of the vaccine.

In addition, vaccines have been recently developed to serve as live viral vectors expressing heterologous host genes. Examples of such live viral vectors include herpesviruses (such as pseudorabies and bovine herpesvirus type 1, and 2), poxviruses (Blancou et al., 1986; Fekadu et al., 1991; Taylor et al., 1991), human adenovirus 5, (Prevec et al., 1990) and flaviviruses (Arroyo et al., 2001; Monath et al., 1999; Pletnev et al., 2001; Pletnev et al., 1992; Pletnev et al., 2000; Pletnev and Men, 1998; Pletnev et al., 2002; Pletnev et al., 2006). The recombination in genomes of chimeric viruses probably plays an important role in the reduction of viral fitness that leads to attenuation. This creates a selective pressure to accumulate mutants that restore viral fitness by adapting heterologous genomic parts to each other, potentially leading to a loss of attenuation. The accumulation of mutations and genetic stability of flaviviruses were previously reported (Dunster et al., 1999; Laassri et al., 2011; Pugachev et al., 2004; Pugachev et al., 2002; Pugachev et al., 2007).

Of paramount importance is the need to demonstrate the genetic stability of the recombinant construct and confirm the fidelity of the heterologous gene inserted into the vector genome (WHO, 1990). An important consideration for the licensure and use of any genetically engineered live vaccine is the stability of the vector and that of the recombinant construct. In addition, genetic stability is an important safety concern, since predictions of vaccine behavior rely heavily on the knowledge of the genetic makeup of the recombinant. If a recombinant vaccine is to be useful, it should undergo no substantial mutation either during production of the vaccine by passage of working seed or after administration to the target species.

Therefore, it is essential to identify genomic loci that are prone to mutations and determine their phenotypes. If mutations in unstable genomic loci increase virulence, then methods to prevent their emergence and control their presence in vaccine preparations must be developed. On the other hand, if the fitness-restoring mutants do not lead to de-attenuation, then it may be desirable to incorporate them into the genetic makeup of the vaccine strain.

Additionally, increased yields of such viruses during vaccine production may help stabilize the genome by relieving selective pressure, thereby preventing random and potentially undesirable mutations from being passively selected through the "passenger effect".

Conventional assays of genetic stability of viruses by combined sequencing and sequence analysis are generally too insensitive to detect small proportion of mutant viruses in a quasispecies, and are laborious. Thus, sensitive high-throughput microarray techniques including microarray for resequencing and sequence heterogeneity (MARSH), microarray analysis of viral recombination (MAVR) assay, and microarray for quantitation of known virulent mutations (MQNVM) have been developed and applied as valuable tools to evaluate the genetic stability of live viral vaccines (Cherkasova et al., 2003; Laassri et al., 2011; Laassri et al., 2005; Laassri et al., 2007).

These microarray approaches allow large-scale full-genome mutational screening of live viral vaccines from various sources including cell culture, humans, monkeys, and mice. They can be used to improve quality control and to accelerate development of safer and more effective vaccines. The study of genetic stability also contributes to our understanding of live viral vaccine evolution.

2. DNA microarray: An overview

DNA microarray is a high-throughput hybridization technology used for quantitative and qualitative assessments of gene-expression, chromosomal aberrations, and mutations in molecular biology and biotechnology. It consists of an arrayed series of tens or thousands of micro-spots of oligonucleotides of specific DNA sequence, known as probes. Probes can be short regions of a gene or other DNA elements used to hybridize DNA or RNA samples (targets) under high-stringency conditions. Probe-target hybridizations are usually detected and quantified using fluorophore-labeled targets to determine the relative quantities of nucleic acid sequences in the target. Since an array can contain tens of thousands of probes, a microarray experiment enables analysis of many genes simultaneously. Therefore, arrays have dramatically accelerated many types of investigations.

In standard microarrays, the probes are attached via their engineered chemical group to a solid surface by a covalent bond to a chemical matrix (via epoxy - silane, amino-silane, lysine, polyacrylamide, or others). The solid surface can be glass, a silicon chip, or microscopic beads.

DNA microarrays can be used to measure changes in expression level, to detect and quantify single nucleotide polymorphisms (SNPs), to genotype and resequence mutant genomes, or to determine recombinant nucleotide sequences. Microarrays vary in fabrication, operating protocols, accuracy, and efficiency. Additional factors affecting microarray experiments are experimental design and methods for analyzing the data.

3. Microarray for resequencing and sequence heterogeneity

Several approaches based on hybridization of viral probes with oligonucleotide microarrays have been applied for rapid analysis of genetic variations during the microevolution of viruses. Microarray for resequencing and sequence heterogeneity (MARSH) was used to identify mutations in vaccine viruses and their derivatives, revealing the degree of their evolutionary divergence and quantifying mutant genome proportions present.

The MARSH assay was based on the hybridization of fluorescently-labeled RNA produced from the virus genome with microarrays of oligonucleotide probes that are complementary to and cover the entire viral genome or specific genes. Quantitative comparison of hybridization data produced for a test sample with the data for homogeneous reference RNA reveal mutations that have emerged and accumulated during the replication of vaccine strains *in vitro* or *in vivo* (Figure 1).

MARSH (Figure 1) microchips were fabricated using a set of short oligonucleotides (T_m ~ 50°C) overlapping at half length, matching genomic sequences of virus strains, and covering a specific viral region of interest in the genome or the entire viral genome. Each oligonucleotide probe was synthesized with an aminolink group at its 5' end for immobilization on a specific platform and purified after automated synthesis. Microarrays were printed on sialylated (aldehyde-coated) glass slides by using a contact microspotting robot equipped with a microspotting pin. Each oligonucleotide probe was spotted several times within a single microarray for redundancy to increase the reliability of results.

For RNA viruses, cDNA was prepared with reverse transcriptase using a specific reverse primer at the 3' end region of the genome. Microarrays of immobilized oligonucleotide probes were hybridized with fluorescently-labeled RNA transcribed by T7 RNA polymerase from PCR-amplified viral cDNA. First, the viral genome was amplified using the specific primers (the reverse primer contains T7 promotor) to produce the needed DNA segments. RNA for hybridization was produced by *in vitro* transcription of the PCR products with a T7 RNA polymerase kit. Each RNA product (~10 µg) can be fluorescently labeled with a Cy3 RNA Labeling Kit. Labeled RNA samples were purified using spin columns.

Microarray hybridization was performed as follow: fluorescently-labeled RNA samples were vacuum-dried prior to hybridization, reconstituted in Hybridization Buffer, and denatured by incubation for one minute at 95°C. The final concentration of each fluorescent target in the hybridization solution should not exceed 0.1 µM. An aliquot of the hybridization mixture (~10 µl) was applied to the microarray area and covered with an individual plastic cover slip. Hybridization was performed in an incubation chamber for one hour at 45°C. Fluorescent images of processed microarray slides were captured using a ScanArray 5000, and the images were analyzed using ScanArray Express software.

In the experiments presented on Figure 1, each microarray contained 4 identical sub-arrays that were simultaneously hybridized in order to assess reproducibility of the hybridization results and to eliminate outlier data points. Hybridization signals from individual sub-array elements that differ from the average value were calculated for all 4 replicates of the oligonucleotide probes by more than two standard deviations were discarded; the number of such invalidated data points should not exceed 0.1%. Next, average values from the 4 sub-arrays were normalized by the total fluorescence signal from the entire array. Finally, normalized signals from the reference sample (homogeneous RNA) were divided by the respective normalized signals from the test samples, and the results were expressed as a fluorescence ratio (Figure 1C). Regions with no mutations should have ratios close to one, while test samples with mutations reduced hybridization with some oligonucleotide probes and therefore, produce ratios greater than one.

The MARSH assay was first used to analyze mutations that accumulate in the region coding for VP1, the most variable capsid protein of poliovirus (Cherkasova et al., 2003; Laassri et al., 2005). Later, this microarray approach was expanded to discriminate between vaccinia strains and to evaluate genetic stability of the vaccinia virus Ankara (MVA) B5R gene following propagation of a cloned isolate of MVA in Vero and MRC-5 cell lines (Laassri et al., 2007), to analyze the variability of the structural region of West Nile (WN) virus (Grinev et al., 2008), and to evaluate stability of the entire genome of a WN/Dengue 4 chimeric virus under study as a new candidate of WN vaccine (Laassri et al., 2011).

The MARSH microarray approach facilitates rapid analysis of viral genes and genomes, and circumvents traditional more laborious methods. With the microarray method, many samples can be analyzed simultaneously within a few hours. Furthermore, test samples do not need to be cloned, thus preserving the natural composition of viral gene populations. This method permits large-scale full genome screening of viral isolates, useful for epidemiological surveillance, vaccine quality control, and analysis of genetic changes in viruses that may occur in response to drug treatment.

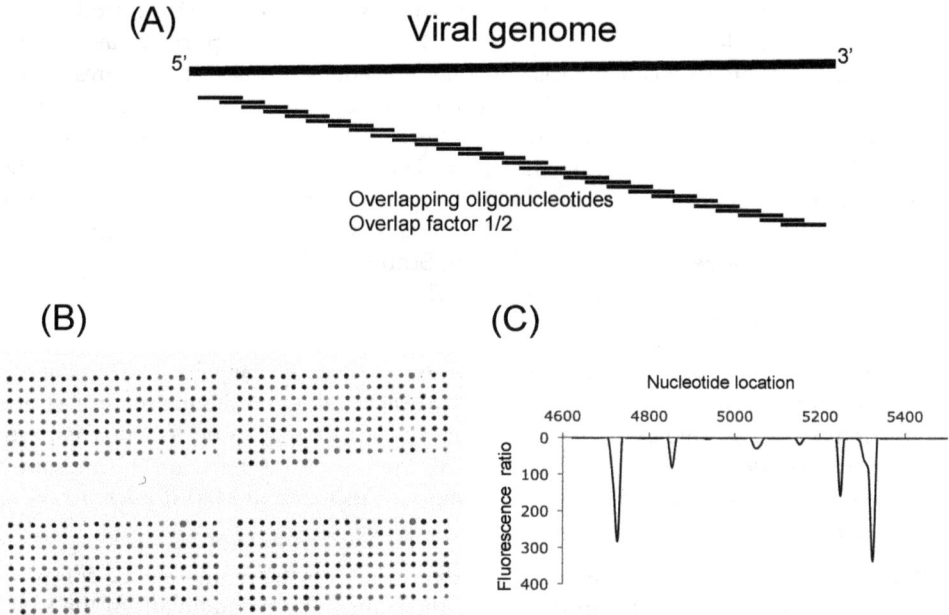

Fig. 1. Schematic overview of the MARSH assay: (A) Oligonucleotide microarray containing short oligonucleotides overlapped at half-length covering the entire viral genome. Each microarray has four identical sub-arrays hybridized at the same time. (B) Images of 4 individual identical sub-arrays hybridized with control sample. (C) Ratio of hybridization signals from reference and test sample preparations was plotted to reveal peaks indicating the presence of mutations. The hybridization images and the ratio plot are cartoons given for illustration purposes only.

4. Microarray analysis of viral recombination assay

Microarray analysis of viral recombination (MAVR) assay is, in essence, an extension of the genotype-specific oligonucleotide microarrays previously used to identify different viruses and bacteria (Laassri et al., 2003; Volokhov et al., 2003). The MAVR assay was developed to detect recombinations between the three serotypes of oral poliovirus vaccine (OPV) (Cherkasova et al., 2003; Laassri et al., 2005). The locations of oligonucleotide spots within the poliovirus MAVR microarray produce an image that graphically reveals genomic recombination patterns and crossover regions (Figure 2).

The MAVR microarray was composed of genotype-specific oligonucleotide probes selected to identify Sabin strains of OPV (GenBank accession nos. AY184219–AY184221); the selected sequences were spaced ~150 bases from each other in the viral genome, they contain a moderate amount of GC (T_m between 41 and 57°C), spots were printed in three rows according to their location in the genome. Each slide accommodates 5 individual microarrays for MAVR microarrays. Microarrays of immobilized oligonucleotide probes were hybridized with fluorescently-labeled viral cDNA prepared with hexanucleotide random primers and Superscript III reverse transcriptase. Each viral cDNA product (10 µg) was fluorescently labeled with a Cy3 RNA Labeling Kit and purified using spin columns. Hybridization between microarray oligonucleotide probes and fluorescently-labeled cDNA was performed as follow: vacuum-dried fluorescently-labeled cDNA samples were reconstituted in Hybridization Buffer and denatured by incubation for one minute at 95°C. A 10-µl aliquot of the hybridization mixture was applied to the microarray area and covered with an individual plastic cover slip. Hybridization was proceeded in an incubation chamber for two hours at 45°C. Fluorescent images of processed microarray slides were captured using ScanArray 5000. Any recombination in the analyzed viruses was detected as a change of fluorescent patterns of spots in the rows (clinical sample, Figure 2).

One potential caveat regarding MAVR analysis is that it can positively identify recombinations only in regions derived from strains represented on the microarray. If one recombination partner is unknown, the microarray reveals a "gap" or an irregular pattern, calling for nucleotide sequencing as a tool of last resort. Alternatively, conserved oligonucleotide probes with broader specificity might be included in MAVR microarrays to tentatively identify the origin of "orphan" genomic segments.

MAVR mapping of poliovirus genomes has an advantage over the more detailed complete nucleotide sequencing because it can determine more than one recombinant in the same samples without a need for cloning and has an extraordinary throughput. Restriction fragment length polymorphism (RFLP), also used for this purpose, is more time-consuming and less informative than MAVR. In addition, MAVR analysis allows the genotyping of naturally heterogeneous populations. For example, the MARV analysis of a clinical sample (coded 18058) (Laassri et al., 2005) revealed a mixture of poliovirus strains composed of at least two and probably several more different types of recombinants. MAVR combined with cDNA preparation, coping of viral genome directly from clinical specimens (Laassri et al., 2005), opened the possibility of studying natural heterogeneity of viral populations *in vivo*.

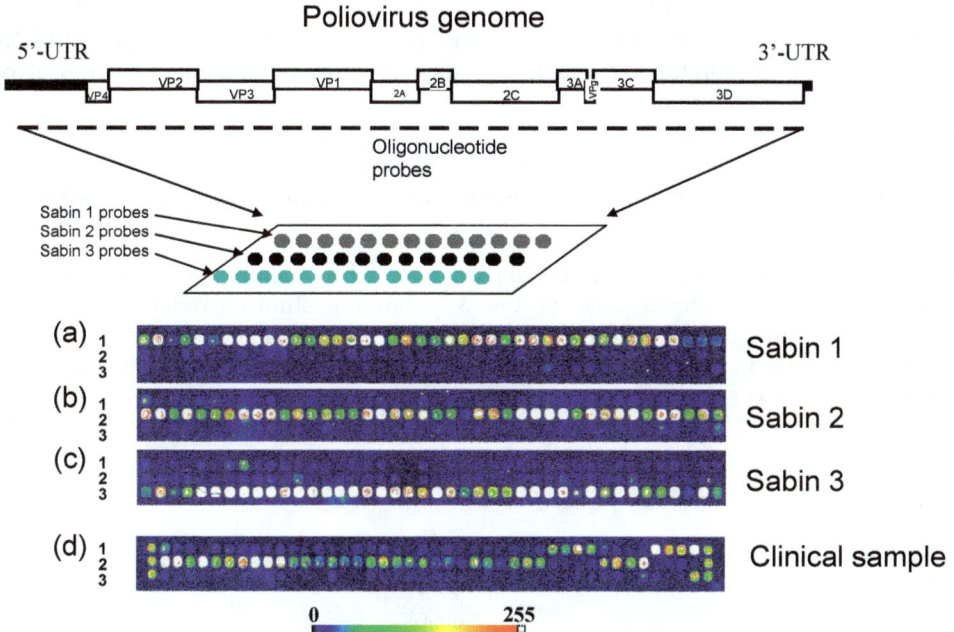

Fig. 2. MAVR analysis of the genome structure of poliovirus. Three rows (coded 1, 2, and 3 in the left of microarrays) of oligonucleotide probes in each microarray are specific to 3 serotypes of poliovirus were spotted according to their location in the genome. Sample names are shown on the right. (a); Hybridization pattern of Sabin 1 genome with the MAVR microarray, (b); Represent the hybridization pattern of Sabin 2 genome with the MAVR microarray, (c); Hybridization pattern of Sabin 3 genome with the MAVR microarray, (c); Hybridization pattern of poliovirus genome extracted from a clinical sample obtained from acute flaccid paralysis (AFP).

5. Microarray assay for quantitation of known virulent mutations

Microarray assay for quantitation of known virulent mutations (MQNVM) was developed to quantify the virulent mutations in the genomes of the three serotypes of oral poliovirus vaccine (OPV) isolated from clinical specimens (Laassri et al., 2005; Laassri et al., 2006).

Sabin strains of OPV mutate rapidly *in vitro* and *in vivo*. Some of these mutations are direct reversions to the alleles of wild-type progenitors of the vaccine strains, whereas others are second-site suppressors of the attenuated phenotype or are incidental changes. Among the best characterized attenuating mutations in the OPV Sabin strains are mutations located in the Internal Ribosome Entry Site (IRES) of the 5'-untranslated region (5'-UTR) (Minor, 1992) (Figure 3). These mutations have been identified in Sabin type 3 poliovirus (472U→C) (Cann et al., 1984), as well as type 2 (481A→G) (Macadam et al., 1993), and type 1 (480G→A and 525U→C) (Otelea et al., 1993); they are believed to selectively affect initiation of translation of viral polyprotein in neuronal cells (Guest et al., 2004; Svitkin et al., 1990). Previously was shown that the content of these revertants was low in vaccine batches that failed the monkey neurovirulence test (Chumakov et al., 1991). Sensitive mutant analysis by PCR and restriction enzyme cleavage (MAPREC) method is used to monitor the quantity of neurovirulent revertants in batches of oral poliovirus vaccine (Chumakov et al., 1991). However, the method is relatively labor-intensive, and is not amenable to analysis of a large number of clinical samples that is needed for studies of genetic stability of vaccine viruses *in vivo*. Therefore, the development of high throughput methods to quantitate revertants in attenuated poliovirus remains a high priority for evaluation of existing and new vacines.

Recently, an MQNVM assay (Figure 4A) was created to identify and quantitate the 4 known reversions in the 5'-UTR of the 3 poliovirus strains (mutations located at nucleotides 480 and 525 for Sabin type 1, nucleotide 481 for Sabin type 2, and nucleotide 472 for Sabin type 3); the assay has been described in detail elsewhere (Laassri et al., 2005; Laassri et al., 2006).

Fig. 3. Locations of the most important primary attenuating mutations of Sabin poliovirus in the IRES region of 5'-UTR of the genome. There are two known attenuating mutations for Sabin 1 (at nucleotides 480 and 525), one for Sabin 2 (at nucleotide 481) and one for Sabin 3 (at nucleotide 525).

To prepare viral genomes directly from stool samples and to quantitate the reversions by MQNVM, 1 g of frozen stool was suspended in 10 ml of Dulbecco's PBS, vortexed, centrifuged, and supernatants aliquoted and stored at -70°C. Viral RNA was isolated from a total of 140 µl of stool supernatant. The extracted RNA was eluted in a final volume of 60 µl of sterile RNase-free water.

For viral cDNA preparation, 10 µl of RNA was added to a reaction mixture containing 1 mM dithiothreitol, 2.5 µg/ml concentrations of each primer (A7-sabin1, 3 and A7-sabin2 (Laassri et al., 2005)), 0.5 mM dNTP mix, and 1x first-strand RT buffer. The final volume of the reaction mix is 50 µl. The mixture was heated for 5 min at 65°C and then quickly chilled on ice. Superscript II reverse transcriptase (12 U/µl) was added to the mixture and incubated for two hours at 42°C, then additional Superscript II reverse transcriptase (4 U/µl) was added, and the mixture incubated for another 3 hours at 42°C.

Full-length poliovirus genome from stool specimens was amplified by PCR (Laassri et al., 2005). The reaction was performed with an XL-PCR kit. The viral full-length amplicons obtained from this PCR amplification were used for MQNVM analysis.

The MQNVM microarrays (Figure 4A) contain 10 spots of oligonucleotide probe for a specific Sabin strain and 10 spots of oligonucleotide probe specific to the revertant virus. They also contain two oligonucleotides specific to a conserved region as control. Each control oligonucleotide was spotted 5 times in the last row. The redundant spotting of oligonucleotide probes was used to improve the quantitation accuracy. Ten individual MQNVM microarrays were spotted on each slide. Hybridization probes are single-stranded DNA (ssDNA) prepared by asymmetric PCR (Laassri et al., 2005). The ssDNA was purified with a PCR purification kit, and diluted in 50 µl of water. Aliquots containing 0.2 µM were labeled with a Cy5 or Cy3 RNA Labeling Kit and purified using spin columns.

Several microarrays spotted on the same slide were simultaneously hybridized for 30 min at 45°C with fluorescently-labeled ssDNA samples prepared from the reference Sabin strain, reference Sabin revertant (or wild-type poliovirus), and one or more test strains. The microarray was then washed for 2 min in 2x standard saline citrate (SSC) with 0.1% sodium dodecyl sulfate (SDS), followed by one min in 2x SSC.

Microarray images were taken with confocal fluorescent scanner ScanArray 5000 equipped with green and red HeNe lasers (543 nm and 632 nm that excite Cy3 and Cy5, respectively). Images were then analyzed using QuantArray software. The values obtained from MQNVM microarrays were normalized, and the percentage of reversion was calculated by dividing the normalized signal from each revertant oligonucleotide probe by the total signal (signal obtained from both revertant and vaccine oligonucleotide probes). The values obtained from 10 replicates of each oligonucleotide probe (vaccine, revertant) were averaged, and the standard deviation calculated.

To study the linearity of MQNVM assay to quantify mutants, another candidate vaccine virus such as West Nile (WN)/Dengue 4 chimeric virus, was used. An MQNVM microarray was developed as described above to analyze the mutation 2337G→C in West Nile/Dengue 4 virus (Laassri et al., 2011), and samples to contain different percentages of the mutant were spiked and analyzed by MQNVM assay, the detected percentages of mutants were plotted against the expected percentages of mutants (Figure 4B). Results demonstrate that mutant quantitation by MQNVM assay is linear, indicating that this assay is suitable to quantitate mutants.

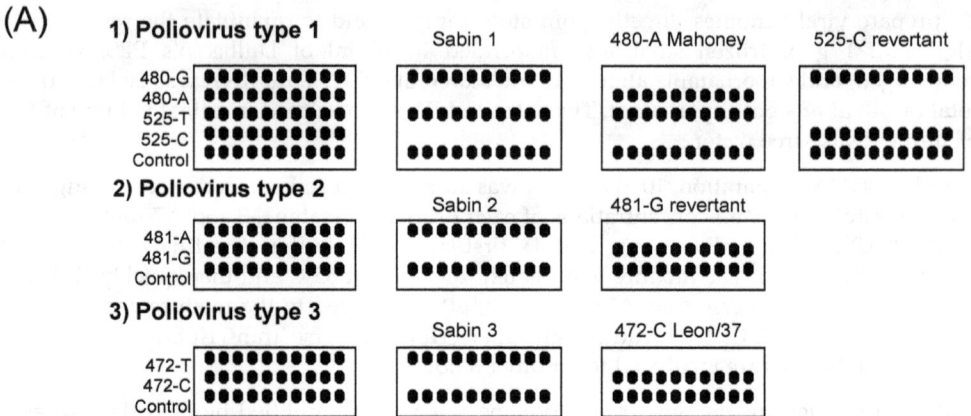

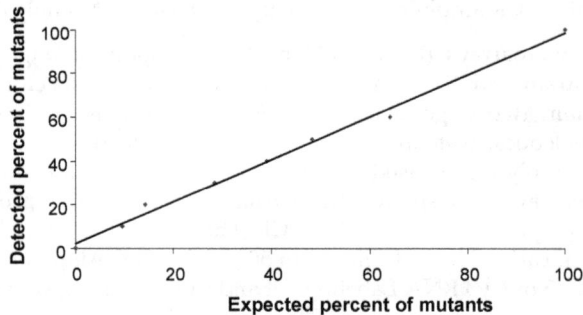

Fig. 4. Layout of a microarray for quantitation of known virulent mutations (MQNVM) in Sabin strains, the hybridization pattern of Sabin strains and their revertants, and the linearity of mutants quantitation with MQNVM assay.

(A): 1) Detection of the 480G→A and 525U→C revertants in the Sabin strain type-1 of poliovirus. The first microarray shows the layout of oligonucleotide probes: 10 spots each of 4 allele-specific oligonucleotide probes were spotted into the top 4 rows; the bottom row contains 5 spots each of a universal oligonucleotide probe and of a Sabin1-specific oligonucleotide probe. The second, third and fourth microarrays show patterns of hybridization of, respectively, Sabin 1, Mahoney (wild-type poliovirus type 1), and revertant strain 11262 (poliovirus type 1). 2) Detection of 481A→G revertants in Sabin 2 strain. The first microarray shows the layout of oligonucleotide probes: 10 spots each of two allele-specific oligonucleotide probes were spotted onto the top 2 rows; the bottom row contains 5 spots each of universal oligonucleotide probe and Sabin 2-specific oligonucleotide probe. The second and third microarrays show patterns of hybridization of, respectively, Sabin 2 strain, and revertant strain 154 (poliovirus type 2). 3) Detection of the 472U→C revertants in Sabin 3 strain. The first microarray shows the layout of oligonucleotide probes: 10 spots each of two allele-specific oligonucleotide probes were spotted into the top two rows; the bottom row contains 5 spots each of universal oligonucleotide probe and Sabin3-specific oligonucleotide probe. The second and third microarrays show patterns of hybridization of, respectively, Sabin 3 strain, and Leon/37 (wild type poliovirus type 3). (B): Evaluation of the linearity of a quantitative MQNVM assay. Samples containing different amounts of WN/Dengue 4 virus 2337G→C mutant were analyzed by MQNVM assay. The results were plotted as observed versus the expected mutant contents. This result shows that mutants quantitation with MQNVM assay is linear with R-squared value (R^2) equal 0.99.

Also, MQNVM assay was used to characterize poliovirus in about 300 stool specimens obtained from children vaccinated with different combinations of OPV and inactivated polio vaccine (IPV) (Laassri et al., 2005; Laassri et al., 2006). The PCR-amplified viral cDNA prepared directly from the stool specimens was used to quantitate reversions in the 5'-UTR of each of the 3 poliovirus serotypes. Fluorescently-labeled ssDNA for hybridization was prepared form each poliovirus serotype as described above and elsewhere (Laassri et al., 2005). Results of our study (Laassri et al., 2006) show that many stool samples from healthy children one week after OPV vaccination contained different percentages of revertants, consistent with earlier observations based on conventional methodology (Cann et al., 1984; Kew et al., 2002; WHO, 2002). The oligonucleotide microarrays simultaneously detected and discriminated between vaccine and revertant sequences and allowed the quantitation of reversions in the 5'-UTRs of all 3 serotypes of poliovirus.

6. Conclusion

Microarray technology is a sensitive and versatile method for genetic analysis that allows screening of mutations in genetic materials and readily detecting single-point mutations. Viral nucleic acid hybridization with immobilized oligonucleotides in microarrays that encompass thousands of individual probes offers a rapid method suitable for simultaneous analysis of a large number of markers distributed over the whole viral genome. The technique generates instant genetic maps of mutant strains and reveals evolutionary divergence and mutational profiles of individual viral stocks.

The simplicity and high throughput of microarray-based analyses might also assist in improving genetic stability of candidate vaccine strains by incorporating mutations conferring better replicative properties. They also facilitate monitoring of molecular consistency in a new viral vaccine during its manufacturing. The same approach can be applied in future development of new live viral vaccines and used as a new paradigm for better quality control tests of vaccines against other pathogens.

The oligonucleotide microarrays described in this chapter have already facilitated the analysis of the genetic diversity of viruses and live virus vaccines at the levels of genomic recombination, nucleotide sequence heterogeneity, and quantitation of single-point mutations. They facilitate rapid analysis of viral genes and genomes, circumventing traditional methods that usually involve much more laborious efforts. Microarray methods can analyze a very large number of samples simultaneously, within few hours. Furthermore, cloning of nucleic acids is not required, thus preserving the natural genomic composition of viral gene populations. Microarray methods open the possibility of a large-scale full-genome screening of viral isolates needed for improved epidemiological surveillance and better vaccine quality control; for example, MQNVM microarrays rapidly, simultaneously, and unambiguously identified viral vaccines and their revertants and quantified the amounts of single-point mutations.

Unlike direct DNA sequencing, the MARSH assay determines only the approximate location of mutations within a single oligonucleotide probe. However, this limitation has the advantage of increasing the sensitivity of detecting genomic changes by microarray, since it reveals several adjacent mutations on different molecules, even mutations present in quantities too low to detect by conventional sequencing (Cherkasova et al., 2003).

Besides having a high throughput capacity the MARV assay easily demonstrates naturally heterogeneous viral populations, even in the same sample, without the need to separate or clone them.

Microarray-based assays for genetic stability of live viral vaccines should greatly assist in evaluating safety. The information obtained from such microarray methods will not only expedite regulatory review of the prospective recombinant vaccines but also provide a method suitable for monitoring consistency of vaccine production as part of routine quality control. Microarray techniques also offer the possibility for a large-scale full-genome screening of viral isolates to improve epidemiological surveillance, and better vaccine quality control.

7. Acknowledgements

We thank Dr. David Asher for his suggestions and critical review of this chapter. Also we thank Dr. Vladimir Chizhikov for his valuable advices concerning microarray development.

8. References

Amexis, G., Oeth, P., Abel, K., Ivshina, A., Pelloquin, F., Cantor, C. R., Braun, A., and Chumakov, K. (2001). Quantitative mutant analysis of viral quasispecies by chip-based matrix-assisted laser desorption/ ionization time-of-flight mass spectrometry. *Proc Natl Acad Sci U S A* 98(21), 12097-102.

Arroyo, J., Guirakhoo, F., Fenner, S., Zhang, Z. X., Monath, T. P., and Chambers, T. J. (2001). Molecular basis for attenuation of neurovirulence of a yellow fever Virus/Japanese encephalitis virus chimera vaccine (ChimeriVax-JE). *J Virol* 75(2), 934-42.

Berkhout, B., Verhoef, K., van Wamel, J. L., and Back, N. K. (1999). Genetic instability of live, attenuated human immunodeficiency virus type 1 vaccine strains. *J Virol* 73(2), 1138-45.

Blancou, J., Kieny, M. P., Lathe, R., Lecocq, J. P., Pastoret, P. P., Soulebot, J. P., and Desmettre, P. (1986). Oral vaccination of the fox against rabies using a live recombinant vaccinia virus. *Nature* 322(6077), 373-5.

Cann, A. J., Stanway, G., Hughes, P. J., Minor, P. D., Evans, D. M., Schild, G. C., and Almond, J. W. (1984). Reversion to neurovirulence of the live-attenuated Sabin type 3 oral poliovirus vaccine. *Nucleic Acids Res* 12(20), 7787-92.

Cherkasova, E., Laassri, M., Chizhikov, V., Korotkova, E., Dragunsky, E., Agol, V. I., and Chumakov, K. (2003). Microarray analysis of evolution of RNA viruses: evidence of circulation of virulent highly divergent vaccine-derived polioviruses. *Proc Natl Acad Sci U S A* 100(16), 9398-403.

Chumakov, K. M., Powers, L. B., Noonan, K. E., Roninson, I. B., and Levenbook, I. S. (1991). Correlation between amount of virus with altered nucleotide sequence and the monkey test for acceptability of oral poliovirus vaccine. *Proc Natl Acad Sci U S A* 88(1), 199-203.

Dietzschold, M. L., Faber, M., Mattis, J. A., Pak, K. Y., Schnell, M. J., and Dietzschold, B. (2004). In vitro growth and stability of recombinant rabies viruses designed for vaccination of wildlife. *Vaccine* 23(4), 518-24.

Domingo, E., Martinez-Salas, E., Sobrino, F., de la Torre, J. C., Portela, A., Ortin, J., Lopez-Galindez, C., Perez-Brena, P., Villanueva, N., Najera, R., and et al. (1985). The quasispecies (extremely heterogeneous) nature of viral RNA genome populations: biological relevance--a review. *Gene* 40(1), 1-8.

Dunster, L. M., Wang, H., Ryman, K. D., Miller, B. R., Watowich, S. J., Minor, P. D., and Barrett, A. D. (1999). Molecular and biological changes associated with HeLa cell attenuation of wild-type yellow fever virus. *Virology* 261(2), 309-18.

Fekadu, M., Shaddock, J. H., Sumner, J. W., Sanderlin, D. W., Knight, J. C., Esposito, J. J., and Baer, G. M. (1991). Oral vaccination of skunks with raccoon poxvirus recombinants expressing the rabies glycoprotein or the nucleoprotein. *J Wildl Dis* 27(4), 681-4.

Grinev, A., Daniel, S., Laassri, M., Chumakov, K., Chizhikov, V., and Rios, M. (2008). Microarray-based assay for the detection of genetic variations of structural genes of West Nile virus. *J Virol Methods* 154(1-2), 27-40.

Guest, S., Pilipenko, E., Sharma, K., Chumakov, K., and Roos, R. P. (2004). Molecular mechanisms of attenuation of the Sabin strain of poliovirus type 3. *J Virol* 78(20), 11097-107.

Hansen, H., Okeke, M. I., Nilssen, O., and Traavik, T. (2004). Recombinant viruses obtained from co-infection in vitro with a live vaccinia-vectored influenza vaccine and a naturally occurring cowpox virus display different plaque phenotypes and loss of the transgene. *Vaccine* 23(4), 499-506.

Kew, O., Morris-Glasgow, V., Landaverde, M., Burns, C., Shaw, J., Garib, Z., Andre, J., Blackman, E., Freeman, C. J., Jorba, J., Sutter, R., Tambini, G., Venczel, L., Pedreira, C., Laender, F., Shimizu, H., Yoneyama, T., Miyamura, T., van Der Avoort, H., Oberste, M. S., Kilpatrick, D., Cochi, S., Pallansch, M., and de Quadros, C. (2002). Outbreak of poliomyelitis in Hispaniola associated with circulating type 1 vaccine-derived poliovirus. *Science* 296(5566), 356-9.

Laassri, M., Bidzhieva, B., Speicher, J., Pletnev, A. G., and Chumakov, K. (2011). Microarray hybridization for assessment of the genetic stability of chimeric west nile/dengue 4 virus. *J Med Virol* 83(5), 910-20.

Laassri, M., Chizhikov, V., Mikheev, M., Shchelkunov, S., and Chumakov, K. (2003). Detection and discrimination of orthopoxviruses using microarrays of immobilized oligonucleotides. *J Virol Methods* 112(1-2), 67-78.

Laassri, M., Dragunsky, E., Enterline, J., Eremeeva, T., Ivanova, O., Lottenbach, K., Belshe, R., and Chumakov, K. (2005). Genomic analysis of vaccine-derived poliovirus strains in stool specimens by combination of full-length PCR and oligonucleotide microarray hybridization. *J Clin Microbiol* 43(6), 2886-94.

Laassri, M., Lottenbach, K., Belshe, R., Rennels, M., Plotkin, S., and Chumakov, K. (2006). Analysis of reversions in the 5'-untranslated region of attenuated poliovirus after sequential administration of inactivated and oral poliovirus vaccines. *J Infect Dis* 193(10), 1344-9.

Laassri, M., Meseda, C. A., Williams, O., Merchlinsky, M., Weir, J. P., and Chumakov, K. (2007). Microarray assay for evaluation of the genetic stability of modified vaccinia virus Ankara B5R gene. *J Med Virol* 79(6), 791-802.

Macadam, A. J., Pollard, S. R., Ferguson, G., Skuce, R., Wood, D., Almond, J. W., and Minor, P. D. (1993). Genetic basis of attenuation of the Sabin type 2 vaccine strain of poliovirus in primates. *Virology* 192(1), 18-26.

Minor, P. D. (1992). The molecular biology of poliovaccines. *J Gen Virol* 73 (Pt 12), 3065-77.

Monath, T. P., Soike, K., Levenbook, I., Zhang, Z. X., Arroyo, J., Delagrave, S., Myers, G., Barrett, A. D., Shope, R. E., Ratterree, M., Chambers, T. J., and Guirakhoo, F. (1999). Recombinant, chimaeric live, attenuated vaccine (ChimeriVax) incorporating the envelope genes of Japanese encephalitis (SA14-14-2) virus and the capsid and nonstructural genes of yellow fever (17D) virus is safe, immunogenic and protective in non-human primates. *Vaccine* 17(15-16), 1869-82.

Otelea, D., Guillot, S., Furione, M., Combiescu, A. A., Balanant, J., Candrea, A., and Crainic, R. (1993). Genomic modifications in naturally occurring neurovirulent revertants of Sabin 1 polioviruses. *Dev Biol Stand* 78, 33-8.

Pletnev, A. G., Bray, M., Hanley, K. A., Speicher, J., and Elkins, R. (2001). Tick-borne Langat/mosquito-borne dengue flavivirus chimera, a candidate live attenuated vaccine for protection against disease caused by members of the tick-borne encephalitis virus complex: evaluation in rhesus monkeys and in mosquitoes. *J Virol* 75(17), 8259-67.

Pletnev, A. G., Bray, M., Huggins, J., and Lai, C. J. (1992). Construction and characterization of chimeric tick-borne encephalitis/dengue type 4 viruses. *Proc Natl Acad Sci U S A* 89(21), 10532-6.

Pletnev, A. G., Karganova, G. G., Dzhivanyan, T. I., Lashkevich, V. A., and Bray, M. (2000). Chimeric Langat/Dengue viruses protect mice from heterologous challenge with the highly virulent strains of tick-borne encephalitis virus. *Virology* 274(1), 26-31.

Pletnev, A. G., and Men, R. (1998). Attenuation of the Langat tick-borne flavivirus by chimerization with mosquito-borne flavivirus dengue type 4. *Proc Natl Acad Sci U S A* 95(4), 1746-51.

Pletnev, A. G., Putnak, R., Speicher, J., Wagar, E. J., and Vaughn, D. W. (2002). West Nile virus/dengue type 4 virus chimeras that are reduced in neurovirulence and peripheral virulence without loss of immunogenicity or protective efficacy. *Proc Natl Acad Sci U S A* 99(5), 3036-41.

Pletnev, A. G., Swayne, D. E., Speicher, J., Rumyantsev, A. A., and Murphy, B. R. (2006). Chimeric West Nile/dengue virus vaccine candidate: preclinical evaluation in mice, geese and monkeys for safety and immunogenicity. *Vaccine* 24(40-41), 6392-404.

Prevec, L., Campbell, J. B., Christie, B. S., Belbeck, L., and Graham, F. L. (1990). A recombinant human adenovirus vaccine against rabies. *J Infect Dis* 161(1), 27-30.

Pugachev, K. V., Guirakhoo, F., Ocran, S. W., Mitchell, F., Parsons, M., Penal, C., Girakhoo, S., Pougatcheva, S. O., Arroyo, J., Trent, D. W., and Monath, T. P. (2004). High fidelity of yellow fever virus RNA polymerase. *J Virol* 78(2), 1032-8.

Pugachev, K. V., Ocran, S. W., Guirakhoo, F., Furby, D., and Monath, T. P. (2002). Heterogeneous nature of the genome of the ARILVAX yellow fever 17D vaccine revealed by consensus sequencing. *Vaccine* 20(7-8), 996-9.

Pugachev, K. V., Schwaiger, J., Brown, N., Zhang, Z. X., Catalan, J., Mitchell, F. S., Ocran, S. W., Rumyantsev, A. A., Khromykh, A. A., Monath, T. P., and Guirakhoo, F. (2007). Construction and biological characterization of artificial recombinants between a wild type flavivirus (Kunjin) and a live chimeric flavivirus vaccine (ChimeriVax-JE). *Vaccine* 25(37-38), 6661-71.

Svitkin, Y. V., Cammack, N., Minor, P. D., and Almond, J. W. (1990). Translation deficiency of the Sabin type 3 poliovirus genome: association with an attenuating mutation C472---U. *Virology* 175(1), 103-9.

Taylor, J., Trimarchi, C., Weinberg, R., Languet, B., Guillemin, F., Desmettre, P., and Paoletti, E. (1991). Efficacy studies on a canarypox-rabies recombinant virus. *Vaccine* 9(3), 190-3.

Volokhov, D., Chizhikov, V., Chumakov, K., and Rasooly, A. (2003). Microarray-based identification of thermophilic Campylobacter jejuni, C. coli, C. lari, and C. upsaliensis. *J Clin Microbiol* 41(9), 4071-80.

WHO (1990). Potential use of live viral and bacterial vectors for vaccines. WHO meeting, Geneva, 19-22 June, 1989. *Vaccine* 8(5), 425-37.

WHO (2002). Paralytic poliomyelitis in Madagascar, 2002. *Wkly Epidemiol Rec* 77(29), 241-2.

9

Application of a Microarray-Based Assay for the Study of Genetic Diversity of West Nile Virus

Andriyan Grinev, Zhong Lu, Vladimir Chizhikov and Maria Rios
Center for Biologics Evaluation and Research,
US Food and Drug Administration
USA

1. Introduction

1.1 Molecular virology and epidemiology of West Nile virus

West Nile virus (family *Flaviviridae*, genus *Flavivirus*, WNV) is a small, enveloped, single stranded, positive RNA genome virus. WNV is a member of the Japanese encephalitis serogroup, which includes St Louis encephalitis virus (SLEV), Japanese encephalitis virus (JEV), Murray Valley encephalitis virus (MVEV), Kunjin virus (KUNV), and Usutu virus (USUV), which have all been shown to cause disease in humans. The virion consists of an envelope and prM-M dimers surrounding an icosahedral capsid of approximately 50 nm in size (Beasley, 2005). The WNV genomic RNA is approximately 11 kb in length, and contains 10 genes within a single open reading frame (ORF) that encodes for a single polyprotein flanked by 5′ and 3′ untranslated regions (UTR). The approximately 3430 amino acid WNV polyprotein is processed by cellular proteases and by the viral NS2B-NS3 protease into 3 structural and 7 non-structural proteins (NS) (Fig. 1).

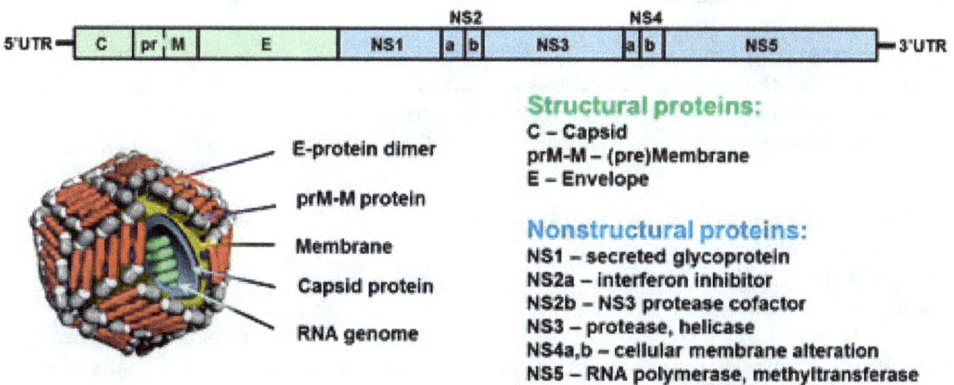

Fig. 1. Scheme of WNV genome and virion composition. The 11 kb positive RNA genome contains a single ORF encoding the 3 structural proteins that form the virus particle and the 7 non-structural proteins required for virus replication and immune evasion.

The structural proteins i.e., capsid (C), premembrane-membrane (prM-M), and envelope (E), interact with the viral genomic RNA and with the host cell membrane to assemble viral particles. The structural proteins are not only essential for virion assembly and release, but they are also the major targets for virus neutralizing antibodies. The seven viral nonstructural proteins (NS1, NS2A, NS2B, NS3, NS4A, NS4B and NS5) are all necessary for genome replication (Khromykh et al., 2000). NS1 is a secreted glycoprotein implicated in immune evasion (Schlesinger, 2006). NS2A plays a role in virus assembly as well as inhibiting IFN-β promoter activation (Leung et al., 2008; Mackenzie et al., 1998). NS3 contains an ATP-dependent helicase, and in conjunction with the NS2B protein, functions as a serine protease, which is required for virus polyprotein processing (Chappell et al., 2005; Clum et al., 1997; Falgout et al., 1991). NS4A is responsible for a rapid expansion and modification of the endoplasmic reticulum that helps establish replication domains (Khromykh et al., 1998; Mackenzie et al., 1998). NS4B blocks the IFN response (Evans et al., 2007; Munoz-Jordan et al., 2005). NS5 is a methyltransferase and RNA-dependent RNA polymerase (Beasley, 2005; Egloff et al., 2002).

The untranslated regions (UTR) are involved in translation and viral RNA replication and likely play an important role in genome packaging. Both the 5' UTR and the 3' UTR in the WNV genome form highly conserved secondary and tertiary structures, some elements of which are similar among mosquito-borne flaviviruses. The cyclization of the flavivirus genome is necessary for viral RNA replication. In addition to base pairing between 5'-3' UAR and 5'-3' CS specific sequences involved in cyclization, a third stretch of nucleotides was identified to form a double-stranded region between the 5' and 3' UTRs (Friebe & Harris, 2010). Different functional regions have been described inside the 5'UTR and 3'UTR of flaviviruses based on such factors as nucleotide content, degree of sequence conservation, occurrence of repeated sequence motifs, and predicted secondary structure (Gritsun & Gould, 2007; Markoff, 2003; Proutski et al., 1997; Tajima et al., 2006). The 5' end of the WNV genomic RNA has a type I cap structure (m^7GpppAmp) mediating cap-dependent translation. The 5'UTR contains two functional elements, the stem-loop A (SLA) and capsid-coding region hairpin (cHP) essential for RNA replication. The 3'UTR is generally divided into three regions based on the differences in the level of conservation: (1) the variable region is located immediately after the ORF; (2) the intermediate region has a moderate level of conservation and contains several hairpin motifs; (3) the conserved 3'-terminal region contains a cyclization sequences and stable stem-loop structure (Bryant et al., 2005; Markoff, 2003). These regions are believed to contain sequences that confer identity of the flaviviruses as demonstrated by attempts to exchange portions of the 3'UTR between WNV and dengue virus (DENV) that resulted in chimeric viruses which were unable to replicate (Yu et al., 2008).

WNV is maintained in nature by transmission between mosquitoes and birds, but it can also infect humans, other mammals (Beasley, 2005; Petersen & Marfin, 2002) and reptiles (Klenk et al., 2004) by mosquito bite (Fig.2). *Culex spp.* mosquitoes are the main vectors of WNV, although the virus has also been found in at least 43 other mosquito species (Granwehr et al., 2004; Higgs et al., 2004; Petersen et al., 2001). WNV can be transmitted vertically and overwinter in hibernating female mosquitoes, providing the mechanisms for viral persistence and reemergence each spring (Nasci et. al, 2001). WNV has spread within many bird species, including crows, magpies, and jays, house sparrows, house finches, grackles, and others representing 63 species, 30 families and 14 orders (Kramer & Bernard, 2001). They are all primarily competent reservoirs for WNV infection. By contrast, mammals including humans and horses are "dead-end" hosts in this enzootic cycle. They do not develop prolonged high-level viremia, so the concentration of the virus in blood is insufficient to infect a feeding mosquito. Most human infections are asymptomatic (~80%). The severity of symptomatic cases ranges from flu-like illness (~20% of infections) to severe neurological disease (~ 1%) (Hayes & Gubler, 2006). Additional modes of transmission were identified in 2002, including human-to-human by blood transfusion, breast-feeding, transplacental transmission, and by organ transplants extending the impact of WNV to blood safety and other areas of public health worldwide (Austgen et al., 2004; Pealer et al., 2003; Sbrana et al., 2005).

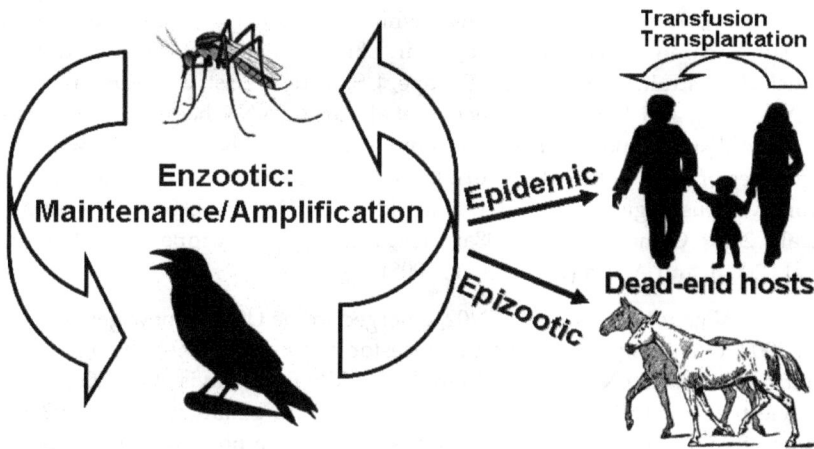

Fig. 2. Scheme of WNV transmission cycle. The maintenance of WNV in nature depends on an enzootic cycle involving many avian and mosquito species. Humans and other animals are incidental hosts that can become infected by WNV-infected mosquito bites. The virus can be also transmitted from human to human by blood transfusion and by solid organ transplantation.

Historically, since its isolation in Uganda in 1937, WNV outbreaks occurred in Africa, Europe, the Middle East, and caused a rare and mild febrile illness in humans and horses. A significant geographical expansion occurred starting from 1999 when the virus was introduced into North America. First detected in the U.S. in 1999, WNV has become endemic, causing yearly summer outbreaks. In 2002, the virus spread westward and the number of reported human cases increased dramatically. The North American epidemics of 2002 and 2003 represent the largest WNV outbreaks ever reported. WNV now is the most widespread arbovirus in the world (Kramer et al., 2008). The spread of the virus and intensity of the outbreak was correlated with the appearance of a new genotype with higher virulence and ability to disseminate in mosquitoes (Beasley, 2005). From the first outbreak in 1999 through 2010, WNV is estimated to have infected ~4 million humans in the US, causing over 30,000 serious illnesses, including 12,729 neuroinvasive disease cases with 1,206 deaths reported to the CDC (http://www.cdc.gov/ncidod/dvbid/westnile/). The virus has also been detected in the continental U.S. and in several areas of Canada, Central and South America, and the Caribbean. The persistence of WNV indicates that it has become endemic in the Western Hemisphere (Kramer et al., 2008). The pattern of yearly reoccurring outbreaks in North America differs from that of sporadic outbreaks observed in Europe and Africa. However, in recent years, WNV epidemics in humans and horses have become more frequent in several Southern European countries, and these epidemics could potentially be associated with an emergence of new viral genotypes (Chevalier et al. 2011). The speed with which the virus spread over the world triggered great interest and prompted a detailed investigation of the genetic evolution of the virus in search of the cause of its rapid adaptability.

Based on phylogenetic analyses, WNV has been initially divided into two major genetic lineages (Lanciotti et al., 2002). Lineage 1 included viruses circulating in Europe, Israel, United States, India, Russia, and Australia, while the Lineage 2 contained strains that circulated in sub-Saharan Africa and Madagascar. Lineage 1 was further divided into 3 sub-clades: 1a (including strains from Africa, Europe, US, Middle East, and Russia), 1b (Kunjin strain from Australia), and 1c (India) (Lanciotti et al., 2002). WNV has now been reported to have at least five distinct lineages based on phylogenetic analysis of all known full WNV genome sequences of viral isolates that correlates well with the geographical points of their isolation from various regions around the world including the U.S. (Davis et al., 2005; Herring et al., 2007; Grinev et al., 2008a; McMullen, 2011), Europe and Mediterranean (Parreira et al., 2007), and Africa (Botha et al., 2008).

In 2001 a new WNV genotype, named WN02, emerged in the US. The new genotype became prevalent in 2002, eventually displacing the ancestor genotype NY99, which is believed to have been introduced to the New World from the Middle East (Davis et al., 2005; Herring et al., 2007; Lanciotti et al., 1999). When compared to the WN99 genotype, the WN02 genotype possesses a few fixed silent nucleotide mutations and one amino acid substitution in the E protein (E-V159A). The highest rate of nucleotide sequence divergence among viruses isolated from 2002-2010 varies in the range of 0.4% - 0.6% (Davis et al., 2005; Ebel et al., 2004; Grinev et al., 2008a; McMullen et al., 2011). It is noteworthy that 80% of the nucleotide changes are observed in the structural regions represented by U<->C transitions; 75% among them are silent mutations (Grinev et al., 2008a). The possible explanation for the rapid displacement of the WN99 genotype by the new dominant genotype WN02 is due to the ability of the new viruses to more efficiently proliferate in domestic mosquitoes (Jerzak et al., 2005; Moudy et al., 2007). Phylogenetic analysis of modern WNV isolates demonstrates the existence of at least two subtypes of WN02 genotype co-circulating in North America. In addition to the common E-V159A amino acid substitution two other substitutions have become fixed in the significant part of North American WNV population: NS4a-A85T and NS5-K318R. Positive selection of these two amino acid substitutions potentially could impact viral fitness, phenotype and virulence (McMullen et al., 2011). As indicated by sequence analysis of new WNV genetic variants isolated in different areas of the US, the virus continues to diverge from the precursor isolate. Thus, changes in the WNV genome and viral proteins have the potential to negatively affect the sensitivity of screening and diagnostic assays currently used for virus detection, and to impact the development of vaccines and potential antiviral therapeutic agents. Therefore, development of new methods able to rapidly detect the emergence of WNV genetic variants is critical for epidemiological surveillance.

1.2 Basics of nucleotide microarrays

The monitoring and surveillance of pathogens is highly dependent on the capability of detection technology to simultaneously monitor multiple genomic signatures specific for different genetic variants of the pathogen. One of the approaches that enable this type of analysis is microarray technology. Generally, nucleotide microarrays are microscopic slides loaded with hundreds or thousands of pathogen-specific probes (DNA fragments or synthetic oligonucleotides) which can specifically hybridize with the target molecules to produce either quantitative (gene expression) or qualitative (diagnostic) data. DNA microarray technology provides an opportunity to perform parallel nucleic acid hybridization with a large number of immobilized oligonucleotides on a small surface area. Microarrays have the unique potential to simultaneously detect and identify defined pathogens, as well as to detect mutations within the complete viral genomes and target areas of a pathogen's genome. It provides a significant advantage for the field of clinical microbiology and molecular epidemiological studies. Printed microarrays historically were the first arrays utilized for detection of mutations in many research laboratories and they are so-called because of the "printing" of the probes onto the surface of a glass microscope slide (Fig. 3).

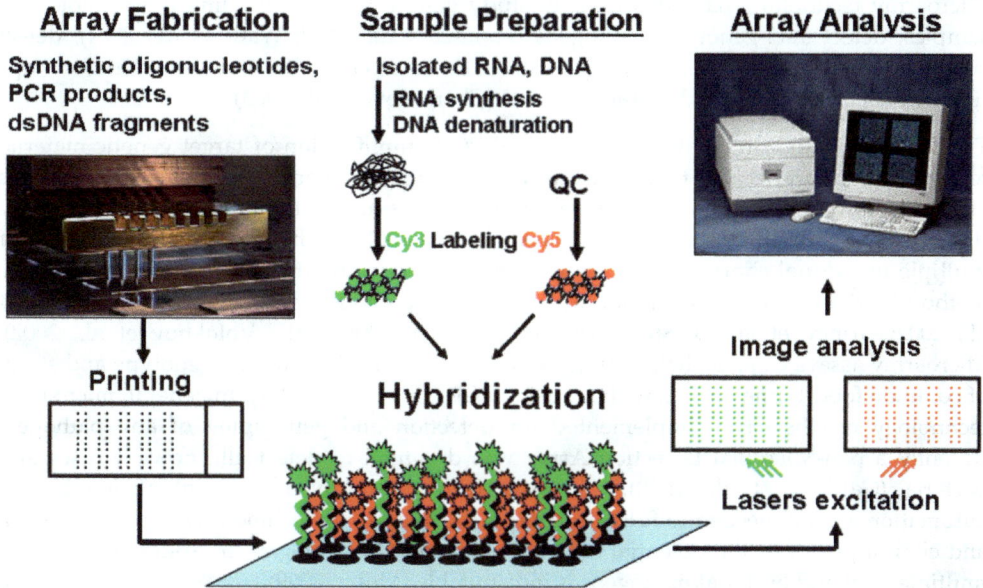

Fig. 3. Schematic outline of a printed microarray experiment. Microarray probes are spotted onto a surface of glass slide. In this example, the target sample and quality control sample (QC) with different fluorescent labeling were mixed and hybridized to the microarray probes. The efficiency of hybridization is monitored through measurement of the fluorescent signal from each spot by using a laser microarray scanner equipped with two lasers: 632 nm and 543 nm for excitation of Cy5 and Cy3 dyes respectively. The obtained fluorescent images are analyzed using specific computer software.

Chemically activated glass slides are commonly used for microarrays because they permit irreversible attachment of microarray probes, allow for efficient hybridization kinetics between probes and analyzed targets, and have a low background fluorescence level (Cheung et al., 1999). Printed arrays can be produced using double-stranded DNA (dsDNA) fragments or oligonucleotides. For dsDNA microarrays, the probes usually consist of PCR products (amplicons) obtained using primers designed from a known genomic sequence or shotgun library clones. The double-stranded probes must be denatured prior to hybridization, either using a special printing buffer or after immobilization on the glass surface (Tomiuk et al., 2001). These microarrays containing relatively long 200-800 bp dsDNA probes usually demonstrate a high sensitivity but are not very useful for detection of minor genetic differences (e.g., single nucleotide mutations) between the probe and target nucleic acid (Hager et al., 2006). In contrast to dsDNA microarray probes, the length of oligonucleotide probes usually ranges from 20 to 80 nucleotides. Use of shorter probe lengths increases the microarray specificity and enables efficient detection of minor genetic changes between the probe and target (Chou et al., 2004). For more efficient attachment of oligonucleotide probes to the functional groups on the surface of chemically activated glass slides (usually aldehyde, epoxy, and succinimide groups), the 5' or 3' end of the probes contain primary amino groups introduced during chemical synthesis.

Microarray technology has been used to study gene expression in clinical and biological samples, detect and genotype pathogens (Honma et al., 2007; Wade et al., 2004), detect single base pair mismatches (Anthony et al., 2003; Hacia et al., 1999), design genomic maps (Roerig et al., 2005), and study viral evolution (Cherkasova et al., 2003).

The methods based on combination of initial PCR amplification of target genetic material followed by hybridization of amplicons with specific microarray oligonucleotide probes allowed for reconstitution of instant mutation profiles and determination of evolutionary divergence of individual viral isolates (Neverov et al., 2006). The microarrays consisting of multiple individual short oligoprobes were shown to be an efficient and sensitive genetic method for detection of single point mutations in viral and bacterial genomes (Chizhikov et al., 2002; Grinev et al., 2008b; Laassri et al., 2003, 2005, 2007; Volokhov et al., 2002). Microarray assays can also help simultaneously detect and identify the genotype and strain of common food-borne viruses without using PCR (Chen et al., 2011). In general, microarray technology can be easily implemented for detection and genotyping of any pathogen. Recently a pan-Microbial Detection Array was designed to detect all known viruses and bacteria (Gardner et al., 2010). Although further improvements, optimizations, and automation are still needed to fully implement the microarray technique in routine research and clinical practices, the potential role of these robust technologies in rapid diagnostics of multiple viral and bacterial pathogens is indisputable (Miller, 2009).

2. Material and methods

2.1 Plasma samples

The microarray development and evaluation study included the analysis of a total of 34 plasma specimens from blood donor units identified as positive for WNV by nucleic acid tests used to screen blood donations (Table 1).

Isolate ID	Year	Location	Passage	GenBank No.
FDA/HU-02	2002	NY	FFP,P1-P3	AY646354
ARC10-02	2002	MI	P1	AY795965
ARC12-02	2002	OH	P1	DQ666453
ARC16-02	2002	IN	P1	DQ666456
BSL5-03	2003	UT	P1	DQ005530
BSL9-03	2003	TX	P1	DQ666458
BSL62-03	2003	SD	P2	DQ666460
RMS1-03	2003	MN	P1	DQ666462
RMS2-03	2003	IN	P1	DQ666463
RMS3-03	2003	IN	P1	DQ666464
RMS4-03	2003	IA	P1	DQ666465
BSL2-04	2004	AZ	P1	DQ666467
BSL4-04	2004	AZ	P2	DQ666468
BSL6-04	2004	AZ	P1	DQ666469
BSL7-04	2004	AZ	P2	DQ666470
BSL8-04	2004	AZ	P2	DQ666471
GCTX1	2005	TX	P1	DQ666449
GCTX2	2005	TX	P1	DQ666450
BSL2-05	2005	SD	P1	DQ666452
BSL6-05	2005	AZ	P1	DQ666472
BSL9-05	2005	TX	P1	DQ666473
BSL10-05	2005	LA	P1	DQ666474
BSL13-05	2005	AZ	P1	DQ666451
ARC140-07	2007	ID	P1	JF957168
BSL2-09	2009	NV	P1	JF957175
BSL5-09	2009	AZ	P1	JF957176
BSL6-09	2009	NV	P1	JF957177
BSL11-09	2009	NV	P1	JF957178
BSL18-09	2009	LA	P1	JF957179
BSL20-09	2009	NV	P1	JF957180
BSL22-09	2009	SD	P1	JF957181
BSL24-09	2009	TX	P1	JF957182
BSL27-09	2009	TX	P1	JF957183
CO7-09	2009	CO	P1	JF957184

Table 1. WNV isolates used for microarray assay validation. FFP indicates fresh frozen plasma sample. Passage P1 indicates first isolation in Vero cells; P2 and P3 indicates subsequent virus passages. Isolates in boldface were used for the full genome array validation.

These samples were collected in different geographic locations of the continental U.S. from the 2002-2009 epidemic seasons under IRB approved informed consent. In addition to 23 previously published isolates, which were used for structural region investigation (Grinev et al., 2008b), 11 WNV isolates from 2007 and 2009 were used to conduct microarray analyses of their full genomes in order to detect emerged genetic differences in comparison with that of the reference WNV strain NY99.

2.2 Viral isolates

West Nile virus isolation from tested plasma samples was performed using Vero cells. Vero cells were plated in T75 flasks and grown to 85% confluence in EMEM (GIBCO BRL, Gaithersburg, MD, USA) supplemented with 5% fetal bovine serum (FBS) (Hyclone, Logan, UT) and 10μg/mL of penicillin/streptomycin (GIBCO). For viral isolation, growth medium was removed, 500 μl of each plasma sample were added to individual flasks and the total volume was adjusted to 5 ml with fresh medium. Vero cells were incubated with the viral inoculum for 2 hours, either at room temperature under gentle rocking or at 37∘C with mixing every 10-15 min. After incubation, 10 ml of fresh medium were added, and the cultures were additionally incubated at 37 °C in 5% CO_2, and observed daily under phase microscopy for gross morphological degeneration i.e., cytopathic effect (CPE). Supernatants were harvested when extensive CPE was observed. Harvested supernatants were centrifuged to remove cell debris and aliquots were frozen at –80 °C until further analysis.

2.3 RNA extraction

Total RNA was extracted from 1-3 ml of plasma samples with Trizol reagent (Invitrogen, Carlsbad, CA), according to the manufacturer protocol with additional step of ethanol precipitation. Viral RNA from 140 μl of infected Vero cell culture supernatants was extracted by using the QiaAMP viral RNA extraction kit (Qiagen, Valencia, CA), according to the manufacturer protocol. Each RNA sample was dissolved in 60 μl of RNase-free water and stored at -80∘C.

2.4 Reverse transcription

Reverse transcription was performed in 20-40 μl reaction volume at 47∘C for 2 h using a mixture of specific reverse primers (Table 2) and SuperScript III (Invitrogen, Carlsbad, CA) reverse transcription system according to the manufacturer's instructions. 2-3 μl of the reaction mixture were used for the subsequent DNA amplification.

2.5 PCR amplification

PCR fragments covering the entire structural region of the FDA-Hu2002 plasma sample were amplified by semi-nested PCR (Figure 4) from cDNA using the Hi-Fidelity PCR system (Invitrogen, Carlsbad, CA) according to the manufacturer's instructions. Primers used for PCR amplification are shown in Table 2.

cDNA was amplified in the first round of PCR in a GeneAmp 9700 thermocycler (Applied Biosystems Inc., Foster City, CA) using the following protocol: denaturation at 94 °C for 30 s., 35 cycles of 30 s at 94 °C, 30 s at 50 °C, and 2 min at 68 °C. The final extension was carried out at 68 °C for 7 min. PCR products were purified using the QIAquick PCR Purification Kit (Qiagen Valencia, CA), according to the manufacturer's protocol. 5 μl of the 1st round PCR product was used for the 2nd round of amplification with reverse primers containing the T7 RNA polymerase promoter sequence at the 5' ends tagged to the WNV sequence using the cycling program described above and the protocol for Hi-Fidelity PCR kit.

Set	Primers name and sequence
F1-1300	F1 5'-AGTAGTTCGCCTGTGTGAGCTGAC (1-24) R1300 5'-TTGGCGCATGTGTCAATGCT (1300-1319) R1300m 5'-GCCTAATACGACTCACTATAGGGTGTCAATGCTTCCTTTGCCA
F980-R2000	F980 5'-CTTGGAATGAGCAACAGAGA (976-995) R2000 5'-GTTAGGTCGTTCAATGAAGC (1996-2015) R2000m 5'-GCCTAATACGACTCACTATAGGGAGGTCGTTCAATGAAGCCAC
F1690-R2490	F1690 5'-GAGACGTTAATGGAGTTTGA (1675-1694) R2670 5'-CTTCACTGCTTCCCACATTTG (2665-2685) R2490m 5'-GCCTAATACGACTCACTATAGGGCTGATGTCTATGGCACACCC
F2340-R3470	F2340 5'-TTCGGAGGCATGTCCTGGAT (2326-2345) R3470 5'-CTGATCTCCATACCATACCAACA (3455-3476) R3470m 5'-GCCTAATACGACTCACTATAGGGCTGATCTCCATACCATACCAACAGCC
F3330-R4120	F3330 5'-GAGAGCTGCGGACACCGTGGACC (3334-3356) R4120 5'-CATAGCAGACTTGCTCCTTTCT (4115-4136) R4120m 5'-GCCTAATACGACTCACTATAGGGCATAGCAGACTTGCTCCTTTCTTT
F4070-R4950	F4070 5'-CTGTTGATGGTCGGAATAGG (4054-4076) R4950 5'-CCTGGTTTCGTCTGGACGTT (4933-4952) R4950m 5'-GCCTAATACGACTCACTATAGGGCCTGGTTTCGTCTGGACGTTCTT
F4810-R5650	F4810 5'-CGCCTGGACCCATACTGG (4801-4818) R5650 5'-CCATTCGTATCCAGAGTTCCA (5644-5664) R5650m 5'-GCCTAATACGACTCACTATAGGGAGACCATTCGTATCCAGAGTTCCAAGC
F5510-R6430	F5510 5'-AGCATTGCAGCAAGAGGTTA (5491-5510) R6430 5'-TAGTGCCTGGTGATCCGAGTACAC (6412-6435) R6430m 5'-GCCTAATACGACTCACTATAGGGAGATGCCTGGTGATCCGAGTACACCTG
F6290-R6770	F6290 5'-CGACCGGAGGTGGTGCTTTGATGG (6288-6311) R6770 5'-CCTGGAACTTCAGCCATCCA (6751-6770) R6770m 5'-GCCTAATACGACTCACTATAGGGAGACCTGGAACTTCAGCCATCCAACA
F6690-R7550	F6690 5'-CCTCCTCATGCAGCGGAA (6675-6692) R7550 5'-GAGCTTGCTCCATTCTCCCA (7543-7562) R7550m 5'-GCCTAATACGACTCACTATAGGGAGAGCTTGCTCCATTCTCCCAAAGCG
F7420-R8260	F7420 5'-CCACACCCATCATGCAGAA (7409-7427) R8260 5'-CGTTGGAGCAGCTCCATCTT (8260-8279) R8260m 5'-GCCTAATACGACTCACTATAGGGAGACGTTGGAGCAGCTCCATCTTCTCT
F8170-R9050	F8170 5'-CATAGGACGATTCGGGTCCT (8155-8174) R9050 5'-CTCTTTCCCATCATGTTGTAAATGC (9035-9059) R9050m 5'-GCCTAATACGACTCACTATAGGGAGACTCTTTCCCATCATGTTGTAAATGCAAG
F8920-R9810	F8920 5'-CAGCTTTGGGTGCCATGTT (8906-8924) R9810 5'-GAACCTGCTGCCAATCATACC (9794-9814) R9810m 5'-GCCTAATACGACTCACTATAGGGAGAACCTGCTGCCAATCATACCATCC
F9750-R10630	F9750 5'-TCCTCAATGCTATGTCAAAGGT (9734-9755) R10630 5'-GGTCCTCCTTCCGAGACGGT (10619-10638) R10630m 5'-GCCTAATACGACTCACTATAGGGAGAGGTCCTCCTTCCGAGACGGTTCTG
F10550-R11029	F10550 5'-TGAGTAGACGGTGCTGCCTG (10537-10556) R11029 5'-AGATCCTGTGTTCTCGCACCACCAG (11005-11029) R11029m 5'-GCCTAATACGACTCACTATAGGGAGATCCTGTGTTCTCGCACCACCAGCCA

Table 2. Forward and reverse primers used for PCR amplification of different regions of WNV genome. Numbers in brackets indicates the primer position in the NY99 genome. Reverse primers marked as *'m'* contained the T7 RNA polymerase promoter sequence.

Viral RNA samples isolated from Vero cell culture supernatants were amplified using forward and T7 tagged reverse primers (Table 2) using the OneStep RT-PCR Kit (Qiagen Valencia, CA), according to the manufacturer's Q-Solution protocol with the 40 cycle program as recommended. PCR products were separated by electrophoresis in 0.8% agarose gel prepared in 1×TAE buffer containing 0.2 mg/ml of ethidium bromide. The stained DNA fragments were excised under UV light. PCR products were purified using the MinElute Gel Extraction Kit (Qiagen Valencia, CA), according to the manufacturer's protocol, and stored at -20 °C. DNA fragments were used for the subsequent RNA synthesis using T7 RNA polymerase.

2.6 Preparation of fluorescently labeled WNV RNA for hybridization

The scheme for preparation of fluorescently labeled RNA samples is shown in Figure 4. Single-stranded RNA (ssRNA) samples used for hybridization were synthesized by T7 polymerase-driven transcription of the PCR products using the MEGA script T7 High Yield Transcription Kit (Ambion, Austin, TX) according to the manufacturer's instructions. The MICROMAX ASAP RNA Labeling Kit (Perkin Elmer, Boston, MA) was used to incorporate Cy3 fluorophore into the ssRNA molecules. Fluorescently labeled ssRNA samples were purified from unincorporated dye using the Centrisep Spin Columns (Princeton Separations, Adelphia, NJ), dried under vacuum, and solubilized in the MICROMAX Hybridization Buffer III at a final concentration of 0.5-1.0 μM. The Cy5 antisense-QC oligonucleotide was prepared by 5'-end labeling with indocarbocyanine (Cy5)-dCTP during synthesis. The Cy5-antisense QC oligonucleotide was purified by high performance liquid chromatography (HPLC).

2.7 Microarray oligoprobe design and microchip fabrication

Oligonucleotide probes (oligoprobes) were designed on the basis of the nucleotide sequence of the reference strain NY99 (GenBank accession no.: AF196835) using the OligoScan software. A total of 1274 oligoprobes overlapping by half-lengths, with melting temperatures around 50°C, were designed for microarray-based detection of single point mutations in the entire WNV genome. Each oligoprobe spotting mixture contained 20 μM specific oligoprobe and 1 μM quality control (QC) oligonucleotide in 1× printing buffer (150 mM sodium phosphate, pH 8.5). The probes were used to print five identical arrays (each array contained triplicate set of oligoprobes) per each amine-binding glass slide (CodeLink, Amersham Biosciences, Piscataway, NJ) using a contact microspotting robot PIXSYS 5500 (Cartesian Technologies, Inc.) and a ChipMaker microspotting device equipped with CMP-7 pins delivering approximately 2 to 3 nl of a spotting mixture per spot (Tele-Chem International Inc.). Normally, the size of spots did not exceed 250 μm in diameter. After printing slides were processed and stored according to the manufacturer's protocol. Thus, each microarray slide could be used for simultaneous analysis of five RNA samples. The array layout is shown in Fig. 4. The quality of each lot of printed slides was tested by hybridizing the last printed slide with Cy5-labeled antisense QC oligonucleotide.

2.8 Hybridization conditions

Hybridization between microarray oligoprobes and fluorescently labeled ssRNA samples was performed for 1 hour at 50°C. Before hybridization, Cy3-labeled ssRNA samples was mixed with the Cy5-QC oligonucleotide at a molar ratio of 10 to 1, followed by denaturing at 95°C for 2 min and rapid chilling on ice. The area of each array on the slide was covered with a separate 20×9-mm glass cover-slip containing two rails of 35μm (±10μm) in height (Erie Scientific, Portsmouth, NH) that created a tiny "hybridization chamber" between the rail-elevated cover-slip and microarray slide. Each hybridization mixture (approximately 5-7 μl) was gently loaded by pipette into this space; the slides were placed in a hybridization cassette (Tele-Chem International, Inc., Sunnyvale, CA) to minimize sample evaporation and hybridized for 1 hour at 50°C. After hybridization, the slides were washed twice for 5 min with 2× SSC with 0.1% SDS pre-warmed to 50°C, then once for 2 min with 0.2× SSC buffer and once for 1 min with 0.1× SSC buffer all at room temperature followed by centrifugation at 1000 rpm for 5 min to remove any traces of the buffer.

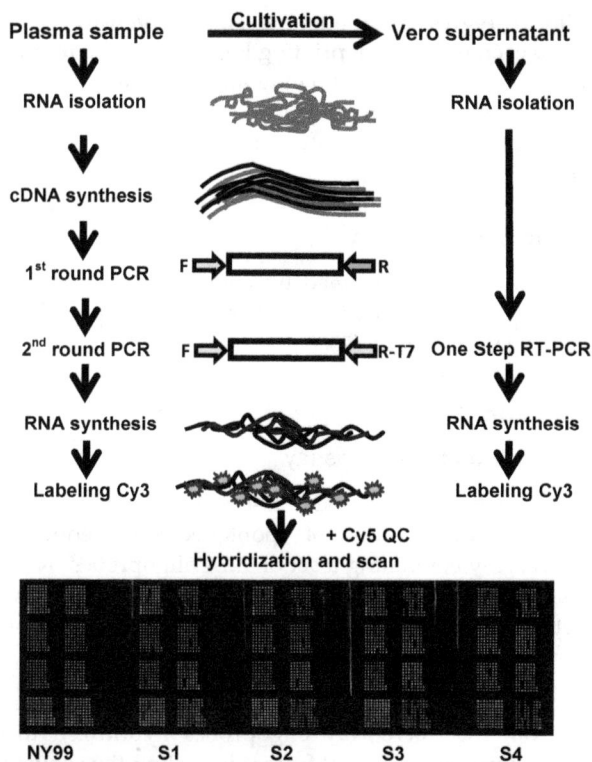

Fig. 4. Scheme of preparation of Cy3 fluorescently labeled RNA samples. The Cy3 image shows the design of array and the results of hybridization experiment. Each microarray slide contained five identical arrays for analysis of five different RNA samples including the reference WNV isolate NY99 and four clinical samples (S1-S4).

2.9 Microarray scanning and image analysis

The fluorescent images of processed microarray slides were generated using GenePix 4100 (Axon Instruments) and ScanArray 5000 (Perkin Elmer) scanners equipped with two lasers operating at 632 nm (for excitation of Cy5 dye) and 543 nm (for excitation of Cy3 dye). Images were analyzed using ScanArray Express 2.1 (Perkin Elmer) and GenePix 3.0 (Axon Instruments) software. Each spot was defined by manual positioning of a grid over the array image. Atypical and empty spots were manually flagged and excluded from further analysis. Background fluorescence readings obtained from the region surrounding each spot were subtracted, and the net value of the Cy3 fluorescence signal from each oligoprobe was divided by the Cy5 signal value from the QC probe of the same spot to minimize the effect of spot size variation on resulting data. Data files generated by ScanArray Express and GenePix software were exported into MS Excel. To identify positions in the WNV genome where mutations occurred, the intensities of fluorescent signals from each array spot obtained for tested WNV isolates were compared with that of the reference NY99 isolate. The fluorescent signal ratio values between the reference NY99 and the tested sample were normalized using a linear regression model. A signal intensity ratio threshold from reference spots, specific to each microarray printing lot, was defined as an average ratio plus two standard deviation values. Any spots showing a ratio greater than the threshold value for a particular printing lot (i.e. considerably lower signal than reference isolate NY99) potentially indicated the presence of a mutation in the genomic area covered by the oligoprobe of that spot.

2.10 Validation of the microarray results

Validation of the microarrays was performed by comparing microarray results with the sequencing data obtained for the tested WNV isolate. Direct sequencing of PCR products was described previously by Grinev et al. (2008a).

3. Results and discussion

3.1 Optimization of the WNV microarray assay

The objective of this study was to develop and assess the feasibility of a DNA microarray approach for rapid throughput detection of spontaneous nucleotide mutations in the genome of WNV. A microarray containing a set of 1274 oligoprobes overlapping by half of their lengths (overlapping factor = 2) was developed and evaluated by assessing the ability of microarray to detect all identified mutations in 34 clinical isolates of WNV that were previously sequenced in our laboratory (Fig. 5).

In fact, optimization of an oligonucleotide microarray assay is a multi-parametric task. Thus, the fluorescent signal from the microarray oligoprobes hybridized to the fluorescently labeled target is known to depend on several factors including the sequence of oligoprobes, the character of mutations, and the position of the mismatched nucleotide in the oligoprobe, as well as the propensity of the RNA hybridization target to form a secondary structure (Relogio et al., 2002; Liu et al., 2005; Naiser et al., 2008). The goal of the optimization process was to determine the hybridization conditions which would enable the sensitive and specific hybridization of the target RNA to the vast majority of oligoprobes composing the WNV microarray. In our study, we optimized the temperature and time of hybridization, the stringency of post-hybridization washing conditions, the image scanning settings (PMT and laser power) to achieve the most efficient discrimination for each probe. It should be noted that the different oligoprobes have slightly different melting temperatures with the hybridization target that also contributes to the difference in sensitivity-specificity balance for each oligoprobe and its targets.

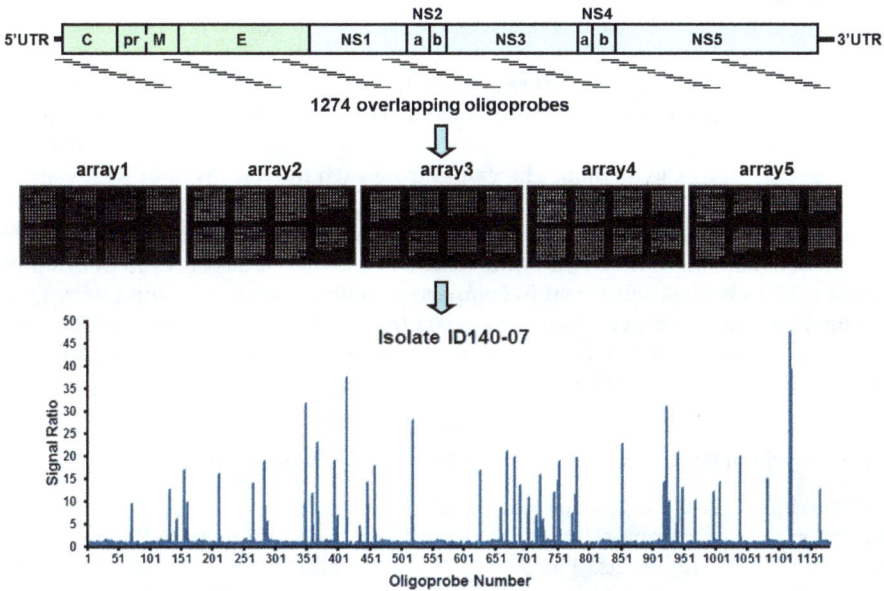

Fig. 5. Scheme of the WNV microarray assay. Five arrays containing 1274 oligoprobes overlapped by half-lengths that covered the entire genome of WNV are shown. The Cy3 hybridization images show the layout of the printed arrays, each oligoprobe was spotted in triplicate. The analysis of the hybridization profile of WNV isolate ID140 from the 2007 U.S. epidemic is shown as a chart of hybridization signal ratios (y-axis) between the reference isolate NY99 and the tested isolate. Oligoprobe numbers are shown on the x-axis.

The oligoprobes were designed to have similar thermodynamic characteristics to ensure uniform hybridization signals from all microarray probes. The design of the microarray probes was performed using the OligoScan software, which is capable of selecting multiple oligonucleotide probes with similar thermodynamic features. The microarray contained overlapping oligoprobes of 15-26 bp in length with melting temperatures around 50°C, which was previously shown to be optimal for the detection of single nucleotide mutations and deletions-insertions (Laassri et al., 2005, 2007). The predicted melting temperatures of the oligoprobes varied from 49.7°C to 52.6°C. Therefore, an optimization of hybridization temperature that would provide efficient hybridization and high specificity for each spot on the array is required. We evaluated three hybridization temperatures: 47, 50 and 53°C. During the optimization process we determined that hybridization and washing of the microarrays at 50°C resulted in better discrimination between perfect matches and mismatches for most of the spots of the array.

The synthetic 5′-aminated oligoprobes were printed on CodeLink Activated slides, previously shown to be suitable slides for detection of single nucleotide mismatches (Laassri et al., 2007). Five arrays containing triplicate sets of oligoprobes were printed on each slide to allow for simultaneous analysis of four target samples in each hybridization experiment. One array on each separate slide was always hybridized with the fluorescently labeled RNA prepared using the reference WN-NY99 strain while other four arrays were used for hybridization with fluorescently labeled RNAs prepared from the WNV isolates to be tested. To assess the reproducibility of microarray fabrication, all hybridizations were repeated twice using different slides. The hybridization temperature and time, post-hybridization washing and the detector setting were optimized. The MS Excel worksheet was developed as a result of analysis of each scan image and data obtained from triplicates of each oligoprobe averaged by the median. The values from each microarray spot were normalized by the signal from the quality controls. The normalized signals from the reference array were then divided by the relevant signals obtained from the sample array. The occurrence of mismatches between oligoprobe and fluorescently labeled RNA target resulted in significant reduction of the hybridization signal in comparison with a perfect matching pair. Thus, the monitoring of the ratios of hybridization signals from unknown samples and the reference samples could be used as a tool for detection of spontaneous mutations in the target region of the WNV genome. The scheme of microarray experiments is shown on Figure 5.

DNA microarray technology provides an opportunity to perform parallel nucleic acid hybridization with a large number of immobilized oligonucleotides on a small surface area.

DNA microarrays containing short oligonucleotide probes (15-25 nt) provide a greater discrimination power compared to microarrays composed of larger oligonucleotides or PCR-amplified DNA fragments. The strongest signal ratios between perfect matched and mismatched sequences were observed when mutations were located near the center of an oligoprobe, and the shortest probes always had better discriminatory power (Urakawa et al., 2003). Although the microarray hybridization method provides limited information about the position of mutations in the analyzed genomic region when compared to sequencing, it has the significant advantage of allowing the identification of "hot spots" where random mutations occur within short (a few nucleotides) areas. It also may allow for the detection of those mutations even when they occur at relatively low levels (up to 1%) as in the case of mixtures of quasispecies that cannot be detected by traditional direct sequencing methods (Cherkasova et al., 2003; Leberre at al., 2007). The efficiency of microarrays for identification and discrimination of closely related bacteria and viruses has been previously demonstrated (Chizhikov et al., 2002; Hsia et al., 2007; Laassri et al., 2003; Nordström et al., 2005; Volokhov et al., 2002; Wade et al., 2004). The use of oligonucleotide microchips for screening of random mutations is based on the ability of microarrays to identify the presence of single-nucleotide mutations in the hybridization template (Hacia et al., 1999; Urakawa et al., 2003).

3.2 Optimization of hybridization probes

In general, single-base-pair discrimination can be achieved by optimization of hybridization or array washing conditions. However, optimization of washing conditions does not guarantee the equal efficiency of mismatch detection for all designed oligoprobes, which is likely to be caused by nature and position of the particular mutation or even formation of the secondary and tertiary structures in the hybridization mixture. The increase of size of the RNA template tends to increase the complexity of the secondary structure of the RNA molecule, and may affect the efficiency of hybridization of the RNA template with the microarray oligoprobes. Therefore, the RNA template may be subdivided into shorter overlapping templates covering the full length of the larger template. The use of a hybridization RNA target of approximately 2500 nucleotides covering the entire structural region of WNV resulted in a twofold reduction of some hybridization signals when compared to the use of a mixture of three overlapping RNA templates in the equivalent molar concentrations. The typical results of this experiment are shown in Figure 6.

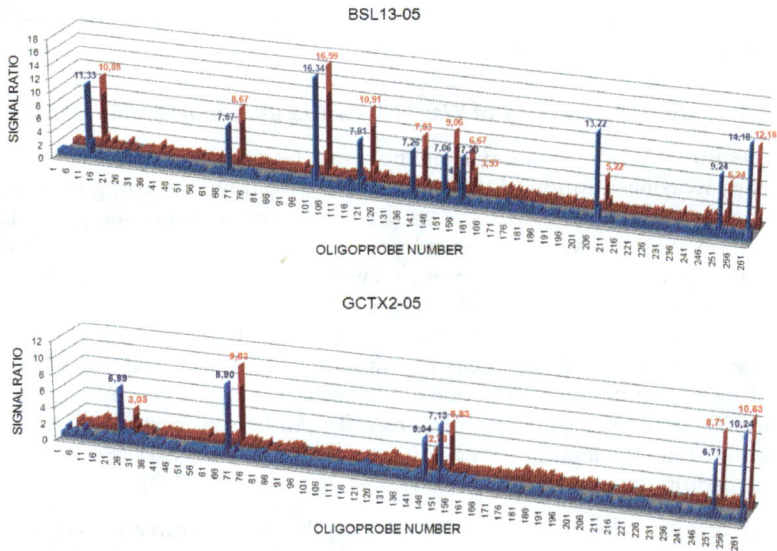

Fig. 6. The analysis of microarray hybridization is shown as a chart that contains the oligoprobe numbers on the x-axis. The y-axis shows the signal ratio values between the isolate NY99 and two isolates from 2005: BSL13-05; and GCTX2-05. The results for hybridization using a mixture of three overlapping RNAs covering the whole structural region of WNV and 5'UTR are shown in blue. The results for hybridization using the one long RNA covering the whole structural region of WNV and 5'UTR are shown in red.

Consequently, mixtures of three RNA templates (instead of one long RNA target) were normally used for hybridization with microarray to ensure the high efficiency of mutation discrimination. The initial use of in-house printed arrays represented a convenient model for the development and optimization of mutation-detecting microarrays. However, the low densities of in-house printed arrays usually force the use of a substantial number of slides for a single hybridization experiment. It is quite laborious and relatively expensive. After optimization of the microarray system including the set of oligoprobes and hybridization conditions, a high density array can be prepared using a well developed system like the widely used GeneChip produced by Affymetrix, which relies on *in situ* synthesis of all oligoprobes covering the entire WNV genome in a single array. In contrast to the printed oligonucleotide arrays described above, the oligonucleotide probes for high-density arrays are synthesized directly on the surface of the microarray, which is usually a small quartz wafer. Because *in situ*-synthesized probes normally form a very tiny spots, multiple overlapping probes for each target may be included to improve sensitivity, specificity, and statistical accuracy. On the other hand, the use of a single array and mixture of RNA probes in a single experiment requires additional optimization of hybridization conditions. We found a loss of some signals while using a template composed of a mixture of 15 overlapping RNAs covering the entire WNV genome, probably due to formation of double stranded molecules between complementary parts of the WNV genome such as the cyclization sequences in the 5′ and 3′ UTRs and tight secondary structure formation. The usage of a number of shorter RNAs produced by multiplex PCR of a target region as a hybridization template might reduce this negative effect and improve the signal, as would chemical defragmentation of the long RNA probe.

3.3 Evaluation of genetic stability of West Nile Virus after isolation in Vero cell culture

Our study required WNV isolation from human plasma samples in Vero cell cultures because of the low concentration of WNV and limited volume of starting material available from many of the specimens. It was shown previously that genetic changes in flaviviruses can be induced by consecutive passages in Vero cell cultures. For example, multiple passage of Dengue virus (DENV), which is closely related to WNV, in Vero cells has resulted in emergence of mutants with amino acid changes in E and occasionally in prM, but not in C. Some nucleotide mutations were detected after the first five passages (Lee et al., 1997). Further studies showed that three nucleotide mutations (two in E and one in NS1 proteins) that resulted in two amino acid alterations (one each in E and NS1) emerged in the DENV genome after 20 continuous passages in Vero cells. Additional passage of the virus for 30 passages caused four nucleotide changes (two each in E and NS1) that resulted in three amino acid substitutions (one in E and two in NS1) (Chen et al., 2003).

In order to investigate whether viral isolation using three consecutive passages in Vero cells resulted in genetic changes in the WNV genome, we cultivated the isolate FDA/HU-02 and compared the sequence and microarray results for the three passages (P1, P2 and P3). Total RNA samples were isolated from aliquots of the original plasma and from each of the passages 1-3. RNA samples were reverse transcribed and amplified by PCR using WNV-specific primers designed to cover the complete structural region of the virus, followed by PCR product purification, direct sequencing, and microarray hybridization. We found no significant changes in fluorescent signal ratio values obtained in one slide microarray hybridization experiment for each original plasma sample and the RNA isolated from each of the Vero cell culture passages. We also performed comparative sequence analysis of the aforementioned viral samples. There was no difference between the genomic sequence obtained from the isolate FDA-Hu2002 (P1) and two passages (P2 and P3) when compared to the genomic sequence obtained from RNA extracted from the original plasma sample. These results are in good concordance with the genetic stability data previously published for the chimeric Dengue and Yellow Fever-Dengue vaccine candidates passaged 10-20 times in Vero cells (Guirakhoo et al., 2004; Butrapet et al., 2006). Therefore we assume that WNV isolates generated from human specimens by a few passages in Vero cells represented the original virus, and that the structural region was not changed during a low number of serial passages in Vero cell cultures.

3.4 Evaluation of oligonucleotide microarray using clinical WNV isolates

The ability of microarray to detect mutations in the target regions was evaluated by testing 34 previously sequenced WNV isolates obtained in the course of the 2002 - 2009 US epidemics.

Gene	prM	ENV			NS1	NS2A	NS3				NS4A	
Strain/nt #	660	1320	1442	2466	3399	4146	4803	6138	6238	6426	6721	6765
NY-99	C	A	T	C	T	A	C	C	C	C	G	T
ARC140-07	T(7.7)	G(7.2)	C(7.1)	T(14.2)	C(23.1)	G(14.4)	T(28.2)	T(8.8)	T(21.2)	T(13.7)	A(16.0)	C(6.1)
BSL2-09	T(13.8)		C(5.4)	T(33.0)		G(3.7)	T(3.9)	T(3.0)	T(8.5)	T(15.0)	A(3.5)	
BSL5-09	T(8.1)	G(25.0)	C(8.7)	T(13.7)	C(5.0)	G(3.8)	T(4.9)	T(8.0)	T(7.3)	T(7.5)	A(3.8)	C(4.2)
BSL6-09	T(29.5)	G(12.4)	C(73.9)	T(10.0)	C(6.1)	G(22.0)	T(30.0)	T(13.6)	T(10.8)	T(10.8)	A(3.6)	C(18.0)
BSL11-09	T(19.8)	G(3.7)	C(4.0)	T(14.5)	C(4.5)	G(7.1)	T(5.7)	T(45.0)	T(8.1)	T(8.9)	A(6.6)	C(6.4)
BSL18-09			C(53.5)	T(7.9)		G(12.2)	T(19.7)	T(12.5)		T(7.0)		
BSL20-09			C(57.8)	T(12.3)		G(3.7)	T(3.7)	T(9.4)		T(4.1)		
BSL22-09			C(50.0)	T(65.8)		G(29.0)	T(3.6)	T(85.7)	T(11.3)	T(3.6)		
BSL24-09	T(63.1)		C(8.9)	T(46.0)		G(9.9)	T(20.0)	T(121.7)	T(21.8)	T(4.0)		
BSL27-09	T(66.3)		C(86.0)	T(8.9)		G(3.1)	T(3.3)	T(12.0)	T(18.1)	T(7.5)		
CO7-09			C(5.3)	T(20.8)		G(12.9)	T(15.5)	T(8.4)		T(29.4)		

Gene	NS4B			NS5								3'UTR
Strain/nt #	6996	7015	7269	7938	8550	8621	8811	9264	9352	9660	10062	10851
NY-99	C	T	T	T	C	A	T	T	C	C	T	A
ARC140-07	T(14.9)	C(19.0)	C(19.8)	C(22.9)	T(14.4)	G(9.9)	C(13.1)	C(12.3)	T(14.4)	T(20.2)	C(15.2)	G(12.7)
BSL2-09	T(12.5)	C(3.7)		C(3.1)	T(4.7)		C(84.0)	C(7.6)	T(5.7)	T(31.0)		G(3.2)
BSL5-09	T(16.2)	C(6.0)	C(7.7)	C(3.3)	T(14.9)	G(4.6)	C(3.7)	C(4.7)	T(4.1)	T(9.9)	C(34.8)	G(7.3)
BSL6-09	T(12.2)	C(5.2)	C(11.5)	C(5.0)	T(46.8)	G(3.9)	C(8.3)	C(3.3)	T(28.4)	T(4.0)	C(5.8)	G(8.3)
BSL11-09	T(13.2)	C(5.0)	C(3.6)	C(3.0)	T(79.4)	G(4.6)	C(11.5)	C(3.1)	T(4.4)	T(3.6)	C(4.4)	G(3.6)
BSL18-09	T(13.5)	C(7.3)		C(3.2)			C(9.9)		T(19.6)			G(4.7)
BSL20-09	T(13.2)	C(4.3)		C(3.4)			C(3.3)		T(4.6)			G(4.8)
BSL22-09	T(31.9)	C(3.3)		C(4.3)			C(7.4)		T(6.4)			G(17.0)
BSL24-09	T(9.6)	C(3.6)		C(5.3)			C(3.5)		T(3.3)			G(4.0)
BSL27-09	T(60.0)	C(7.0)		C(3.9)			C(30.0)		T(16.9)			G(18.1)
CO7-09	T(53.0)	C(3.4)		C(7.8)			C(3.0)		T(16.0)			G(11.0)

Table 3. Fixed nucleotide mutations and ratios of hybridization signals normalized against the reference isolate NY99. Unique mutations are not shown.

23 WNV isolates from 2002-2005 epidemic seasons we used to evaluate the first array containing oligoprobes covering the structural region. Fluorescent signal ratios for all mutations ranged from 4.4 to 85.5 depending on the position of the mismatch within the oligoprobe and the character of the mismatched nucleotide (Grinev et al., 2008b). Table 3 shows the positions of identified nucleotide mutations and ratios of hybridization signals from the respective hybridization templates normalized against the reference isolate NY99 for 24 fixed mutations determined in the complete genomes of isolates from 2007 and 2009 epidemics. All 11 completely sequenced isolates from this study shared 12 nucleotide mutations including a non-silent mutation in Env T1442C. Ten more common mutations were detected in 4 isolates ARC140-07, BSL5-09, BSL6-09 and BSL11-09. The signal ratios for these mutations varied in the range 3.1-73.9. The fluorescence signal ratios produced by single nucleotide mutations in all hybridization experiments exceeded the experimentally determined cut-off threshold value ranging from 0.2 to 2.1 when compared to that of completely matched pairs. The ratio value over 2.1 can be considered as an indication on the potential mutation in a specific genomic region pointing to the need for additional analysis of this region by sequence analysis. This approach has the advantage of substantially reducing the numbers and length of sequences required for proper surveillance studies. The results of our study showed that the WNV microarray was able to unambiguously detect all mutations in the viral genome previously identified by routine sequencing analysis.

4. Conclusion

Viral adaptation through fixation of spontaneous mutations is an important factor potentially associated with reoccurrence of WNV outbreaks in the New World. The emergence of new genetic variants of WNV raise issues of public health importance because they may affect the sensitivity of both screening and diagnostic assays, as well as the development of vaccines and drugs. We have developed and optimized a WNV microarray assay, which enabled simple monitoring of WNV genetic variability and rapid detection of any nucleotide mutation within the entire viral genome. Our microarray system potentially can serve as a high throughput, rapid and effective approach for the identification of WNV mutations, and characterization of circulating WNV genetic variants.

5. Acknowledgments

We would like to thank Drs Caren Chancey, German Anez Gutierrez and Robert Duncan for helpful discussion and review of the manuscript, Dr Majid Laassri for technical assistance, and Dr Konstantin Chumakov for OligoScan software.

6. References

Anthony, R.M.; Schuitema, A.R.; Chan, A.B.; Boender, P.J.; Klatser, P.R. & Oskam, L. (2003). Effect of secondary structure on single nucleotide polymorphism detection with a porous microarray matrix; implications for probe selection. *Biotechniques*, Vol.34, No.5, (May 2003), pp. 1082–1086, 1088–1089, ISSN 0736-6205

Austgen, L.E.; Bowen, R.A.; Bunning, M.L.; Davis, B.S.; Mitchell, C.J. & Chang, GJ. (2004). Experimental infection of cats and dogs with West Nile virus. *Emerging Infectious Diseases*, Vol.10, No.1, (January 2004), pp. 82-86, ISSN 1080-6040

Beasley, D.W. (2005). Recent advances in the molecular biology of West Nile virus. *Current Molecular Medicine*, Vol.5, No.8, (December 2005), pp. 835-850, ISSN 1566-5240

Botha, E.M.; Markotter, W.; Wolfaardt, M.; Paweska, J.T.; Swanepoel, R.; Palacios, G.; Nel, L.H. & Venter, M. (2008). Genetic determinants of virulence in pathogenic lineage 2 West Nile virus strains. *Emerging Infectious Diseases*, Vol.14, No.2, (February 2008), pp. 222-230, ISSN 1080-6040

Bryant, J.E.; Vasconcelos, P.F.; Rijnbrand, R.C.; Mutebi, J.P.; Higgs, S. & Barrett, D.T. (2005). Size heterogeneity in the 3' noncoding region of South American isolates of yellow fever virus. *Journal of Virology*, Vol.79, No.6, (March 2005), pp. 3807–3821, ISSN 0022-538X

Butrapet, S.; Kinney, R.M. & Huang, C.Y. (2006). Determining genetic stabilities of chimeric dengue vaccine candidates based on dengue 2 PDK-53 virus by sequencing and quantitative TaqMAMA. *Journal of Virological Methods*, Vol.131, No.1, (January 2006), pp. 1-9, ISSN 0166-0934

Chappell, K. J.; Nall, T. A.; Stoermer, M. J.; Fang, N. X.; Tyndall, J. D.; Fairlie, D. P. & Young, P.R. (2005). Site-directed mutagenesis and kinetic studies of the West Nile Virus NS3 protease identify key enzyme–substrate interactions. *Journal of Biological Chemistry*, Vol. 280, No.4, (January 2005), pp. 2896–2903, ISSN 0021-9258

Chen, H.; Mammel, M.; Kulka, M.; Patel, I.; Jackson, S. & Goswami,B.B. (2011). Detection and identification of common food-borne viruses with a tiling microarray. *Open Virology Journal*, No.5, (May 2011), pp. 52-59, ISSN 1874-3579

Chen, W.J.; Wu, H.R. & Chiou, S.S. (2003). E/NS1 modifications of dengue 2 virus after serial passages in mammalian and/or mosquito cells. *Intervirology*, Vol. 46, No.5, (May 2003), pp. 289-295, ISSN 0300-5526

Cherkasova, E.; Laassri, M.; Chizhikov, V.; Korotkova, E.; Dragunsky, E.; Agol, V.I. & Chumakov, K. (2003). Microarray analysis of evolution of RNA viruses: Evidence of circulation of virulent highly divergent vaccine-derived polioviruses. *Proceedings of the National Academy of Sciences, USA*, Vol.100, No.16, (August 2003), pp. 9398-9403, ISSN 0027-8424

Cheung, V. G.; Morley, M.; Aguilar, F.; Massimi, A.; Kucherlapati, R. & Childs, G. (1999). Making and reading microarrays. *Nature Genetics*, Vol.21, No.1(Suppl), (January 1999), pp. 15–19, ISSN 1061-4036

Chevalier, V.; Lecollinet, S. & Durand, B. (2011). West Nile Virus in Europe: A Comparison of Surveillance System Designs in a Changing Epidemiological Context. *Vector-Borne and Zoonotic Diseases*, (May 2011) [Epub ahead of print] ISSN 1530-3667

Chizhikov, V.; Wagner, M.; Ivshina, A.; Hoshino, Y.; Kapikian, A.Z. & Chumakov, K. (2002). Detection and genotyping of human group A rotaviruses by oligonucleotide microarray hybridization. *Journal of Clinical Microbiology*, Vol.40, No.7, (July 2002), pp. 2398-2407, ISSN 0095-1137

Clum, S.; Ebner, K.E. & Padmanabhan, R. (1997). Co-translational membrane insertion of the serine proteinase precursor NS2B–NS3(Pro) of dengue virus type 2 is required for efficient in vitro processing and is mediated through the hydrophobic regions of NS2B. *Journal of Biological Chemistry*, Vol.272, No.49, (December 1997), pp. 30715–30723, ISSN 0021-9258

Chou, C.C.; Chen, C.H.; Lee, T.T. & Peck, K. (2004). Optimization of probe length and the number of probes per gene for optimal microarray analysis of gene expression. *Nucleic Acids Research*, Vol.32, No.12, (July 2004), pp. e99, ISSN 0305-1048

Davis, C.T.; Ebel, G.D.; Lanciotti, R.S.; Brault, A.C.; Guzman, H.; Siirin, M.; Lambert, A.; Parsons, R.E.; Beasley, D.W.; Novak, R.J.; Elizondo-Quiroga, D.; Green, E.N.; Young, D.S.; Stark, L.M.; Drebot, M.A.; Artsob, H.; Tesh, R.B.; Kramer, L.D. & Barrett A.D. (2005). Phylogenetic analysis of North American West Nile virus isolates, 2001-2004: evidence for the emergence of a dominant genotype. *Virology*, Vol.342, No.2, (November 2005), pp. 252-265, ISSN 0042-6822

Ebel, G.D.; Carricaburu, J.; Young, D.; Bernard, K.A. & Kramer, L.D. (2004). Genetic and phenotypic variation of West Nile virus in New York, 2000-2003. *American Journal of Tropical Medicine and Hygiene*, Vol.71, No.4, (October 2004), pp. 493-500, ISSN 0002-9637

Egloff, M.P.; Benarroch, D.; Selisko, B.; Romette, J.L. & Canard B. (2002). An RNA cap (nucleoside-2'-O-)-methyltransferase in the flavivirus RNA polymerase NS5: crystal structure and functional characterization. *EMBO Journal*, Vol.21, No.11, (June 2002), pp. 2757–2768, ISSN 0261-4189

Evans, J.D. & Seeger, C. (2007). Differential effects of mutations in NS4B on West Nile virus replication and inhibition of interferon signaling. *Journal of Virology*, Vol.81, No.21, (November 2007), pp. 11809–11816, ISSN 0022-538X

Falgout, B.; Pethel, M.; Zhang, Y.M. & Lai C. J. (1991). Both nonstructural proteins NS2B and NS3 are required for the proteolytic processing of Dengue virus nonstructural proteins *Journal of Virology*, Vol.65, No.5, (May 1991), pp. 2467–2475, ISSN 0022-538X

Friebe, P. & Harris, E. (2010). Interplay of RNA elements in the dengue virus 5' and 3' ends required for viral RNA replication. *Journal of Virology*, Vol.84, No.12, (June 2010), pp. 6103-6118, ISSN 0022-538X

Gardner, S.N.; Jaing, C.J.; McLoughlin, K.S. & Slezak, T.R. (2010). A microbial detection array (MDA) for viral and bacterial detection. *BMC Genomics*, No.11, (November 2010), pp. 668, ISSN 1471-2164

Granwehr, B.P.; Lillibridge, K.M.; Higgs, S.; Mason, P.W.; Aronson, J.F.; Campbell, G.A. & Barrett, A.D. (2004). West Nile virus: where are we now? *Lancet Infectious Diseases*, Vol.4, No.9, (September 2004), pp. 547–556, ISSN 1473-3099

Grinev, A.; Daniel, S.; Stramer, S.; Rossmann, S.; Caglioti, S. & Rios, M. (2008a). Genetic variability ofWest Nile virus in US blood donors, 2002–2005. *Emerging Infectious Diseases*, Vol.14, No.3, (March 2008), pp. 436-444, ISSN 1080-6040

Grinev, A.; Daniel, S.; Laassri, M.; Chumakov, K.; Chizhikov, V. & Rios M. (2008b). Microarray-based assay for the detection of genetic variations of structural genes of West Nile virus. *Journal of Virological Methods*, Vol.154, No.1-2, (December 2008), pp. 27-40, ISSN 0166-0934

Gritsun, T.S. & Gould, E.A. (2007). Direct repeats in the flavivirus 3' untranslated region; a strategy for survival in the environment? *Virology*, Vol.358, No.2, (February 2007), pp. 258-265, ISSN 0042-6822

Guirakhoo, F.; Pugachev, K.; Zhang, Z.; Myers, G.; Levenbook, I.; Draper, K.; Lang, J.; Ocran, S.; Mitchell, F.; Parsons, M.; Brown, N.; Brandler, S.; Fournier, C.; Barrere, B.; Rizvi, F.; Travassos, A.; Nichols, R.; Trent, D. & Monath, T. (2004). Safety and efficacy of chimeric yellow Fever-dengue virus tetravalent vaccine formulations in nonhuman primates. *Journal of Virology*, Vol.78, No.9, (May 2004), pp. 4761-4775, ISSN 0022-538X

Hacia, J.G.; Fan, J.B.; Ryder, O.; Jin, L.; Edgemon, K.; Ghandour, G.; Mayer, R.A.; Sun, B.; Hsie, L.; Robbins, C.M.; Brody, L.C.; Wang, D.; Lander, E.S.; Lipshutz, R.; Fodor, S.P. & Collins, F.S. (1999). Determination of ancestral alleles for human single-nucleotide polymorphisms using high-density oligonucleotide arrays. *Nature Genetics*, Vol.22, No.2, (June 1999), pp. 164–167, ISSN 1061-4036

Hager, J. (2006). Making and using spotted DNA microarrays in an academic core laboratory. *Methods in Enzymology*, Vol.410 (2006), pp. 135–168, ISSN 0076-6879

Hayes, E.B. & Gubler, D.J. (2006). West Nile virus: epidemiology and clinical features of an emerging epidemic in the United States. *Annual Review of Medicine*, Vol.57, (2006), pp. 181-194, ISSN 0066-4219

Herring, B.L.; Bernardin, F.; Caglioti, S.; Stramer, S.; Tobler, L.; Andrews, W.; Cheng, L.; Rampersad, S.; Cameron, C.; Saldanha, J.; Busch, M.P. & Delwart, E. (2007). Phylogenetic analysis of WNV in North American blood donors during the 2003-2004 epidemic seasons. *Virology*, Vol.363, No.1, (June 2007), pp. 220-228, ISSN 0042-6822

Higgs, S.; Snow, K. & Gould, E. (2004). The potential for West Nile virus to establish outside of its natural range: a consideration of potential mosquito vectors in the United

Kingdom. *Transactions of the Royal Society of Tropical Medicine and Hygiene*, Vol.98, No.2, (February 2004), pp. 82–87, ISSN 0035-9203

Honma, S.; Chizhikov, V.; Santos, N.; Tatsumi, M.; Timenetsky Mdo, C.; Linhares, A.C.; Mascarenhas, J.D.; Ushijima, H.; Armah, G.E.; Gentsch, J.R.& Hoshino, Y. (2007). Development and validation of DNA microarray for genotyping group A rotavirus VP4 (P(4), P(6), P(8), P(9), and P(14)) and VP7 (G1 to G6, G8 to G10, and G12) genes. *Journal of Clinical Microbiology,*Vol.45, No.8, (August 2007), pp.2641-2648, ISSN 0095-1137

Hsia, C.C.; Chizhikov, V.E.; Yang, A.X.; Selvapandiyan, A.; Hewlett, I.; Duncan, R.; Puri, R.K.; Nakhasi, H.L. & Kaplan, G.G. (2007). Microarray multiplex assay for the simultaneous detection and discrimination of hepatitis B, hepatitis C, and human immunodeficiency type-1 viruses in human blood samples. *Biochemical and Biophysical Research Communications*, Vol.356, No.4, (May 2007) pp. 1017-1023, ISSN 0006-291X

Jerzak, G.; Bernard, K.A.; Kramer, L.D. & Ebel, G.D. (2005). Genetic variation in West Nile virus from naturally infected mosquitoes and birds suggests quasispecies structure and strong purifying selection.*Journal of General Virology*, Vol.86, No.8 (August 2005), pp. 2175–2183, ISSN 0022-1317

Klenk, K.; Snow, J.; Morgan, K.; Bowen, R.; Stephens, M.; Foster, F.; Gordy, P.; Beckett, S.; Komar, N.; Gubler, D. & Bunning, M. (2004). Alligators as West Nile virus amplifiers. *Emerging Infectious Diseases*, Vol.10, No.12, (December 2004), pp. 2150-2155, ISSN 1080-6040

Kramer L.D. & Bernard, K.A. (2001). West Nile virus infection in birds and mammals. *Annals of the New York Academy of Sciences*, Vol.951, (December 2001), pp. 84-93, ISSN: 0077-8923

Kramer, L.D.; Styer, L.M. & Ebel, G.D. (2008). A Global Perspective on the Epidemiology of West Nile Virus. *Annual Review of Entomology*, Vol.53, (2008), pp. 61-81, ISSN 0066-4170

Khromykh, A.A.; Kenney, M.T. & Westaway, E.G. (1998). trans-Complementation of flavivirus RNA polymerase gene NS5 by using Kunjin virus replicon-expressing BHK cells. *Journal of Virology*, Vol.72, No.9, (September 1998), pp. 7270–7279, ISSN 0022-538X

Khromykh, A.A.; Sedlak, P.L. & Westaway, E.G. (2000). cis- and trans-acting elements in flavivirus RNA replication. *Journal of Virology*, Vol.74, No.7, (July 2000), pp. 3253–3263, ISSN 0022-538X

Laassri, M.; Chizhikov, V.; Mikheev, M.; Shchelkunov, S. & Chumakov, K. (2003). Detection and discrimination of orthopoxviruses using microarrays of immobilized oligonucleotides. *Journal of Virological Methods*, Vol.112, No.1-2, (September 2003), pp. 67–78, ISSN 0166-0934

Laassri, M.; Dragunsky, E.; Enterline, J.; Eremeeva, T.; Ivanova, O.; Lottenbach, K.; Belshe, R. & Chumakov, K. (2005). Genomic analysis of vaccine-derived poliovirus strains in stool specimens by combination of full-length PCR and oligonucleotide microarray hybridization. *Journal of Clinical Microbiology*, Vol.43, No.6, (June 2005), pp. 2886–2894, ISSN 0095-1137

Laassri, M.; Meseda, C.A.; Williams, O.; Merchlinsky, M.; Weir, J.P. & Chumakov, K. (2007). Microarray assay for evaluation of the genetic stability of modified vaccinia virus

Ankara B5R gene. *Journal of Medical Virology,* Vol.79, No.6, (June 2007), pp. 791-802, ISSN 0146-6615

Lanciotti, R.S.; Roehrig, J.T.; Deubel, V.; Smith, J.; Parker, M.; Steele, K.; Crise, B.; Volpe, K.E.; Crabtree, M.B.; Scherret, J.H.; Hall, R.A.; MacKenzie, J.S.; Cropp, C.B.; Panigrahy, B.; Ostlund, E.; Schmitt, B.; Malkinson, M.; Banet, C.; Weissman, J.; Komar, N.; Savage, H.M.; Stone, W.; McNamara, T. & Gubler, D.J. (1999). Origin of the West Nile virus responsible for an outbreak of encephalitis in the northeastern United States. *Science,* Vol.286, No.5448, pp. 2333-2337, ISSN 0036-8075

Lanciotti, R.S.; Ebel, G.D.; Deubel, V.; Kerst, A.J.; Murri, S.; Meyer, R.; Bowen, M.; McKinney, N.; Morrill, W.E.; Crabtree, M.B.; Kramer, L.D. & Roehrig, J.T. (2002). Complete genome sequences and phylogenetic analysis of West Nile virus strains isolated from the United States, Europe, and the Middle East. *Virology,* Vol.298, No.1, (June 2002), pp. 96-105, ISSN 0042-6822

Leberre, V.; Baranowski, E.; Deplanche, M.; Trouilh, L. & François, J.M. (2007). Detection of minority variants within bovine respiratory syncytial virus populations using oligonucleotide-based microarrays. *Journal of Virological Methods,* Vol.148, No.1-2, (March 2007), pp. 271-276, ISSN 0166-0934

Lee, E.; Weir, R.C. & Dalgarno, L. (1997). Changes in the dengue virus major envelope protein on passaging and their localization on the three-dimensional structure of the protein. *Virology,* Vol.232, No.2, pp. 281-290, ISSN 0042-6822

Leung, J.Y.; Pijlman, G.P.; Kondratieva, N.; Hyde, J.; Mackenzie, J.M. & Khromykh, A.A. (2008). Role of nonstructural protein NS2A in flavivirus assembly. *Journal of Virology,;* Vol.82, No.10, (October 2008), pp. 4731–4741, ISSN 0022-538X

Liu, S.; Li, Y.; Fu, X.; Qiu, M.; Jiang, B.; Wu, H.; Li, R.; Mao, Y. & Xie Y. (2005). Analysis of the factors affecting the accuracy of detection for single base alterations by oligonucleotide microarray. *Experimental and Molecular Medicine* , Vol.37, No.2, (April 2005), pp. 71-77, ISSN 1226-3613

Mackenzie, J.M.; Khromykh, A.A.; Jones, M.K. & Westaway, E.G. (1998). Subcellular localization and some biochemical properties of the flavivirus Kunjin nonstructural proteins NS2A and NS4A. *Virology,* (June 1998), Vol.245, No.2, pp. 203–215, ISSN 0042-6822

Markoff L. (2003). 5'- and 3'-noncoding regions in flavivirus RNA. *Advances in virus research,* Vol.59, pp. 177-228, ISSN 0065-3527

McMullen, A.R.; May, F.J.; Li, L.; Guzman, H.; Bueno, R. Jr; Dennett, J.A.; Tesh, R.B. & Barrett, A.D. (2011). Evolution of new genotype of west nile virus in north America. *Emerging Infectious Diseases,;*Vol.17, No.5, (May 2011), pp. 785-93, ISSN 1080-6040

Miller, M. B. (2009). Solid and liquid phase array technologies. In *Molecular microbiology: diagnostic principles and practice,* 2nd ed., ASM Press, Washington, DC, ISBN 978-155-5814-7-7

Moudy, R.M.; Meola, M.A.; Morin, L.L.; Ebel, G.D. & Kramer, L.D. A newly emergent genotype of West Nile virus is transmitted earlier and more efficiently by Culex mosquitoes. *American Journal of Tropical Medicine and Hygiene,* Vol.77, No.2, (August 2007), pp. 365–370, ISSN 0002-9637

Muñoz-Jordán, J.L.; Laurent-Rolle, M.; Ashour, J.; Martínez-Sobrido, L.; Ashok, M.; Lipkin, W.I. & García-Sastre, A. (2005). Inhibition of alpha/beta interferon signaling by the

NS4B protein of flaviviruses. *Journal of Virology,* Vol.79, No.13, (July 2005), pp. 8004–8013, ISSN 0022-538X

Naiser, T.; Ehler, O.; Kayser, J.; Mai, T.; Michel, W. & Ott, A. (2008). Impact of point-mutations on the hybridization affinity of surface-bound DNA/DNA and RNA/DNA oligonucleotide-duplexes: comparison of single base mismatches and base bulges. *BMC Biotechnology,* Vol.8, (May 2008), pp. 48, ISSN 1472-6750

Nasci, R.S.; Savage, H.M.; White, D.J.; Miller, J.R.; Cropp, B.C.; Godsey, M.S.; Kerst, A.J.; Bennett, P.; Gottfried, K. & Lanciotti RS. (2001). West Nile virus in overwintering Culex mosquitoes, New York City, 2000. *Emerging Infectious Diseases,* Vol.7, No.4, (July-August 2001), pp. 742–744, ISSN 1080-6040

Neverov, A.A.; Riddell, M.A.; Moss, W.J.; Volokhov, D.V.; Rota, P.A.; Lowe, L.E.; Chibo, D.; Smit, S.B.; Griffin, D.E.; Chumakov, K.M. & Chizhikov, V.E. (2006). Genotyping of measles virus in clinical specimens on the basis of oligonucleotide microarray hybridization patterns. *Journal of Clinical Microbiology,* Vol.44, No.10, (October 2006), pp. 3752-3759, ISSN 0095-1137

Nordström, H.; Falk, K.I.; Lindegren, G.; Mouzavi-Jazi, M.; Waldén, A.; Elgh, F.; Nilsson, P.& Lundkvist, A. (2005). DNA microarray technique for detection and identification of seven flaviviruses pathogenic for man. *Journal of Clinical Microbiology,* Vol.77, No.4, (December 2005), pp. 528-540, ISSN 0095-1137

Parreira, R.; Severino, P.; Freitas, F.; Piedade, J.; Almeida, A.P. & Esteves A. (2007). Two distinct introductions of the West Nile virus in Portugal disclosed by phylogenetic analysis of genomic sequences. *Vector-Borne and Zoonotic Diseases,* Vol.7, No.3, (Fall 2007), pp. 344-352, ISSN: 1530-3667

Pealer, L.N.; Marfin, A.A.; Petersen, L.R.; Lanciotti, R.S.; Page, P.L.; Stramer, S.L.; Stobierski, M.G.; Signs, K.; Newman, B.; Kapoor, H.; Goodman, J.L. & Chamberland, M.E. (2003).West Nile Virus Transmission Investigation Team. Transmission of West Nile virus through blood transfusion in the United States in 2002. *The New England Journal of Medicine,* Vol.349, No.13, pp. 1205-1206. ISSN 0028-4793

Petersen, L.R. & Roehrig, J.T. (2001).West Nile Virus: a reemerging global pathogen. *Emerging Infectious Diseases,* Vol.7, No.4, (July-August 2001), pp. 611–614, ISSN 1080-6040

Petersen, L.R. & Marfin, A.A. (2002). West Nile virus: a primer for the clinician. *Annals of Internal Medicine,* Vol.137, No.3, (August 2002), pp. 173-179, ISSN 0003-4819

Proutski, V.; Gould, E.A. & Holmes, E.C. (1997). Secondary structure of the 3' untranslated region of flaviviruses: similarities and differences. *Nucleic Acids Research,* Vol.25, No.6, (March 1997), pp. 1194-1202, ISSN 0305-1048

Relógio, A.; Schwager, C.; Richter, A.; Ansorge, W. & Valcárcel, J. (2002). Optimization of oligonucleotide-based DNA microarrays. *Nucleic Acids Research,* Vol.30, No.11, (June 2002), pp. e51, ISSN 0305-1048

Roerig, P.; Nessling, M.; Radlwimmer, B.; Joos, S.; Wrobel, G.; Schwaenen, C.; Reifenberger, G. & Lichter, P. (2005). Molecular classification of human gliomas using matrix-based comparative genomic hybridization. *International Journal of Cancer,* Vol.117, No.1, (October 2005), pp. 2095-2103, ISSN 0020-7136

Sbrana, E.; Tonry, J.H.; Xiao, S.Y.; da Rosa, A.P.; Higgs, S. & Tesh, R.B. (2005). Oral transmission of West Nile virus in a hamster model. *The American Journal of Tropical Medicine and Hygiene,* Vol.72, No.3, (March 2005), pp. 325-329, ISSN 0002-9637

Schlesinger, J.J. (2006). Flavivirus nonstructural protein NS1: complementary surprises. *Proceedings of the National Academy of Sciences of the USA*, Vol.103, No.50, (December 206), pp. 18879–18880, ISSN 0027-8424

Tajima, S.; Nukui, Y.; Ito, M.; Takasaki, T. & Kurane, I. (2006). Nineteen nucleotides in the variable region of 3' non-translated region are dispensable for the replication of dengue type 1 virus in vitro. *Virus Research*, Vol.116, No.1-2, (March 2006), pp. 38-44, ISSN 0168-1702

Tomiuk, S. & Hofmann, K. (2001). Microarray probe selection strategies. *Briefings in bioinformatics*, Vol.2, No.4, (December 2001), pp. 329–340, ISSN 1467- 5463

Urakawa, H.; El Fantroussi, S.; Smidt, H.; Smoot, J.C.; Tribou, E.H.; Kelly, J.J.; Noble, P.A. & Stahl, D.A. (2003). Optimization of single-base-pair mismatch discrimination in oligonucleotide microarrays. *Applied and Environmental Microbiology*, Vol.69, No.5, (May 2003), pp. 2848-2856, ISSN 0099-2240

Volokhov, D.; Rasooly, A.; Chumakov, K. & Chizhikov, V. (2002). Identification of Listeria species by microarray-based assay. *Journal of Clinical Microbiology*, Vol.44, No.12, (December 2005), pp. 4720–4728, ISSN 0095-1137

Wade, M.M.; Volokhov, D.; Peredelchuk, M.; Chizhikov, V. & Zhang, Y. (2004). Accurate mapping of mutations of pyrazinamide-resistant Mycobacterium tuberculosis strains with a scanning-frame oligonucleotide microarray. *Diagnostic Microbiology and Infectious Disease*, Vol.49, No.2, (June 2004), pp. 89–97, ISSN: 0732-8893

Yu, L.; Nomaguchi, M.; Padmanabhan, R. & Markoff, L. (2008). Specific requirements for elements of the 5' and 3' terminal regions in flavivirus RNA synthesis and viral replicatio. *Virology*, Vol.374, No.1, (April 2008), pp. 170-185, ISSN 0042-6822

10

Inter and Intra-Host Evolution of Dengue Viruses and the Inference to the Pathogenesis

Day-Yu Chao
National Chung Hsing University
Taiwan

1. Introduction

Dengue viruses, like many RNA viruses are highly mutagenic, which have a potential to generate approximately one nucleotide mutation per round of genome replication (Domingo, Escarmis et al. 1996). The extent of genetic diversity differs among the sylvatic/urban cycles, chronologically isolates, serotypes and genotypes of Dengue viruses. All four serotypes of Dengue viruses (DENV-1 to DENV-4) evolved independently in their particular ecologic niche and could further classified into different genotypes among each serotypes based on nucleotide homology. Currently DENV-1 is classified into five different genotypes, DENV-2 into six different genotypes, DENV-3 into five different genotypes and DENV-4 into four different genotypes (Holmes and Twiddy 2003; Klungthong, Zhang et al. 2004; King, Chao et al. 2008; Vasilakis and Weaver 2008).

Genotype switch was correlated with DHF epidemic in certain region in recent years, which implied a more virulent strain may be evolved from the Southeast Asian and transmit to a new susceptible indigenous population to cause severe disease outcome (Thant, Morita et al. 1996; Sittisombut, Sistayanarain et al. 1997; Cologna and Rico-Hesse 2003). Our previous studies showed that the larger of quasispecies among structure proteins than non-structural proteins, which probably implied the more selection immune pressure on the structure proteins instead of merely randomly changing during replication (Chao, King et al. 2005). This has been further validated by the data in E gene where the Domain I, the antigenic sites had larger sequences diversity than other domains. An increasing question raised upon this observation has focused on the concept that whether quasispecies distribution of dengue viruses as a reservoir of virus variants plays an essential role in diversification and selection of variants which replicates better in human population in the context of the high incidence density and clustering of dengue epidemic. Therefore, to closely examine the inter- and intra-host evolution of dengue viruses among full-genomic sequence will be required to elucidate the relationship between host immune responses, viral strain infected and disease outcomes.

The recent epidemics of dengue in Taiwan started when dengue virus serotype 2 (DENV-2) was first introduced into the southern off-islet of Hsiao-Liu-Chiu in 1981 after its absence for 38 years since World War II (WWII) (Tsishe 1932; Wu 1986). Later in 1987-1988, the epidemic of DF, caused by DENV-1 mainly in Kaohsiung and Pingtung in southern Taiwan. Tainan City, located just at north of Kaohsiung, had not had a dengue epidemic since l942-43 until three outbreaks occurred recently (Ko 1989; Chen, King et al. 1997). The first of three DF epidemics in Tainan, was caused by DENV-1, resulted in 38 confirmed DF cases in 1994 with no DHF cases observed. Three years later in 1997, a more localized DENV-2 outbreak of DF occurred, involving only 14 confirmed cases. The third epidemic of dengue, attributed to DENV-3, began in late 1998 and continued into January, 1999. During this 1998-1999 epidemic, 142 confirmed dengue cases including at least 14 DHF cases were officially reported (Chao, Lin et al. 2004).

During the epidemic, two interesting epidemiological phenomena were found based on our previous publications(Chao, Lin et al. 2004). First, the DHF/DF ratio increased with time, from 11% during the first time interval, to 20% and 30% during the second and third time intervals, respectively. Second, the majority (73.3%, 88/120) of the dengue cases were primary infections, including DHF, which showed no significant association with secondary infections [13 DHF cases had primary infection and 8 DHF cases had secondary infection, odds ratio=1.92 (95%CI 0.64-5.76), p=0.19]. Therefore, we hypothesize that intense transmission of dengue virus within closed environment may drive emergence of DENV-3 strain with higher propensity of causing sever disease. Thus, viruses, isolated in first passaged-C6/36 cell culture, from three well-characterized family clusters were chosen for molecular genetic study.

Since previous study suggested that there were a genetic marker in DENV-2 viruses, which differentiated American genotype with the southeast Asian genotype and caused wind-sweeping epidemic in the central and south America since 1981 epidemic in Cuba(Leitmeyer, Vaughn et al. 1999; Sariol, Pelegrino et al. 1999; Rodriguez, Alvarez et al. 2005; Rodriguez, Alvarez et al. 2005), it will be interesting to look at the DENV-3 viruses whether such genetic marker or virulence marker exists or not. We started by the consensus-direct sequencing of full-length genome to identify the genetic markers associated with virulence and the most probable quasispecies regions. Candidate regions were then amplified, cloned, and randomly selected multiple clones were sequenced to better understand the population dynamic of quasispecies variation among family pairs. The clonal sequencing result was also conducted in selected regions by obtaining amplicons directly from viremic plasma and *Aedes aegytie* passage one virus.

2. Materials and methods

2.1 Case definition

A confirmed dengue case was defined as a person with the illness that fulfilled any of the following laboratory diagnostic criteria: (1) isolation of dengue virus from serum; or (2) identification of dengue-specific cDNA fragment from plasma or serum by reverese-transcriptase polymerase chain reaction (RT-PCR)(Lanciotti, Calisher et al. 1992); or (3) seroconversion of dengue-specific IgM from negative to positive but seronegative for Japanese encephalitis (JE)-specific IgM by IgM-enzyme-linked immuno-sorbent assay (IgM-ELISA)(Shu, Chen et al. 2003); or (4) a 4-fold or greater titer rise in dengue-specific IgG antibody in paired serum samples(Shu, Chen et al. 2002). The clinical diagnosis of DHF was based upon revised WHO's criteria in 1997(World Health Organization 1997), as follows: (1)

fever, (2) hemorrhagic manifestations, including a positive tourniquet test result, (3) thrombocytopenia (100,000/mm³ or less), and (4) evidence of plasma leakage manifested by at least one of the following: hemoconcentration, presence of pleural effusion or ascites (documented by radiography, ultrasound, or computed tomographic scan) or hypoproteinaemia. Hemoconcentration, which was defined as a 20% increase in hematocrit compared with stabilized hematocrit at hospital discharge or revisit after discharge, was calculated as the ratio of the difference of maximum and minimal hematocrit values, divided by the minimal value. In consideration of references used in most hospitals in Taiwan, hypoproteinaemia was defined as a serum albumin level less than 3 gm/dL. Those confirmed dengue cases were classified as primary, secondary, or indeterminate infections, depending on the ratio of DENVVgue-specific IgM/IgG as measured by the capture IgM and IgG ELISA test(Vaughn, Nisalak et al. 1999; Shu, Chen et al. 2003).

2.2 Family cluster chosen

In choosing family clusters during 1998 epidemics, minimum of one confirmed dengue patient within the same household unit were selected and it ended up with 12 family clusters identified. The definite disease classification, disease onset date, and detail demographic data were recorded for all dengue patients in the same family clusters. Virus isolates for molecular genetic characterizations from three family clusters were selected, based on the following criteria: (1) with DF and DHF patients in the same family clusters; (2) the duration of disease onset between DF and DHF patients in the same family clusters is longer than 3 day and shorter than 10 days. This is based on the assumption of that if the dengue virus is transmitted within the same household from DF patient to DHF patient and the duration of disease onset between two cases as indicated, there is a high propensity that virus may transmit by mosquito mechanically, influenced by the multiple feeding behavior of *Aedes aegytie*. The bottleneck transmission by the mechanical transmission may create the opportunity of transmitting the higher virulent virus population from DF patient to naïve individual who may have higher odds ratio of developing DHF. The detail description of basic information including onset date, sex, age, viral load and immune status for three family clusters was summarized in Table 1.

Cluster	ID	Disease status	Onset date	Age	Sex	Immune status	Viral load (RNA copies/ml plasma)
1	364	DF	11/20	38	F	Primary	107,000
1	368	DHF	11/27	27	M	Primary	3,890,000
2	390	DF	12/1	69	M	Secondary	<600
2	388	DHF	12/5	57	F	Primary	1,540,000
3	414	DF	12/7	36	F	Primary	<600
3	407	DHF	12/12	63	M	Secondary	1,360,000

Table 1. The Demographic Data and Viral Load From Each Patient Within the Family Clusters.

2.3 Virus isolation

Acute-phase serum or plasma samples were collected from patients within seven days after the onset of fever and stored in -70ºC freezer until tested. Plasma sample aliquots were used to infect C6/36 *Aedes albopictus* mosquito cell lines as described previously(Kuno, Gubler et al. 1985) and were identified as DENVV-3 by indirect fluorescent antibody tests with serotype-specific monoclonal antibodies (DENV-1:H47, DENV-2:H46, DENV-3:H49, DENV-4:H48). Briefly, C6/36 cells were seeded into 75-cm² tissue culture flasks at 5x10⁵ cells per flask in Mitsuhashi & Maramorosch insect medium (MM) (Sigma, St. Louis, MO) and Dulbecco's minimum essential medium (DMEM) (Invitrogen, San Diego, CA) with 1:1 ratio containing 10% fetal calf serum and 100% antibiotics-antimycotics (Invitrogen, San Diego, CA). As cells reached 80% confluent after seeding, the medium was removed and only 1ml left. 40 ul of patient's plasma was added to each flask and the virus was allowed to infect cells by rolling the flasks every 15 minutes for 2 hours. After absorption, fresh maintenance medium containing only 2% fetal calf serum was added and incubated in 37ºC incubator. The culture supernatant was collected at day 7 and 14 post infection, and used to re-infect BHK-21 cells for plaque assay to determine virus titer as previously described. For the molecular sequencing, the C6/36 cell one-passaged virus was used for full-length consensus sequencing to identify potential heterogeneous regions for clonal sequencing as described later.

2.4 Preparation of viral RNA and RT-PCR amplification

Dengue viral RNA was isolated either from viremic plasma or C6/36 passaged one supernatant by using QIAamp viral RNA mini kit (Qiagen, Germany) following the manufacturer's protocol. The eluted RNA was subjected to Titan™ one tube RT-PCR System (Boehringer Mannheim) to amplify overlapping regions of DENV-3 sequence by virus specific synthetic oligonucleotide primers. The oligonucleotide primer pairs were designed based on published full-length DENV-3 sequence data for H87 and 80-2 obtained from Genebank at the National Center for Biotechnology Information (NCBI) and some unpublished DENV-3 sequences (personal communication, Chang et. al.; Centers for Disease Control and Prevention, Fort Collins, CO., U. S. A.). Ten overlapping fragments were generated which spanned genomic regions 1 to 1181, 530 to 1694, 1259 to 1694, 2171 to 3417, 3142 to 4677, 4123 to 5686, 5443 to 7477, 7246 to 8750, 8501 to 10316, 9991 to 10688 as previously described(Chao, King et al. 2005).

2.5 PCR product cloning, purification and sequencing of PCR fragments

PCR product was purified by using the QIAquick PCR Purification Kit Protocol following manufacture suggested protocol (Qiagen, Germany). The purified double-stranded DNA fragments were subjected to sequence analysis the cycle-sequencing dye terminator method using the Big Dye Terminator Cycle Sequencing Ready Reaction kit (Perkin-Elmer, Applied Biosystems, Foster City, Ca.). We estimated the amount of DNA for each sequence reaction by comparing the band intensity of a 1:10 diluted product (1 µL product + 9µL DEPC dH20) with 2 and 4µL of the high DNA mass ladder (Invitrogen, San Diego, CA) after gel electrophoresis using EtBr containing 1% agarose gel. For each sequencing reaction, approximately 50 to 100 ng purified DNA was combined with 3.2 pmol of sequencing primer (3.2 µL of 10uM primer concentration), 8.0 µL of reaction cocktail (containing dNTP, dye-labeled-ddNTP terminators and Taq polymerase) and deionized distill water to bring the final volume to 20ul. Start on sequencing cycling program on the thermocycler as suggested in the manufacturer's protocol (30 cycles of 96°for 10sec; 50°C for 5sec; 60°C for 4 min) and hold at 4°C forever.

The reaction mixture was column purified by home-made Sephadex™ G50, fine DNA grade (Amersham Pharmacia, Biotech, AB, Sweden) filled column, and the DNA was dried in a vacuum centrifuge for 20 minutes. Finally, the DNA pellet was resuspended in Hi-D formamide (denaturing reagent), transferred to a 96-well plate, heated for 2 min at 95°C and kept on ice prior to run the 3100 automate sequencer (Perkin-Elmer, Aplied Biosystems). We used a short capillary (47 cm by 50 um diameter) and Performance Optimized Polymer 6 for the run.

The PCR product from the potential heterogeneous regions ligated with the T/A cloning vector, PCRII-TOPO, was used to transform to E. coli TOP10 competent cells (Invitrogen, San Diego, CA). At least 30 recombinant clones were randomly selected, and completely sequenced by using insert flanking primers, T7 and cSP6.

2.6 Mosquito feeding and inoculation

To determine the sequence diversity inside the mosquitoes, 70 5-day old adult female mosquitoes of each Ae. aegypti and Ae. albopictus were starving for 2 days prior to oral feed on virus spiked rabbit blood. Mosquitoes of each species were enclosed in two fine mesh net covered cans. A mixture of 5 ml fresh-prepared rabbit blood (with heparin), 1 ml 1% sucrose and 2 ml virus stock (10^6PFU/ml) was spreading over the net at room temperature for one hour for mosquito to feed on virus-spiked blood mixture. One ml each of the blood mixture was collected before feeding, feeding 0.5 hour and after feeding, centrifuged at 3000rpm for 10 minutes to collect the supernatants. The virus titer in each collection was determined by the plaque assay. Mosquitoe pools were relocated to the insectory room for 14 days after blood feeding. Ten other female Ae. aegypti mosquitoes were inoculated with the 1:5 dilution of viral stock by intrathoracic inoculation techniques.

Virus infected mosquitoes were hold at 32°C for 7 days and 14 days at which the salivary glands were dissected from surviving females, and the presence or absence of viral antigen in these tissues was determined by the indirect fluorescent antibody techniques (IFA). The salivary glands from oral feeding or intrathoracically inoculated mosquitoes were placed inside the 1.5ml eppendorf with 200ul PBS, and frozen at –70°C before RNA extraction.

2.7 Nucleotide and Amino acid sequence analysis

Overlapping chromatogram files retrieved from the automate sequencer were analyzed and edited using the SeqMan program in the Lasergene software package (DNASTAR, inc. Madison, Wis.). The derived consensus sequences after excluding the sequences of primers were aligned using GCG package (Genetic Computer Group, Wis.). For full-length genomic sequences we paid special attention for the regions consistently presented mixed-chromatographic picks. These regions were identified and selected for the clonal sequence analysis. Pairwise comparisons of both nucleotide and amino acid sequences between isolates and clonal sequences were performed using the program MEGA v2.1 (Molecular Evolutionary Genetics Analysis, Pennsylvania State University, PA) to determine the mean and range of proportion of difference (hamming and p distance)(Kumar, Tamura et al. 2004). The obtained nucleotide sequences were aligned with the sequences of available DENV-3 strains and the DENV-2 Jamaica strain obtained from the Genebank at NCBI, using the multiple sequences alignment program PILEUP with the default gap penalties. The

PHYLIP package, that utilized the neighbor-joining method to calculate nucleotide evolutionary distances, was used to generate a phylogenetic tree.(Felsenstein 1993)

2.8 Statistical analysis

All the data from the questionnaires and laboratory results were entered into the database and analyzed by SAS (Statistical Analytical System, Wisconsin, 6.12 version). Chi-square test was performed to compare differences in two groups for discrete data. The exact p-values by Fisher exact test were calculated if the expected number was smaller than 5.

3. Results

3.1 Geographical distribution and chronological spread of genotypes

DENV-3 was the only serotype isolated during the 1998 epidemic, which was identified in 38 (26.7%) of the confirmed dengue cases. The phylogenetic tree analysis showed that the virus strain belongs to DENV-3 genotype 2, which comprises Thailand strains according to the classification of Lanciotti et al (Fig 1)(Lanciotti, Lewis et al. 1994). Further examining the phylogenetic analysis results based on full-length genomes by neighbor-joining method for DENV-3 viruses including 4 different genotypes and other serotypes of dengue viruses suggested the similar results with previous reports by Wittke et al(Wittke, Robb et al. 2002), which separated the viruses into 4 main groups except genotype 4 because of lacking full-length sequence. The proximity of different genotypes of DENV-3 was that genotype I and IV were closer than genotype type II and III. After comparing the full-length sequences of the DF and DHF isolates, no genetic makers can be found to differentiate the disease severity, which might suggest the genomic virulence determination site does not exist.

Next, we constructed a contingency table to analyze all strains by geographical regions and genotypes (Table 2). Based on the geographical distribution of isolates by time, the Philippines-Indonesia-Malaysia region has all genotypes except traditional American genotype (genotype IV). The oldest genotype V represented the DENV-3 prototype H87, which rarely being isolated except one in China in 1980 and the other in Malaysia in 1981. It is very possible genotype I evolved from genotype V and first appeared also in the Philippines-Indonesia-Malaysia region since 1974 in Malaysia and 1973 in Indonesia. It later spread into other southeast Asian countries, including Taiwan and further into south Pacific Island, including Fiji in 1992 and Tahiti in 1989. Genotype II however, first evolved from Thailand in 1962 and later spread into Malaysia and replaced the original genotype I to cause epidemic locally in 1993-4. Genotype III was the wide-spread genotype, which evolved probably also from the Philippines-Indonesia-Malaysia region and in the 1980's this genotype spread west to India, Sri Lanka and Africa and east to Taiwan. Genotype III was the only genotype isolated in the Latin America since 1994, which the introduced Asian genotype replaced the American genotype caused a great DF/DHF epidemics in many countries. Genotype IV was the traditional American genotype rarely isolated in other countries. Its distribution was limited in Puerto Rico in 1963 and 1977 and in Tahiti in 1965, which suggested virus exchange between west-Pacific region and the Central- America region during 1960s.

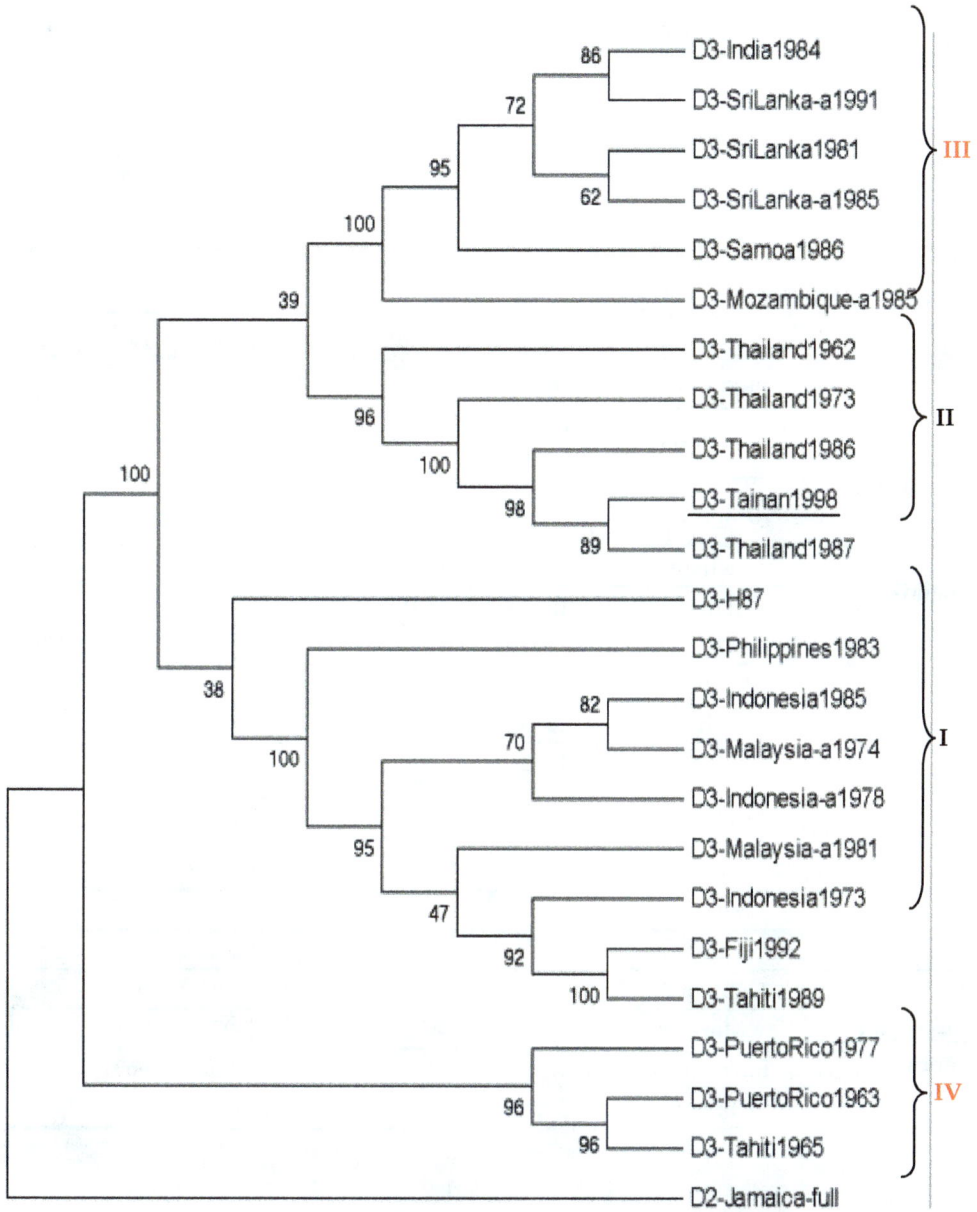

Fig. 1. Phylogram Generated by Parsimony Analysis of Nucleic Acid Sequences from PrM/M and E Genes of 24 DENV-3 Viruses Isolates.The numbers displayed above the horizontal lines correspond to the percentage of bootstrap analysis. Parsimony analysis was performed by the heuristic branch swapping alogorithm of MEGA-2.

Geographical region	Country	Genotypes				
		I	II	III	IV	V
Philippines-Indonesia-Malaysia region	Philippines	1983				1956
	Malaysia	1974, 1981,1997	1992,1993,1994			1981
	Indonesia	1973,19788, 1985,1989	1976			
Thailand-Vietnam	Vietnam		1994,1996			
	Thailand	1988	1973,1986-1998	(1962)		
	China					1980
Pacific islands	Polynesia				1964,1969	
	Tahiti	1989			1965	
	Fiji	1992				
south Asia	Sri Lanka			1981,1985, 1989,1991		
	India			1984		
Africa	Somoa			1986		
	Mozambique			1985		
Central/south America	Guatemala			1996-98	1977	
	Mexico			1995-97		
	Puerto Rico				1963	
	Brazil			2000		

Table 2. Geographical Distribution of DENV-3 Genotypes.

3.2 Nucleotide and amino acid sequence diversity among different regions of full-genome

If DHF patients were not caused by secondary infection or more virulent strain of DENV-3 viruses during 1998 epidemic, would it be possible that there existed a sub-variant in the quasispecies of DENV-3 viruses which caused more severe form of disease after dengue viral infection? Three family clusters were chosen as states in Material and Methods section. More than 20 clones containing the PCR products of different genes of dengue viruses from 6 dengue patients were completed sequenced, aligned and analyzed by excluding the primer sequences. To examine the extent of sequence variation, we determined the mean pairwise p-distance, which is the number of substitution divided by total nucleotide (amino acids) sequenced for each pair of clones. The results were summarized in Table 3. In genearal, non-structural protein such as NS3 and NS5 had the least sequence diversity than structural protein such as capsid or envelope protein. Among non-structural protein, NS3 also presented the least sequence diversity with mean p-distance of nucleotide ranged from 0.09-0.24% and that of amino acid ranged from 0.2-0.5%. Among structural protein, envelope protein presented the largest sequence diversity with mean p-distance of nucleotide ranged from 0.2-0.4% and that of amino acid ranged from 0.4-0.8%. The difference of mean pairwise p-distance among different genes was statistically significant (p<0.01).

ID	C-PrM			E			NS3			NS5		
	No of clones	nucleo-tide	Amino acid	No of clones	nucleo-tide	Amino acid	No of clones	nucleo-tide	Amino acid	No of clones	nucleo-tide	Amino acid
364	23	0.002671	0.006671	26	0.00223	0.004316	19	0.000936	0.003287	13	0.00174	0.004133
368	29	0.00245	0.004863	13	0.003815	0.006638	27	0.001182	0.003053	17	0.002416	0.00348
388	21	0.002027	0.004001	25	0.003878	0.006436	26	0.001247	0.00323	18	0.003005	0.006095
390	22	0.002165	0.004183	13	0.003179	0.00761	25	0.002437	0.005662	25	0.002459	0.003603
407	13	0.001807	0.004416	23	0.003828	0.006235	23	0.001592	0.001968	16	0.004251	0.006818
414	24	0.002549	0.004901	20	0.002555	0.004773	18	0.001013	0.001639	26	0.003697	0.006487
Mean	22	0.002278	0.004839	20	0.003697	0.006487	23	0.001401	0.00314	19	0.003122	0.005915

Table 3. Sequence Diversity (Mean p-Distance) Among Different Genome Regions of dengue Viruses.

We next examined the relationship between the extents of sequence variation among different genes. As shown in Fig. 2, there was a trend of increase in the mean pairwise p-distance of nucleotides of envelope protein as that of the NS3 protein increase (simple linear regression, r=0.6, p=0.01). Similarly, a linear relationship was also observed when comparing the mean pairwise p-distance of amino acid between these two genes (Fig. 3).

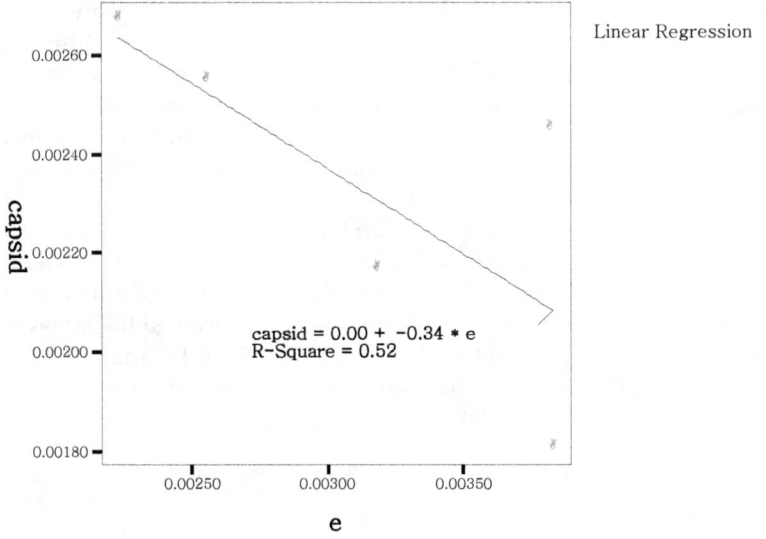

Linear Regression

capsid = 0.00 + −0.34 * e
R−Square = 0.52

(a)

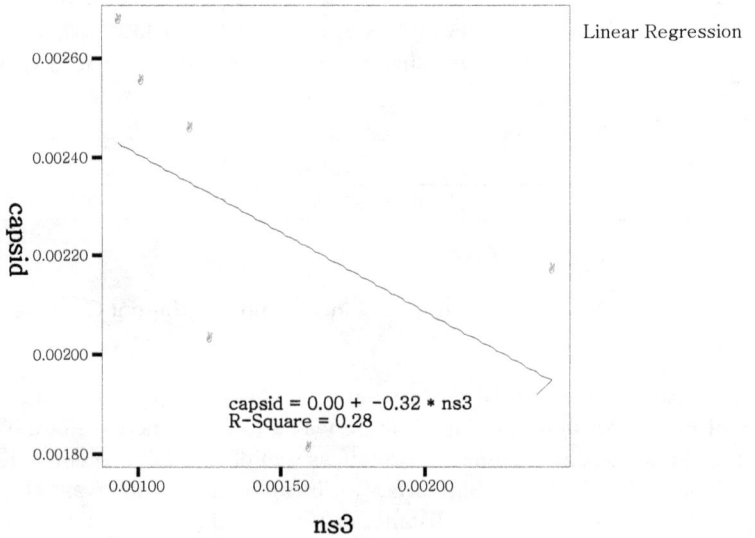

Linear Regression

capsid = 0.00 + −0.32 * ns3
R−Square = 0.28

(b)

Linear Regression

e = 0.00 + 0.45 * ns3
R-Square = 0.14

(c)

ns3

capsid = 0.00 + -0.22 * ns5
R-Square = 0.37

(d)

ns5

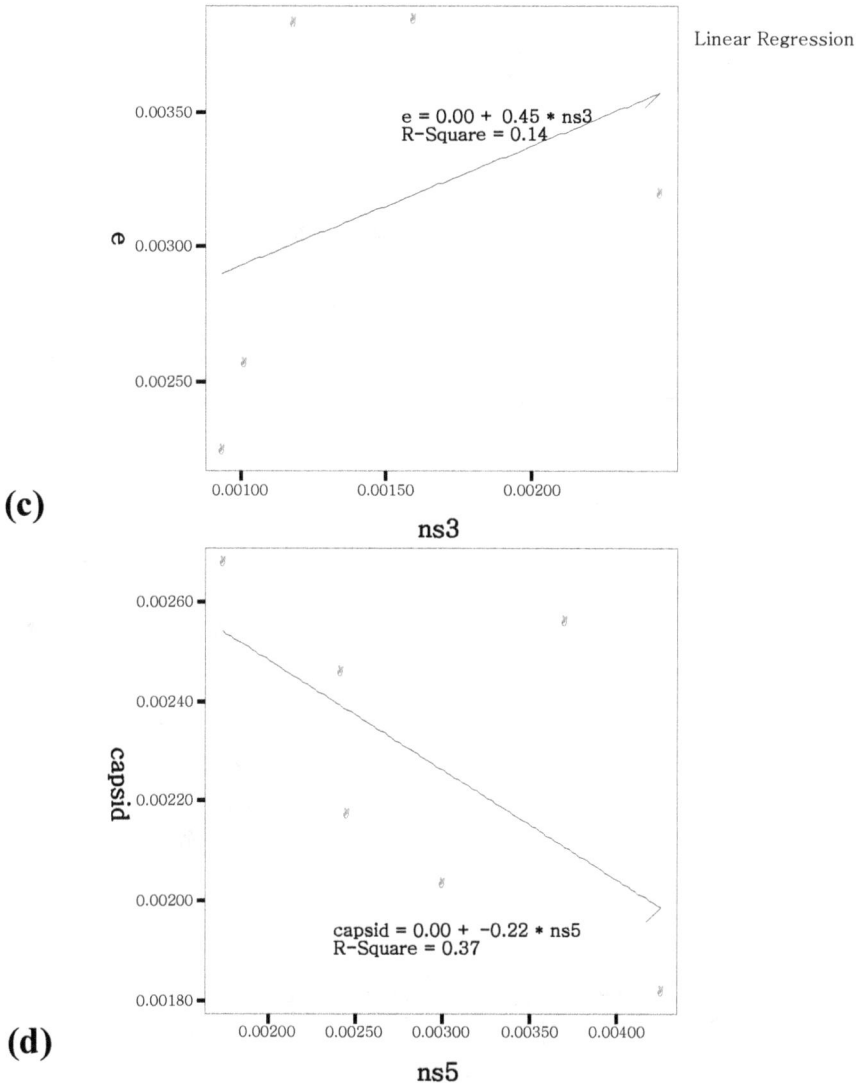

Fig. 2. The Relationship of Intrahost Nucleotide Sequence Diversity Between Different Genomes. Y-axis and X-axis represented the mean p-distance of nucleotides of indicated protein from different patients' isolates. The line was fitted regression line generated by SPSS software and the upward line indicated the positive correlation; the downward line indicated the negative correlation. (a) showed the relationship of mean p-distance between capsid and envelope protein; (b) showed the relationship between capsid and NS3 protein; (c) showed the relationship between NS3 and envelope protein; (d) showed the relationship between capsid and NS5 protein. Only the regression line in (c) showed statistical significance, meaning the higher sequence diversity in envelope protein correlated with the higher sequence diversity in NS3 protein.

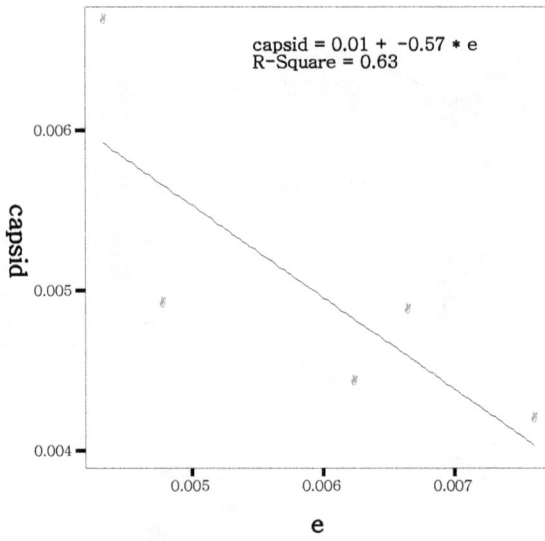

(a)

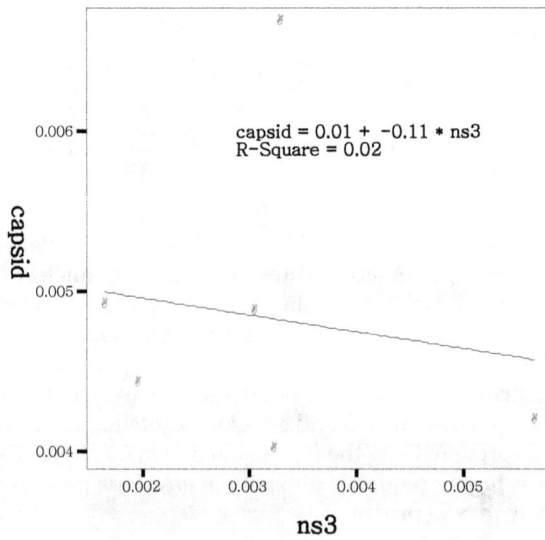

(b)

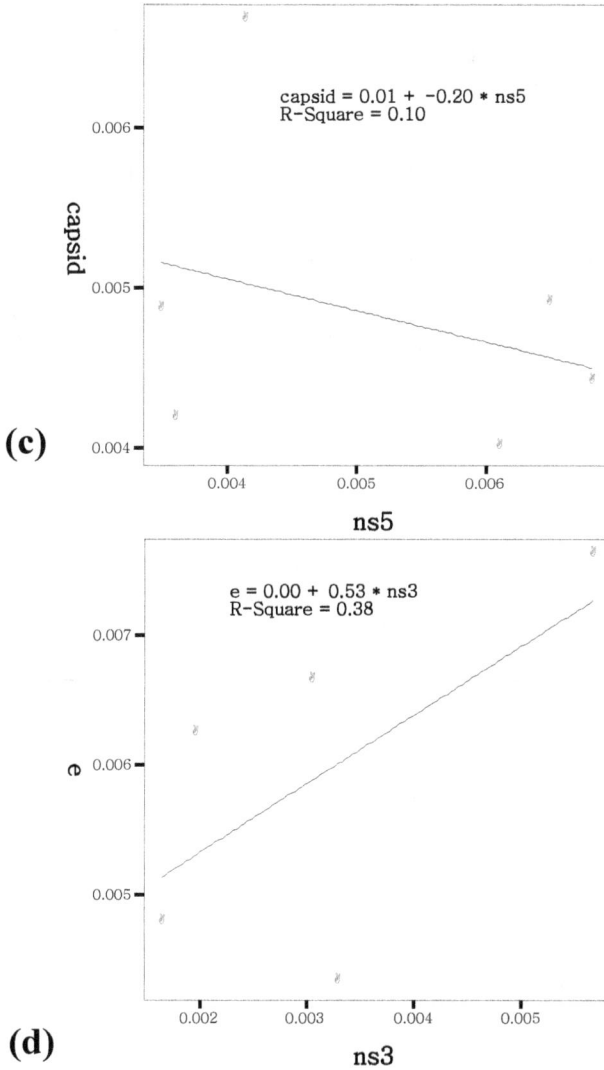

Fig. 3. The Relationship of Intrahost Amino Acid Sequence Diversity Between Different Genomes. Y-axis and X-axis represented the mean p-distance of amino acid of indicated protein from different patients' isolates. The line was fitted regression line generated by SPSS software and the upward line indicated the positive correlation; the downward line indicated the negative correlation. (a) showed the relationship of mean p-distance between capsid and envelope protein; (b) showed the relationship between capsid and NS3 protein; (c) showed the relationship between NS3 and envelope protein; (d) showed the relationship between capsid and NS5 protein. Only the regression line in (d) showed statistical significance, meaning the higher sequence diversity in envelope protein correlated with the higher sequence diversity in NS3 protein.

3.3 Correlation between sequence diversity and phenotypic change

The relationship between genotype and phenotype has been an interest among evolutionary biologists and virologist(Clarke, Duarte et al. 1993; Bielefeldr-Ohmann and Barclay 1998; Arias, Lazaro et al. 2001). Several examples of viral clones isolated from mutant spectra showed altered biological properties, such as HIV mutants with resistant to antiviral inhibitors(Farci, Shimoda et al. 2000; Delwart, Magierowska et al. 2002). The correlation between quasispecies mutant clones with DHF phenotypic change has not been discussed before. With extensive searching, there was no such genetic marker of subvariants from clonal sequencing were found among those family clusters. However, by comparing intra-host variation, there was concomitant increase sequence diversity along with the decrease of identical clones from the first case to the other one in the same family cluster. As shown in Fig. 4, three family clusters present consistent increase sequence diversity from 2.3, 2.5, 3.0 in first case in each family cluster to 4.8, 4.0, 3.3 in the second case in each family cluster.

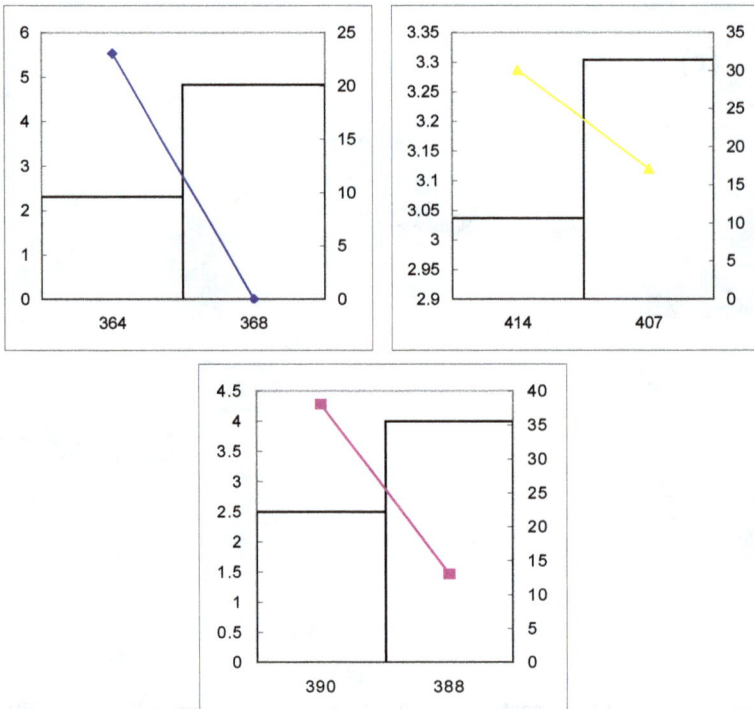

Fig. 4. Genetic Distance Among the Variants and Percent of Identical Clones Between Patients With Different Disease Manifestations Within Individual Family Cluster. Y-Axis on the left indicated the mean hamming distance and on the right indicated the percentage of identical clones, measured by calculating the number of identical clones among total clones sequenced. The box represented the value of mean hamming distance of indicated patient ID shown on the bottom and the line represented the value of identical clones percentage of indicated patient ID shown on the bottom. (a)-(c) indicated the three family clusters with DF and DHF in the same family cluster. The increase of mean hamming distance correlated with the decrease numbers of identical clones.

Consistent with sequence diversity, the percentage of identical clones decreased from 23%, 38%, 30% in first case in each family cluster to 0%, 13%, 17% in the second case in each family cluster. This kind of consistency was correlated with viral load, but not with primary or secondary infection, sampling date, age or sex.

3.4 Comparison of sequence diversity among original plasma, cultured viruses and mosquito inoculation

Little is known about the sequence diversity within mosquitoes, original plasma of patients and cell line passaged viral stocks. The sequence diversity among three family clusters was reconfirmed in the patient's original plasma and mosquito inoculation by one passage viral stock. 30 clones were picked after PCR direct cloning and analyzed after direct sequencing. There was also consistent change with the extent of quasispecies and the number of identical clones. It was found that the sequence diversity of nucleotide (hamming distance) was the lowest in original plasma (1.4) than that in one passage viral stock (2.5) or mosquito inoculation (1.7) at patient ID 368. Although we did not do mosquito inoculation for patient sample ID 388 and 407, the consistent trend was observed when comparing original plasma and one passaged virus stocks (Table 4). Consistent with our general impression is the decrease of sequence diversity when virus was inoculated into the mosquitoes. The hamming distance of nucleotide dropped from 2.46 to 1.7 and the number of identical clones was also increased from 15% to 31%. As shown in Fig. 5, there was a decrease in sequence diversity of original plasma compared with the C6/36 one passage cultured virus, and also increase the percentage of identical clones of original plasma compared with also the cultured virus.

ID	No of clones	nucleotide	Amino acid
P368*	45	1.405556	1.061111
368**	13	2.461538	1.692308
368i^	28	1.70154	1.47077
P388*	20	2.7898	2.33
388**	25	3.50370	2.87
P407*	48	1.039894	0.583333
407**	23	1.905138	1.296443

*indicate sequence diversity from original plasma
**indicate sequence diversity from one passage in c6/36 cell lines of original plasma
^indicate sequence diversity from mosquito injection after one passage in c6/36 passage

Table 4. Sequence Diversity (Mean Hamming Distance) of Envelope Gene Among Different Passage Histories of Dengue Viruses.

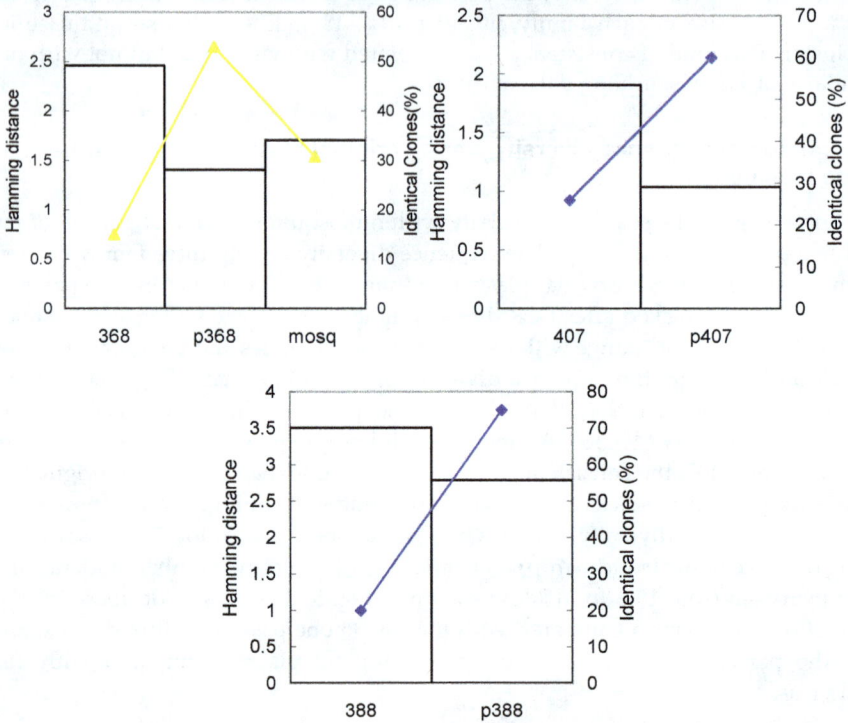

Fig. 5. Comparison of the Sequence Diversity (Hamming distance) of Nucleotide Among Original Plasma, Cultured Viruses and Mosquito Inoculation. Y-axis on the left indicated the value of hamming distance and on the right indicated the percentage of identical clones measured by the number of identical clones among total clones sequences from the isolates ID indicated at the bottom of x-axis. The value of hamming distance for individual isolates are represented as box and the value of percentage of identical clones are shown as dot and connected between isolates by solid line. (a) – (c) indicated the hamming distance and percentage of identical clones from original plasma of DHF patient, which designated by p368, p388, p407, and one passaged in C6/36 virus isolates of patients' plasma, which designated by 368, 388, 407. (a) shows the hamming distance of virus isolates from salivary gland of mosquito-*Ae. Aegypti* after intrathoracically inoculation of C6/36 passaged isolates ID368, which designated by mosq.

4. Discussion

Although several reports have described the uniqueness of 1998 DENV-3 epidemic where different serotypes of dengue viruses co-circulated(de Melecio, R. Barea et al. 1998; Harris, Videa et al. 2000; Rahman, Rahman et al. 2002), the epidemiology of the 1998 epidemic in non-endemic countries, including Republic of China in Taiwan has not been well-described. Without the interference of other serotypes of dengue viruses, this study allowed detailed examination of population genetics in human and mosquito hosts during intensive transmission, which led to a better understanding of the dynamic transmission of viruses during the course of epidemic.

Most molecular epidemiological studies conducted so far have relied on consensus sequences of a virus population or an isolate. When more and more studies on the great potential for variation and phenotypic diversity of some important RNA virus pathogen, such as HIV, HCV and poliovirus have been done, it is necessary to give an appropriate set of micro-environment conditions to observe how virus quasispeices change through transmission during epidemics. Before we started to research on quasispecies of dengue viruses, it would be important to ask which region of full-genome is the most representative one for studying heterogeneity. Many viruses proposed their hyper-variable region for studying quasispecies including structure protein, like HIV V3 region of envelope protein(Delwart, Magierowska et al. 2002), HCV E2 gene (Farci, Shimoda et al. 2000; Curran, Jameson et al. 2002) and non-structure protein (NS5A) in HCV(Blight, Kolykhalov et al. 2000). What about dengue viruses? After picking up five most probable heterogeneous regions for cloning and sequencing, it was found the structural proteins especially envelope protein are more seuquence-diversed than other non-structural proteins, especially NS5 region. The structural proteins we studies here including the capsid and envelope protein. The C protein present in virions as a structure component is a small and highly positively charged protein, including N-terminal hydrophilic region, central hydrophobic region and C-terminal hydrophobic domain. The envelope protein is also the structural protein of virions plays a role in a number of biological activities including virion assembly, receptor binding and membrane fusion and is the major target for neutralization antibodies. On the other hand, the NS3 protein is the second largest viral protein, which encodes protease and helicase bi-functional protein. The C-terminal helicase protein is the region sequenced in this study. The NS5 protein undoubtly is the largest protein which role acts as the viral RNA polymerase and is the most conserved of the flavivirus protein(Chambers, Hahn et al. 1990; LinDENVVbach and Rice 2001). The larger of quasispecies among structural proteins than non-structural proteins probably implies the more selection immune pressure on the structural proteins instead of merely randomly changing during replication. This can be further validated by the data in E gene where the Domain I, the antigenic sites had larger sequences diversity than other domains. An increasing question raised upon this observation has focused on the concept that whether quasispecies distribution of dengue viruses as a reservoir of virus variants plays an essential role in diversification and selection and contribute to the dengue virus evolution.

Studies of sequence heterogeneity like our studies need to take precautions to ensure that artifacts are not introduced during the amplification of virus genomes. In this study, we used C6/36 one-passaged cultured viruses for all viruses isolates among clusters and the approach of thermostable RT-PCR kit and molecular cloning, which has been shown to be a simple and valuable method for characterization of mutant spectra of virus quasispecies. Even though this does not absolutely devoid of RT-PCR error, the relative comparison among different genomes is less biased and the linearity between E gene and other proteins should be able to trust. Furthermore, according to the study by Arias et al, the biological and molecular clones provided statistically indistinguishable definitions of the mutant spectrum with regard to the types and distributions of mutations, mutational hot-spots and mutation frequencies(Arias, Lazaro et al. 2001). Therefore, the molecular cloning procedure employed in this study provides a simple and easy protocol for the characterization of mutant spectra of viruses.

Whether dengue viruses, like other RNA viruses, exist as a quasispecies was first proved experimentally by Wang et al with nucleotide sequence diversity of the envelope gene ranging from 0.1% to 0.84% and p-distance ranging within 0.21-1.67%(Wang, Lin et al. 2002). Instead of using mean diversity represented by using the number of substitutions divided by the total number of nucleotides sequenced, our data on envelope gene used mean pairwise hamming distance ranging from 2.3 to 4.8 which is similar as being used in HCV's study(Farci, Shimoda et al. 2000). The E gene sequenced in this study contained total 1239 nucleotides and 394 amino acids, which was longer in length than previous study but similar diversity with narrower range (p-distance ranging within 0.22-0.38%). From our highly characterized family clusters, it presented consistent increase in sequence diversity from 2.3, 2.5, 3.0 in the first case (DF) to 4.8, 4.0, 3.3 in the second case (DHF) of each family cluster. Consistent with sequence diversity, the percentage of identical clones decreased from 23%, 38%, 30% in the first case to 0%, 13%, 17% in the second case in each family cluster. Further exclusion of those clones with only one single nucleotide mutation and inclusion of clones with more than two mutations, the result was the same. The percentage of identical clones decreased from 11%, 7%, 10% in the first case to 10%, 0%, 4% in the second case in each family cluster.

We hypothesize that intense transmission of dengue virus within closed environment may drive emergence of DENV-3 strain with higher propensity of causing severe diseases. It is plausible there exists at least three virus variants at any stage in mosquito or human during virus replications(Fig 6): variant M: replicate efficiently in mosquito, variant H: replicate efficiently in human, and variant N: replicate equal well in human and mosquito. We will use abbreviations at the following description. Variant N makes up the majority virus population in the quasispecies spectrum either in human, mosquito or tissue culture. Relative percentage of variant M increases when virus replicated in the mosquito; however, relative percentage of variant H increases when virus transmitted to human by mosquitoes. If virus can maintain an efficient Transovary-transmission in mosquito indefinitely, variant M will increase graduately, which might occur in the sylvatic cycle of dengue viruses and during inter-epidemic period. Variant H will increase and progeny virus or newly derived H-variant may have a higher replication capacity, thus higher virus load and higher DHF potential if human virus is transmitted mechanically from human to human through mosquito probing. Variant N possess quasispecies memory both originated from variant M and variant H. The evolution of dengue viruses comes from the random mutation accumulated in the variant N, which forms the bottleneck transmission of dengue viruses during transmission from low viremic human to mosquitoes. The possible explanation for our results is that a minor virus subpopulation with increased virulence gains rapid advantage in a direct transmission condition (ie, within family) from a certain mammalian host with peak viremia.

Several researches have been done to relate quasispecies with adaptability and host range. The alternating host cycle in arboviruses may constrain the evolution and sequence diversity among viral population. Single-host-cell adaptation by serial passage of alphavirus in mammalian cell line (BHK cell) resulted in more mutations than alternating in mammalian and mosquito cell passages(Weaver, Brault et al. 1999). So is the genetic diversity in RNA virus quasispecies, which is controlled by host-virus interaction(Schneider and Roossinck 2001). Our data also showed similar result when comparing genetic diversity among mosquito inoculates, original plasma and single passage viruses. Thus, human-mosquito-human transmission acts as a bottleneck transmission with profound fitness stability. There is also evidence showing that virus transfer might take place with high frequency between human where the donor is at the peak of viremia(Clarke, Duarte et al. 1993; Dockter, Evans et al. 1996). Thus, the chance of a minor, more virulent virus subpopulation being transferred could have increased. The most explosive outbreak of Ross River virus-induced epidemic polyarthritis in Polynesia in the 1970s was most likely caused by the arrival of a single viraemia traveler combining with a larger susceptible population and no appropriate intermediate hosts(Bielefeldr-Ohmann and Barclay 1998). In contrast, in endemic areas pr during periods of low mosquito activity or low viraemia titers, the virus is propagated in a bottleneck transmission mode, which repeatedly selects against the variants most virulent for human.

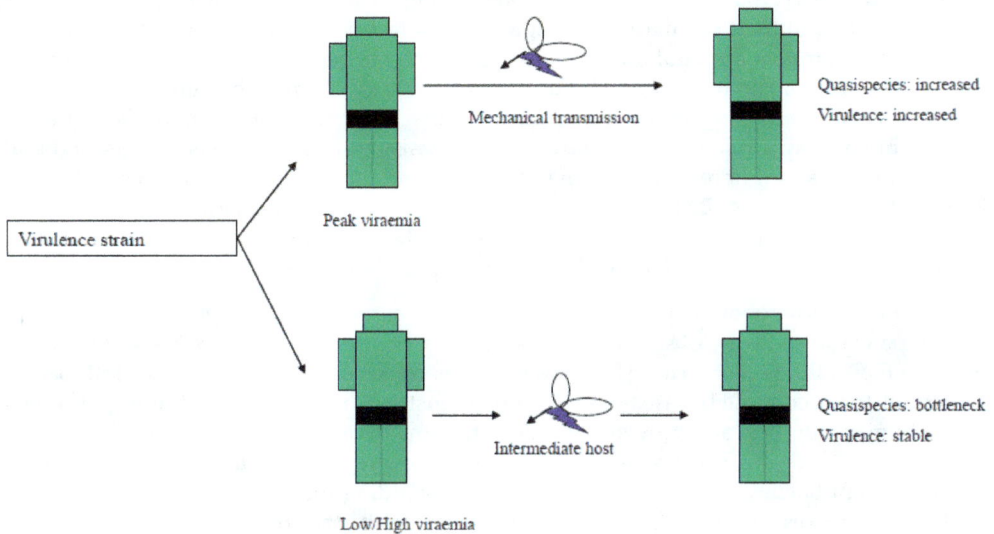

Fig. 6. Proposed model of transmission-replication mode of dengue virus and likely consequences for the quasispecies. A minor virus subpopulation with better replication capability resulting high viremia might rapidly gain selection advantage in a condition of possible direct mechanical transmission condition (ie, within a family) with peak viremia and thus provides chances to increase the viral virulence. the alternating transmission cycles of arboviruses in vertebrates and invertebrates may constrain the evolution and sequence diversity among viral populations.

Our quasispecies-based scenarios reserves a lot questions remained to be answered. First, what is the "virulence-factors" of the minor population might be? As we stated before, there is no such genetic marker in consensus sequenced or clonal sequences found in our study or in other studies. However, there exists several genetic markers repeated appearing in different isolates. Because the frequency is not high enough to say "hot-spot", those mutants reflected genomes that were dominant in the prior evolutionary history of the virus (previous passage), defining as quasispecies memory and keep to be transmitted during the epidemic(Domingo, Ruiz-Jarabo et al. 2002). Changes in virus genome may affect virus binding or replication by increasing the number of host cell receptor-specificities or binding affinities or enhance virus protein synthesis in target tissue and thus viral cytopathogenicity. Or those mutants inside quasispecies hit some viral epitopes rendering them highly inflammagenic in some genetically predisposed individuals, without following discernible evolutionary genetic changes. With the lack of evidences of drawing conclusions based on inadequacy of animal models, the biological mechanism for changes in virulence by a replication-advantage of the mutant could be multi-factorial, which ultimately cause an imbalance in the production of components for virus entry, accumulation or assembly(Yao, Strauss et al. 1996). Also, what is the role of defective interfering (DI) particles in the pathogenesis of dengue virus infection? The involvement of DI particles has been well-studied in many RNA viruses, such as LCMV and alphavirus, which leads to virus persistence and increased cytopathology and inflammation. Our study, consistent with previous studies(Wang, Lin et al. 2002; Wang, Sung et al. 2002), also suggested the DI particles exist in human plasma with dengue viral infection. Combinations of DI particles with standard virus in the viral quasispecies leading to curing of the infection or increasing virus replication remain to be studied. Furthermore, what is the relationship of quasispecies in different compartments of the host, such as the liver, the plasma and the different immune organ play a role in dengue viral pathogenesis? In HIV, it has been suggested that different mutants in different compartments of HIV patients were associated with different tissue pathology(Marras, Bruggeman et al. 2002), as as to dengue virus which has been suggested to be able to replicate in many kinds of cells and body comparments, including liver cell, endothelial cell, dendritic cell, monocyte/ macrophage and CNS.

Finally, our study provided the most important association between genotype and phenotype made so far by observing the consistent increase in sequence diversity from the first case (DF) to the second case (DHF), which was consistent with the higher viral load in the sera of the second DHF cases of each family cluster. Thus, by preventing high viremia titer in human infection and prevent high density transmission of dengue virus, which thus decrease the viral qusispecies size and prevent severe disease manifestation will be important in public health prevention. Better understanding the evolution and quasispecies of dengue viruses with biological determinant(s) and the role of pre-existing sub-neutralizing antibody, will be crucial for future dengue vaccine.

5. References

Arias, A., E. Lazaro, et al. (2001). "Molecular intermediates of fitness gain of an RNA virus: characterization of a mutant spectrum by biological and molecular cloning." *J. Gen. Virol.* 82: 1049-1060.

Bielefeldr-Ohmann, H. and J. Barclay (1998). "Pathogenesis of Ross River virus-induced diseases: a role for viral quasispecies and persistence." *Microb. Pathog.* 24: 373-383.

Blight, K. J., A. A. Kolykhalov, et al. (2000). "Efficient initiation of HCV RNA replication in cell culture." *Science* 290: 1972-1974.

Chambers, T. J., C. S. Hahn, et al. (1990). "Flavivirus genome: organization, expression and replication." *Annu. Rev. Microbiol.* 44: 649-688.

Chao, D. Y., C. C. King, et al. (2005). "Strategically examining the full-genome of dengue virus type 3 in clinical isolates reveals its mutation spectra." *Virol J* 2: 72.

Chao, D. Y., T. H. Lin, et al. (2004). "1998 dengue hemorrhagic fever epidemic in Taiwan." *Emerg Infect Dis* 10(3): 552-554.

Chen, W.-J., C. C. King, et al. (1997). "Changing prevalence of antibody to dengue virus in paired sera in the two years following an epidemic in Taiwan." *Epidemiol. Infect.* 119: 277-279.

Clarke, D. K., E. A. Duarte, et al. (1993). "Genetic bottlenecks and population passages cause profound fitness differeneces in RNA viruses." *J. Virol.* 67(1): 222-228.

Cologna, R. and R. Rico-Hesse (2003). "American genotype structures decrease dengue virus output from human monocytes and DENVVdritic cells." *J Virol* 77(7): 3929-3938.

Curran, R., C. L. Jameson, et al. (2002). "Evolutionary trends of the first hypervariable region of the hepatitis C virus E2 protein in individuals with differing liver disease severity." *J. Gen. Virol.* 83: 11-23.

de Melecio, C. F., H. H. R. Barea, et al. (1998). "Dengue outbreak associated with multiple serotypes -- Puerto Rico, 1998." *Morb Mortal Wkly Rep (MMWR)* 47(44): 952-956.

Delwart, E., M. Magierowska, et al. (2002). "Homogeneous quasispecies in 16 out of 17 individuals during very early HIV-1 primary infection." *AIDS* 16: 189-195.

Dockter, J., C. F. Evans, et al. (1996). "Competitive selection *in vivo* by a cell for one variant over another: implications for RNA virus quasispecies *in vivo*." *J. Virol.* 70(3): 1799-1803.

Domingo, E., C. Escarmis, et al. (1996). "Basic concepts in RNA virus evolution." *FASEB J.* 10: 859-564.

Domingo, E., C. M. Ruiz-Jarabo, et al. (2002). "Emergence and selection of RNA virus variants: memory and extinction." *Virus Res* 82: 39-44.

Farci, P., A. Shimoda, et al. (2000). "The outcome of acute hepatitis C predicted by the evolution of the viral quasispecies." *Science* 288: 339-344.

Felsenstein, J. (1993). "PHYLIP: Phylogeny Inference Package (Univ. of Wahsington, Seattle), Version 3.5c."

Harris, E., E. Videa, et al. (2000). "Clinical, epidemiologic, and virologic features of dengue in the 1998 epidemic in Nicaragua." *Am J Trop Med Hyg* 63(1-2): 5-11.

Holmes, E. C. and S. S. Twiddy (2003). "The origin, emergence and evolutionary genetics of dengue virus." *Infect Genet Evol* 3(1): 19-28.

King, C. C., D. Y. Chao, et al. (2008). "Comparative analysis of full genomic sequences among different genotypes of dengue virus type 3." *Virol J* 5: 63.

Klungthong, C., C. Zhang, et al. (2004). "The molecular epidemiology of dengue virus serotype 4 in Bangkok, Thailand." *Virology* 329(1): 168-179.

Ko, Y. C. (1989). "Epidemiology of dengue fever in Taiwan." *Kaohsiung J Med Sci* 5: 1-11.

Kumar, S., K. Tamura, et al. (2004). "MEGA3: Integrated software for Molecular Evolutionary Genetics Analysis and sequence alignment." *Briefings in Bioinformatics* 5(2): 150-163.

Kuno, G., D. J. Gubler, et al. (1985). "Comparative sensitivity of three mosquito cell lines for isolation of dengue viruses." *Bull World Health Organ* 63: 279-286.

Lanciotti, R. S., C. H. Calisher, et al. (1992). "Rapid detection and typing of dengue viruses from clinical samples by using reverse transcriptase-polymerase chain reaction." *J Clin Microbiol* 30(3): 545-551.

Lanciotti, R. S., J. G. Lewis, et al. (1994). "Molecular evolution and epidemiology of dengue-3 viruses." *J Gen Virol* 75: 65-75.

Leitmeyer, K. C., D. W. Vaughn, et al. (1999). "Dengue virus structural differences that correlate with pathogenesis." *J. Virol.* 73(6): 4738-4747.

LinDENVVbach, B. D. and C. M. Rice (2001). "*Flaviviridae*: The viruses and their replication, p.991-1110. *In* D. M. Knipe and P. M. Howley (ed.) Fields' Virology, 4th ed. Lippincott Williams & Wilkins, Philadelphia, PA.".

Marras, A., L. A. Bruggeman, et al. (2002). "Replication and compartmentalization of HIV-1 in kidney epithelium of patients with HIV-associated nephropathy." *Nature Medicine* 8(5): 522-526.

Rahman, M., K. Rahman, et al. (2002). "First outbreak of dengue hemorrhagic fever, Bangladesh." *Emerg Infect Dis* 8(7): 738-740.

Rodriguez, R., M. Alvarez, et al. (2005). "Virus evolution during a severe dengue epidemic in Cuba, 1997." *Virol* 334: 154-159.

Rodriguez, R., M. Alvarez, et al. (2005). "Dengue virus type 3, Cuba, 2000-2002." *Emerging Infectious Diseases* 11(5): 773-774.

Sariol, C., J. L. Pelegrino, et al. (1999). "Detection and genetic relationship of dengue virus sequences in seventeen-year-old paraffin-embedded samples from Cuba." *Am J Trop Med Hyg* 61(6): 994-1000.

Schneider, W. and M. J. Roossinck (2001). "Genetic diversity in RNA virus quasispecies is controlled by host-virus interaction." *J. Virol.* 75(14): 6566-6571.

Shu, P., L. Chen, et al. (2002). "Potential application of nonstructural protein NS1 serotype-specific immunoglobulin G enzyme-linked immunosorbent assay in the seroepidemiologic study of dengue virus infection: correlation of results with those of the plaque reduction neutralization test." *J Clin Microbiol.* 40(5): 1840-1844.

Shu, P., L. Chen, et al. (2003). "Comparison of capture IgM and IgG ELISA and nonstructural protein NS1 serotype-specific IgG ELISA in the differentiation of primary and secondary dengue virus infections." *Clin Diagn Lab Immunol* 10(4): 622-630.

Sittisombut, N., A. Sistayanarain, et al. (1997). "Possible occurrence of a genetic bottleneck in dengue serotype 2 viruses between the 1980 and 1987 epidemic seasons in Bangkok, Thailand." *Am J Trop Med Hyg* 57(1): 100-108.

Thant, K.-Z., K. Morita, et al. (1996). "Detection of the disease severity-related molecular differences among new Thai dengue-2 isolates in 1993, based on their structural proteins and major non-structural protein NS1 sequences." *Microbiol. Immunol* 40(3): 205-216.

Tsishe, H. C. (1932). "Dengue epidemic in Tainan area in Cho-Ho 6th year." *Formosa J Med* 31: 767-771.

Vasilakis, N. and S. C. Weaver (2008). "The history and evolution of human dengue emergence." *Adv Virus Res* 72: 1-76.

Vaughn, D. W., A. Nisalak, et al. (1999). "Rapid serologic diagnosis of dengue virus infection using a commercial capture ELISA that distinguishes primary and secondary infections." *Am J Trop Med Hyg* 60(4): 693-698.

Wang, W.-K., S.-R. Lin, et al. (2002). "Dengue type 3 virus in plasma is a population of closely related genomes:quasispecies." *J Virol* 76(9): 4662-4665.

Wang, W.-K., T.-L. Sung, et al. (2002). "Sequence diversity of the capsid gene and the nonstructural gene NS2B of dengue-3 virus *in vivo*." *Virology* 303: 181-191.

Weaver, S. C., A. C. Brault, et al. (1999). "Genetic and fitness changes accompanying adaptation of an arbovirus to vertebrate and invertebrate cells." *J. Virol.* 73(5): 4316-4326.

Wittke, V., T. E. Robb, et al. (2002). "Extinction and rapid emergence of strains of dengue 3 virus during an interepidemic period." *Virology* 301: 148-156.

World Health Organization (1997). "Dengue haemorrhagic fever: diagnosis, treatment, prevention and control. 2nd edition. Geneva." *Wld Hth Org*: 12-23.

Wu, Y. C. (1986). "Epidemic dengue 2 on Liouchyou Shiang, Pingtung County in 1981." *Chinese J Microbiol Immunol* 19: 203-211.

Yao, J. S., E. G. Strauss, et al. (1996). "Interactions between PE2, E1 and 6K required for assembly of alphaviruses with chimeric viruses." *J Virol* 70: 7910-7920.

Part 4

Host-Virus Interactions

11

Vaccines and Antiviral Drugs for Diseases Associated with the Epstein-Barr Virus

Limin Chen, Ning Li and Cheng Luo*
Drug Discovery and Design Center,
Shanghai Institute of Materia Medica,
Chinese Academy of Sciences, Shanghai
China

1. Introduction

The Epstein-Barr virus (EBV) is a globally prevalent γ-herpesvirus that infects over 90% of humans and persists for the lifetime of the person. The EB virion is composed of a linear dsDNA molecule, an icosahedral capsid, an amorphous tegument, and an envelope containing viral glycoprotein spikes on its surface (figure 1A). The EBV primarily infects human B-lymphocytes, establishes a latent phase that persists as an incomplete virus, and then induces the transformation as well as proliferation of the infected cells. Under certain circumstances, latent EBV infection can be reactivated, subsequently giving rise to the production of infectious progeny that reinfects cells of the same type. The reactivated virus can also be transmitted to another individual. [1]

EBV-infected B lymphocytes harbor the latent EBV genome as a multicopy episome. There is compelling evidence that most EBV-associated malignancies have escaped this potent virus-specific cytotoxic T-lymphocytes (CTL) response by restricting viral gene expression. EBV expresses more than 80 lytic antigens, whereas latent EBV does not produce progeny virions. However, latent EBV expresses a limited set of viral gene products that maintain the viral genome as well as promote host-cell survival and proliferation. Latent EBV-infected cells express up to nine proteins (figure 1B) and several non-translated RNAs.[2] Among the nine latent EBV proteins, six are EBV nuclear antigens (EBNA1, 2, 3A, 3B, 3C, and LP) and three are latent membrane proteins (LMP1, 2A, and 2B). Based on the latent viral gene expression pattern, the latency is characterized as three main types, namely, types I, II, and III.[3] A type I infected cell only expresses EBNA1; EBNA1, LMP1, and LMP2 are found in type II infected cells. A type III infected cell expresses the full spectrum of latent EBV proteins.

The role of the immune system in the defense against EBV-associated diseases has recently become a popular topic. One hypothesis suggests that numerous neoplasms express viral antigens that should potentially enable them to be recognised and destroyed by the immune system. The EBV infection of B cells is mainly controlled by CD8+ T cells, in addition to

---

*Corresponding Author

natural killer (NK) cells. CD4$^+$ T cell response is also probably important as a source of either effector cells or cytokine help for the massive CD8$^+$ T cell.[4]

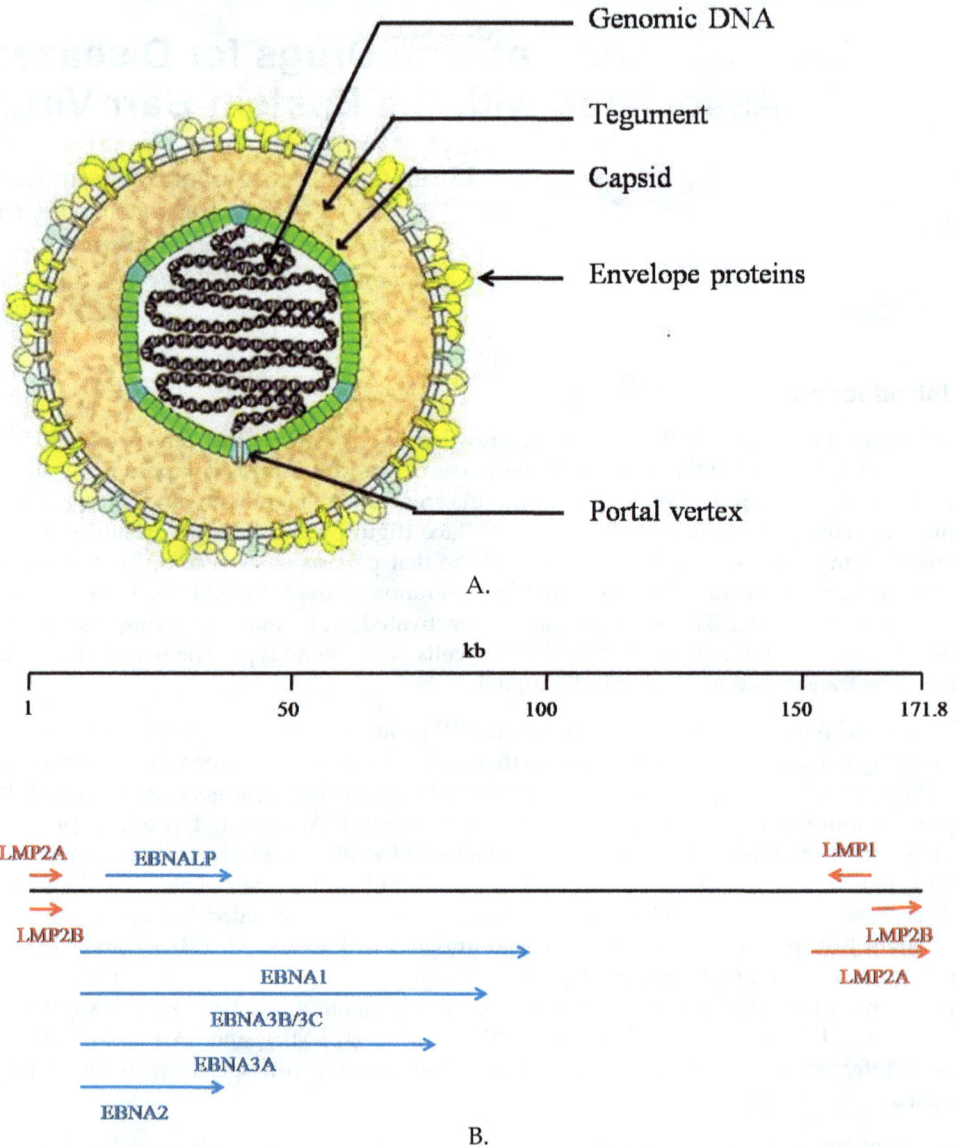

Fig. 1. A) Structure of EB virion. B) Schematic diagram of the EBV genome and location of nine genes expressing latent proteins. The location and polarity of the EBV nuclear antigens (EBNAs) encoding region are shown with blue arrows, latent membrane proteins (LMPs) encoding region are shown with red arrows.

2. EBV-associated diseases

EBV is best known as the aetiological agent of infection mononucleosis (IM), which is most common among adolescents and young adults. IM, a self-limiting disease, is characterised by the appearance of heterophile antibodies in the serum and an atypical lymphocytosis.[5] In developing countries, people are exposed to the virus in their early childhood when they are unlikely to produce noticeable symptoms. In developed countries, such as the United States, the age of first exposure may be delayed to late childhood and young adulthood age when symptoms are more likely to manifest.

2.1 EBV-associated cancers

EBV is associated with an increasing number of lymphoproliferative processes and epithelial neoplasias not only in immunodepressed or immunodeficient patients, but also in immunocompetent persons. EBV-related tumors are characterised by the active expression patterns of viral gene products (Table 1).

	IM	BL	HL	NPC	PTLD
EBNA1	+	+	+	+	+
EBNA2	+	−	−	−	+
EBNA3A	+	−	−	−	+
EBNA3B	+	−	−	−	+
EBNA3C	+	−	−	−	+
EBNALP	+	−	−	−	+
LMP1	+	−	+	+	+
LMP2A	+	−	+	+	+
LMP2B	+	−	+	+	+
EBERs	+	+	+	+	+
BARTs	+	+	+	+	+
Latency	III	I	II	II	III

Table 1. Expression pattern of latent EBV infected genes in EBV- related diseases. IM: infection mononucleosis; BL: Burkitt's lymphoma; HL: Hodgkin's lymphoma; NPC: Nasopharygeal carcinoma; PTLD: Posttransplantation lymphoproliferative disorder.

Burkitt's lymphoma (BL), first described by Denis Parsons Burkitt in 1956, are classified into three forms: endemic, sporadic, and AIDS-associated BL.[6] Endemic BL, initially found in

Africa, is the most common childhood lymphoma in western countries and accounts for approximately 5% of all adult lymphomas.[7] About 97% of endemic BL patients are EBV-positive, suggesting the strong association of endemic BL with EBV. On the other hand, EBV DNA can be detected in 20%-40% of sporadic and AIDS-associated BL.[6] EBNA1 is the only EBV antigen expressed in BL.

Hodgkin's lymphoma (HL) has first been described by Thomas Hodgkin in 1832.[8] Approximately 25%-50% of classical HL cases are associated with the presence of EBV in Reed-Sternberg (RS) mononuclear, and multinuclear cells,[9] which are major components of the tumor.[10] RS cells produce cytokines and chemokines, including TGF-B, IL-10, and TARC. These cytokines and chemokines possibly enable RS cells to modulate the immune response and escape CTL detection.[11] The human tumor-associated antigen RCAS1 expressed in RS cells induces the apoptosis of activated cytotoxic T cells and natural killer (NK) cells. EBV-positive RS cells expressing RCAS1may evade the host immune response.[12]

Nasopharygeal carcinoma (NPC), the most common tumor that develops in the nasopharynx, is extremely common in Southeast Asia and Africa.[13] The EBV genome has been coincidentally found in all NPC specimens,[14] i.e. NPC shows a 100% association with EBV. The latent EBV gene expression pattern in NPC is generally very similar with that detected in most EBV-related HL cases. However, there is no detectable IL-10 and TARC expression in NPC tumor cells, suggesting that the mechanism for escaping immune recognition and destroying NPC tumor cells is different from HL.[11] LMP1 is known to have oncogenic properties during latent infection in NPC, and is thought to be a key modulator in the pathogenesis of NPC.[13] LMP1 triggers the NF-κB, AP-1, and STAT signaling pathways in NPC. Ultimately, all signaling cascades triggered by LMP1 lead to the disruption of the cell cycle, inducing cell transformation.[13]

Posttransplantation lymphoproliferative disorder (PTLD) is an uncontrolled proliferation of B lymphocytes occurring in immunocompromised patients following organ transplant with immunosuppressant medication.[15] The relationship between EBV and PTLD has first been noted by Crawford et al. in 1980.[16] PTLD has a complex clonal diversity ranging from polymorphic B-lymphocyte hyperplasia to malignant monoclonal lymphoma.[5] The B-lymphoma cells of PTLD patients express a full spectrum of latent EBV genes.[17]

In HIV-associated lymphomas, the HIV-induced immunodeficiency may increase the traffic of EBV-infected B-cells, leading to a wide variety of AIDS-related lympholmas.[18] the incidence of non-HL (NHL) in AIDS has increased. Primary cerebral lymphoma (or primary central nervous system lymphoma), a form of NHL, is strongly related to EBV because EBV DNA is present in cerebrospinal fluid.

2.2 EBV-associated autoimmune diseases

There is increasing evidence that EBV is a possible triggering factor of many human-autoimmune diseases.[19]

Multiple sclerosis (MS) is a neurological disease characterised by chronic inflammation and demyelination within the central nervous system. A higher frequency of EBV seropositivity and a higher prevalence of high anti-EBV antibody titres exist in patients with MS compared with controls.[20] About 99% of MS patients are EBV-seropositive.[21] To determine whether antibodies to EBV are elevated before the onset of MS, Levin et al. have conducted a study on the blood samples of more than three million US military personnel. The results demonstrate that the presence of high EBV antibody titres in human increases the risk for developing MS by 34-fold.[22] Although there is no sufficient evidence to conclude that EBV virus causes MS, in some cases, the first attack of MS occurrs at the time of primary EBV infection.[23] Apparently, T cells controlling EBV-infected B cells in MS patients are impaired.[24]

Systemic lupus erythematosus (SLE), an autoimmune chronic inflammatory disease that generates a multi-systemic rheumatic disorder, which ultimately causes organ failure.[25] Compared with controls, SLE patients have increased EBV viral load,[26] anti-EBV antibody levels[27], and numbers of latently infected peripheral B cells. The functional T cell responses of SLE patients are also impaired, and they are positive for the presence of EBV DNA. About 99% of young SLE patients are EBV seropositive.[26] Verdolini et al.[28] have reported a 22-year-old woman who immediately developed SLE after contracting EBV- induced IM. The data obtained from this case suggest that EBV infection can work as a trigger of SLE in some cases, particularly if the patient is genetically susceptible. The T cells controlling the EBV-infected B cells in SLE become defectived that cannot control the numbers of EBV-infected B cells.[29]

Rheumatoid arthritis (RA) is a widespread autoimmune disease characterized by the infiltration of CD4+ T cells and NK cells into synovial joints.[30] Compared with controls, RA patients exhibit increased viral load, anti-EBV antibody titres, and frequency of circulating EBV-infected B cells.[31] The high frequency of EBV-infected B cells in RA patients may be explained by the impaired control of infected B cells by EBV-specific T cells.[32]

Sjögren's syndrome is an autoimmune disorder characterised by lymphoid infiltrates in the salivary gland. Patients with this disorder have elevated levels of anti-EBV antibodies[33] and decreased EBV-specific T cell cytotoxicity.[34]

Other autoimmune disorders, such as autoimmune thyroid disease,[35] scleroderma,[36] autoimmune liver disease (primary biliary cirrhosis and autoimmune hepatitis),[37] inflammatory bowel disease (ulcerative colitis and Crohn's disease),[38] as well as cryptogenic fibrosing alveolitis[39] are also associated with EBV. Patients with any of these disorders all have increased EBV DNA loads and increased serum levels of anti-EBV antibodies.

3. Vaccines

EBV is known to be associated with a large number of human malignancies in immunocompetent and immunosuppressed individuals. Prophylactic vaccines against some pathogenic viruses are excellent public health interventions in terms of safety and effectiveness. Accordingly, there is a great demand for effective vaccines against EBV. Interest in formulating an effective vaccine against EBV is increasing, but onlt a few clinical trials have been conducted. No candidate vaccine has yet been proven sufficiently effective as to warrant commercialisation.

Vaccines should be able to either block primary EBV infection or significantly reduce the EBV load during primary EBV infection. Almost all, if not all, EBV-associated malignancies develop years after the primary EBV infection. Given that immunisation with whole viral proteins does not elicit an efficient CTL response, focus has been directed towards developing peptide vaccines based on defined epitope sequences. Two broad approaches being considered to design effective vaccines for controlling EBV-associated diseases are discussed in the following subsections.

3.1 EBV structural antigens as target antigens

The EBV is enveloped by a membrane composed of four major virus-specific proteins, namely, gp350, gp220, gp85, and p140.[40] The EBV mainly binds to B cells, via the interaction of the gp220 present in the envelope of the virus with cell receptor CD21. This interaction fosters infection. Later, the EBV produces a latent infection mainly in B cells.[41] Most strategies for developing EBV vaccines have focused on the virus membrane antigen, which consists of at least three glycoproteins.

Prophylactic vaccines are known to function primarily via the induction of virus-neutralising antibodies. Gp350 contains the main neutralisation epitope and is the primary target of the virus-neutralising antibody response. These features suggest that gp350 is a primary potential vaccine candidate. In the past several decades, there have been several efforts of developing vaccines mainly focused on the use of a subunit preparation of gp350 (recombinant and affinity purified). Abundant recombinant formulations of gp350 presenting as a subunit antigen or expressing from recombinant viral vectors, generated to induce high load neutralising antibodies, have shown significant protection against EBV-induced B-cell lymphomas in cotton-top tamarins. The recombinant gp350 vaccine is able to elicit neutralising antibodies in a phase I/II trial,[42] has a good safety profile, and is well tolerated.[43] The vaccine is proven effective in preventing the development of EBV-induced IM, but has no efficacy in preventing asymptomatic EBV infection.[44] Indeed, highly purified gp350 induces high levels of neutralising antibodies and inhibits tumor formation in cotton-top tamarins when administered subcutaneously administered with adjuvants such as muramyl dipeptide or immune-stimulation complexes,.[45] A number of recombinant vectors, including vaccinia-gp350 and adenovirus 5-gp350, have also been successfully used in these animals to block tumor outgrowth.[46] Nevertheless, the development of neutralising antibody titres in vaccinated animals does not always correlate with protection.[47] Yao et al. have demonstrated that very low levels of neutralising anti-gp350 antibodies are present in the saliva of healthy EBV-immune donors. This finding suggests that such antibodies are unlikely to be the basis of long-term immunity in healthy seropositive individuals.[48] Apparently, a vaccine solely based on gp350 does not completely prevent the infection of every single B lymphocyte or epithelial cell.

Wolf et al.[49] have expressed poly-antigens containing several antigenic determinants of gp220 and gp350. These proteins are useful in the prophylaxis and therapy of EBV-related diseases because they are able to modulate the immune responses of patients suffering from diseases such as NPC, IM, or EBV-related BL. Mond et al.[50] have enhanced B-cell activation and immunoglobulin secretion by co-stimulation of the receptor for antigen gp350/220.

gp85 is also a potential target for vaccine design.[51] Burrows et al.[52] have successfully identified CTL epitopes within the EBV structural antigen gp85. Using ex vivo primary effectors, strong reactivity to gp85 peptides is observed. An animal model system further reveals that gp85 epitopes are capable of generating structural antigen-specific CTL responses and reducing infections with the virus expressing gp85. Queensland Institute of Medical Research has developed a vaccine including several CTL epitopes that provides protection to more than 90% of the Caucasian population.[53] In 1995, the recombinant vaccinia virus expressing the major virus membrane antigen has first been used in humans.[54]

3.2 Latent antigens as potential vaccine candidates

EBV structural antigens are not expressed in latently infected B-lymphocytes. Hence, therapeutic EBV vaccine efforts have been focused on latency antigens expressed in EBV-associated diseases. EBNA1 has been identified as a vaccine antigen. In a specific embodiment, a purified protein corresponding to EBNA1 elicits a strong CD4+ T cell response.[55] Another vaccine for EBNA2 with the aim of treating and/or preventing PTDL has been developed.[56] LMP1 and LMP2 are the only target antigens available for expanding CTL responses in patients with HD and NPC. Duraiswamy et al.[57] have generated a recombinant poxvirus vaccine that encodes a polyepitope protein derived from LMP1. Human cells infected with the vaccine are efficiently recognised by LMP1-specific CTLs from HLA A2 healthy individuals. The outgrowth of LMP1-expressing tumors in HLA A2/K^b mice is also reversed by the vaccine.

EBNA1 is a protein expressed during both the latent and lytic phases of the EBV. EBNA1 is the only viral protein expressed in all EBV-positive proliferating cells in healthy EBV carriers and in all EBV-associated malignancies.[58] Therefore, a possible vaccine would include EBNA1 added to another latent or lytic gene. A group has developed a vaccine comprising a synthetic polypeptide with a plurality of different segments of parent EBV polypeptides, including EBNA1, LMP1, and LMP2.[59] The vaccine is mainly aimed at treating NPC, HL, and PTLD. Taylor et al. [60] have generated a modified vaccinia virus Ankara recombinant, MVA-EL, which expresses the CD4+ epitope-rich C-terminal domain of EBNA1 fused to full-length LMP2. LMP2 is the source of subdominant CD8+ T cell epitopes. MVA-EL has immunogenicity to both CD4+ and CD8+ T cells.

4. Antiviral drugs

4.1 Targeting lytic DNA replication/EBV-encoded DNA polymerase

Lytic phase EBV causes a cell-to-cell infection in the same host or transmits the virus to another individual. Until now, the most successful therapeutic interventions used against EBV infection and its associated diseases target the lytic replication of EBV.

DNA polymerase performs a key step in DNA replication. The polymerase 'reads' an intact DNA strand as a template and uses it to synthesise the new strand. During the lytic phase of the EBV life cycle, EBV DNA polymerase mediates viral DNA replication. Compounds that target EBV DNA polymerase are used to treat diseases associated with lytic EBV infection, and are widely used in various clinical settings. Drugs that may be possible candidates for targeting viral DNA polymerase are categorised into two groups, namely, nucleoside analogues and non-nucleoside DNA polymerase inhibitors.

4.1.1 Nucleoside analogues

Nucleosidic antivirals have been used in the clinical treatment of EBV-associated diseases since the late 1970s. Acyclovir (ACV; 9-(2-hydroxyethoxymethyl) guanine), a synthetic acyclic nucleoside compound, has been initially shown to have a potent inhibitory activity against herpes simplex virus (HSV) infected cells.[61] Subsequently, ACV has been proven as an effective inhibitor of viral DNA replication in lytic EBV-infected cells, but without the same function in latently infected ones.[62] Given that ACV is only effective in the lytic phase by selectively inhibiting EBV DNA polymerase, efficacious compounds urgently need to be developed. Nucleoside analogues are prodrugs that require phosphorylation by viral thymidine kinase to become active. Inspired by ACV, nucleoside analogues such as ganciclovir (GCV; 9-(1,3-dihydroxy-2-propoxymethyl) guanine)[63] and penciclovir (PCV; 9-(4-hydroxy-3-hydroxymethylbut-1-yl) guanine),[64] as well as nucleotide analogues including cidofovir (CDF; (S)-1-(3-hydroxy-2-phosphonylmethoxypropyl) cytosine)[65] and adefovir (9-(2-phosphonylmethoxyethyl) adenine, PMEA)[66] have been developed. GCV reduces the risk of EBV-associated PTLD in renal transplant recipients, and may be more efficacious than ACV.[67] The inhibitive activity of PCV to EBV has also been evaluatedin assays, wherein infectious virus production, viral antigen expression, and viral DNA synthesis are measured. The obtained data suggest that PCV is a selective inhibitor of EBV in cell cultures.[68] CDF, an acyclic nucleoside phosphonate analogue, decreases EBV oncopreteins and enhances radiosensitivity in EBV-associated diseases.[69] In vitro, adefovir is a potent inhibitor against a few viruses including EBV.[70]Nevertheless, efficiencies of these compounds as inhibitors of EBV are limited. To improve bioavailability, the orally available prodrugs valaciclovir (VACV),[71] valganciclovir (VGC; the valine ester of GCV)[72] and famciclovir (FCV)[73] have been introduced in the mid-1990s. VACV, the L-valyl ester of ACV, is rapidly and almost completely converted to ACV in vivo, as well as provided three to five times increase in ACV bioavailability.[74] The pharmacokinetics of the orally administered VGC, the valine ester of GCV, has been studied compared with the pharmacokinetics of oral and intravenous GCV. VGC results in the improved oral absorption of GCV in liver transplant recipients.[75] FCV, the oral form of PCV, is converted to PCV in vivo.[76] Despite the impressive efficiency of these nucleoside analogues in the treatment of herpes simplex infection, all these compounds suffer from the same drawbacks, including toxic side effects, poor oral bioavailability, and potential mutagenesis. Nearly all clinically effective nucleoside analogues also target the same active sites on viral DNA polymerase molecules, such that mutant viruses resistant to one drug are commonly resistant to others.[77]

4.1.2 Non-nucleoside inhibitors

Given the success of ACV and its analogues, additional inhibitors of DNA polymerases have been expectedly identified. For example, foscarnet (the trisodium salt of phosphonoformic acid),[78] apparently inhibits EBV replication within the range of 2μM to 3μM, which is nontoxic to normal cellular growth. The inhibitory effects of foscarnet are exerted at the pyrophosphate binding site of DNA polymerase. Given that foscarnet is not activated by viral kinases, it is often used as an alternative treatment for EBV, and forACV- or GCV-resistant patients. However, foscarnet is more toxic than ACV, has profound metabolic side effects, and must be intravenously administered.[77] A novel class of non-nucleoside inhibitors against DNA polymerases, 4-oxo-dihydroquinolines (represented as PHA-529311 and PHA-570886), has great inhibitory activity against multiple herpesviruses. These inhibitors also show activity against ACV-resistant HSV and varicella-zoster virus isolates, as well as GCV- or foscarnet-resistant cytomegalovirus isolates.[79]

4.2 Targeting latent infections/EBV-encoded latent proteins

Most EBV-associated tumors harbor the latent viral genome as a multicopy episome in the nucleus of the transformed cells. During latent infection, the EBV does not produce progeny virions, but expresses a limited set of viral gene products that promote host-cell survival and proliferation. The EBV-encoded proteins involved in latency that have received the most attention are EBNA1, EBNA2, EBNA3A, EBNA3C, LMP1, and LMP2A. These latent proteins can induce the immortalisation and proliferation of infected cells, and are involeed in immune response evasion, which are essential for neoplasias.

4.2.1 LMP1 as a target protein

LMP1 is an integral membrane protein containing a short N-terminal cytoplasmic tail of 17 amino acids, 6 hydrophobic transmembrane-spanning domains, and a large cytoplasmic C-terminal domain of 200 amino acids.[80] LMP1, is the main transforming protein of EBV, is identified as the principal oncoprotein because it can transform rodent fibroblasts and is essential for the immortalisation of B cells.[81] LMP1 is a functional homologue of the TNF receptor CD40, which that can deliver a signal to rescue cells from apoptosis and drive proliferation.[82] LMP1 mimics CD40 in activating multiple downstream signaling pathways, such as the NF-κB and JNK pathways, Subsequently, LMP1 up-regulates the expression of cellular genes involved in cell proliferation, cytokine secretion, angiogenesis, and tumor metastasis.[83] The expression of LMP1 induces EBV-associated lymphomas in transgenic mice.[84] Based on these characteristics, LMP1 is a potential target for EBV-associated diseases.

Antisense oligonucleotides (AODs) are effective in inhibiting gene expression in a sequence-specific manner.[85] A number of research groups have used antisense molecules for silencing LMP1. This process is performed with the notion of modulating the course of EBV-associated lymphoproliferative disorders because the modulation is vital for B-cells transformation. As expected, silencing the expression of LMP1 rendered the EBV-positive lymphoblastoid cell lines susceptible to chemotherapeutic agents by abrogating Bcl-2 upregulation and consequently enhancing apoptosis.[86] Galletti et al.[87] have examined the efficacy of liposomes, dendrimers or transferrin–polylysine-conjugated oligonecleotides (ONs) for antisense molecules. The data have indicated that only the delivery system exploiting the transferrin receptor pathway internalised active molecules for silence LMP1 expression. Intracellular single-chain antibodies (sFvs), the smallest domain region of an antibody that retains the binding specificity of the parental antibody, could selectively knockout viral or cellular oncoproteins. Piche et al.[88] have reported that an anti-LMP1 sFv increases the sensitivity of EBV-transformed B lymphocytes to drug-induced cell death. The authors suggest that an anti-LM1 sFv used in combination with conventional chemotherapy may be useful for the therapy of EBV-related lymphomas in immunocompromised patients.

4.2.2 LMP2A as a target protein

LMP2A can promote the survival of latently infected cells and prevent EBV reactivation from the latent phase to the lytic phase.[89] LMP2A signaling does not cause B cells to grow, but delivers a critical signal that is essential for the survival of all B cells.[82] In in vitro infected B cells, the LMP2A N-terminal cytoplasmic domain blocks B cell antigen receptor (BCR) signal transduction, preventing the change from the latent to the lytic cycle, thereby maintaining latency.[90] This domain interacts with Syk and Lyn protein tyrosine kinases via multiple phosphotyrosines arranged in ITAM- and SH2-protein binding motifs. Hence, Syk and Lyn are prevented from binding to the cytoplasmic B cells. Syk and Lyn binding to cytoplasmic B cells are able to induce the lytic cycle.

Monroe et al.[91] have used peptide homologues (synthetic ITAM analogues) to inhibit the interaction of proteins and the ITAM-protein binding motif of viral proteins. They have shown that the blocking association of LMP2A ITAM with cellular molecules and the blocking of LMP2A ITAM-mediated signaling are effective strategies for the treatment and prevention of metastases of EBV-induced malignancies.

4.2.3 EBNA1 as a target protein

EBNA1 is an extremely attractive target for preventing EBV infection and treating EBV-related malignancies. EBNA1 is the only viral protein expressed in all EBV-positive proliferating cells in healthy EBV carriers and in all EBV-related malignancies.[92] EBNA1 is essential for the persistence of the EBV episome, and is anti-apoptotic in contributing to infected-cell survival.[93] EBNA1 also has well-defined biochemical and structural properties. It consists of several functional domains, including a well-defined carboxyl-terminal DNA binding domain.[94] This domain is essential for interacting with the viral oriP. OriP consisting of a series of 30 bp repeats acts in cis to permit linked DNAs to replicate as plasmids in cells containing EBV DNA. EBNA1 regulates the function of oriP to which EBNA1 binds an 18 bp palindromic-sequence as a homodimer.[95] The DNA binding and dimerisation interface have been solved by high resolution X-ray crystallography in the apo- and DNA-bound forms.[96]

The approach of using an AOD to target a single selected viral gene product is promising for the treatment of EBV infections.[97] The treatment of EBV-transformed B cells with EBNA1 antisense ONs inhibits the proliferation of EBV-immortalised cells by at least 50% compared with scrambled antisense sequences.[98] In contrast to primary B cells, EBV-transformed B lymphoblastoid cell lines express alpha-v integrins, the adenovirus internalisation receptor, and are also susceptible to adenovirus-mediated gene delivery. The adenovirus delivery of a specific EBNA1 ribozyme to lymphoblastoid cell lines as well as suppressed EBNA1 mRNA, and protein expression, significantly reduce the number of EBV genomes.[99] Recently, Sun et al.[100] have demonstrated that Hsp90 inhibitors can be used to inhibit EBNA1 expression and translation. This effect requires the EBNA1 Gly-Ala repeat domain. Hsp90 inhibitors induce the death of established, EBV-transformed lymphoblastiod cell lines at doses that are nontoxic to normal cells. Hsp90 inhibitors prevent the EBV transformation of primary B cells and strongly inhibit the growth of EBV-induced lymphoproliferative disease in severe combined immunodeficiency (SCID) mice. The authors suggest that Hsp90 inhibitors may be particularly effective for treating EBV-induced diseases requiring the continued presence of the viral genome. Li et al.[101] have identified a new class of small molecule compound inhibitors of EBV latent infection based on their ability to inhibit the DNA binding function of EBNA1. The molecules have been discovered via high throughput in silico virtual screening and further validated by biochemical as well as cell-based assays. Four compounds are identified to have biochemical activity, and two of which have activity in cell-based assays.

4.2.4 EBNA2 and EBNA3 as target proteins

EBNA2 is related to the differentiation and transformation of B cells. EBNA2 acts as a trans-activator molecule that binds to cellular sequence-specific DNA-binding proteins, such as the Jkappa recombination signal binding protein (CBF1/RBP Jkappa). Consequently, the cellular genes CD23 and CD21, as well as the viral genes LMP1 and LMP2A are transactivated.[102] However, EBNA3A and 3C can inhibit EBNA2 activation of transcription by interacting with RBP Jkappa.[103] EBNA3A and 3C, other than EBNA3B, are critical to this B-lymphocyte growth transformation.[104] Farrell et al.[105] have synthesised a 10-aa peptide from the CBF1 interaction domain of EBNA2 as a fusion with the protein transduction domain of HIV-1 TAT (transcriptional transactivator). Treatment of an EBV-immortalised lympfoid cell lines (LCLs) with the EBNA2-TAT peptide stops cell growth and reduces cell viability. EBNA2-TAT peptide treatment also down-regulats the viral LMP1 and LMP2 genes as well as cellular CD23 expression while up-regulating the expression of the cyclin-dependent kinase inhibitor p21. As another form of treatment, Kempkes[106] has provided a mutant RBP-J DNA binding protein capable of binding the Notch protein but unable to bind to EBNA2. The RBP-J DNA binding protein presents an amino acid sequence with at least one mutation in the EBNA2 binding domain, thereby preventing immortalisation. EBNA3C regulates cell cycles by targeting critical cellular complexes such as cyclin A/cdk2, SCFSkp2, and Rb. Knight et al.[107] have used a 20-aa EBNA3C-derived peptide fused to an HIV TAT-tag to disrupt the EBNA3C-mediated cell cycle. The peptide has inhibited a hyperproliferation of EBV-infected B cell lines and reduced in vitro immortalization of primary B lymphocytes by EBV. The peptide also inhibited lymphoblastoid outgrowth from the blood of an EBV-positive transplant patient in vitro.These experiments suggest that inhibitors targeted against EBNA2 and EBNA3C may be have potential novel anti-EBV therapeutics.

5. Therapies

Chemotherapies based on chemical products play important roles in the treatment of EBV-associated diseases. Immunotherapies using antibodies, such as the anti-CD30 and anti-CD20 antibodies (rituximab) are used to treat EBV-related malignancies. Rituximab has been combined with standard chemotherapy for EBV-associated diseases, with promising results.[9, 108] Other therapies such as adoptive immunotherapy, gene therapy and small interfering RNA (siRNA) therapy have also been developed.

5.1 Adoptive immunotherapy

The adoptive transfer of antigen-specific cytotoxic T lymphocytes offers a safe and effective therapy for certain viral infections and could prove useful in the eradication of tumor cells. Helen et al.[109] have reported the long-term detection of gene-marked EBV-specific CTLs in immunocompromised patients at risk for the development of EBV lymphoproliferative disease. Infusions of T cell lines have not only restored cellular immune responses against EBV, but have also established populations of CTL precursors that could respond to in vivo or ex vivo challenge with the virus for as long as 18 months. The adoptive transfer of EBV-

CTLs has been successfully applied in the treatment of PTLD. In 2010, Helen et al.[110] have tried to address the long-term efficacy, safety, and practicality of EBV-specific CTL immunotherapy. They have studied 114 patients who received infusions of EBV-specific CTLs to prevent or treat PTLD. None of the 101 patients who received CTL prophylaxis has developed EBV-positive PTLD, whereas 11 of the 13 patients treated with CTLs for biopsy-proven or probable PTLD have achieved sustained complete remissions. A gene-marking component is used to demonstrate the persistence of functional CTLs for up to 9 years. The conclusion is that CTL lines provide a safe and effective prophylaxis or treatment for PTLD. However, Subklewe et al[111] compared dendritic cells (DCs) with LCLs for T cell stimulation against dominant and subdominant EBV antigens. DCs expand tenfold more EBNA3A and LMP2 specific T cells than LCLs, and expand EBV-specific T cell responses more efficiently than LCLs. In a specific embodiment, a vaccine using DCs charged with EBNA1 elicits a strong T cell response.[112] Kuzushima et al.[113] have introduced EBNA1 and LMP1 mRNAs into APCs. These modified cells can induce EBV-specific CTLs, inhibit the outgrowth of EBV-infected B lymphocytes, and then lyse EBV-infected NK lymphomas and NK cells.

5.2 Gene therapy

Gene therapy strategies for introducing novel compounds or cytotoxic gene products (e.g., HSV1-TK gene into EBV-infected tumor cells followed by GCV therapy) are being actively developed. Such strategies involve the inhibition of EBV oncoproteins or cellular genes that are critical for virus-associated oncogenesis.

Liu et al.[114] have administered a nucleic acid molecule that can limit tumor cell growth and/or cause tumor cell death. The molecule comprises an EBNA1 responsive promoter region operatively linked to a gene necessary for viral replication. This method can be used to treat and prevent EBV-associated tumors. Franken et al.[115] have introduced a suicide gene regulated by the expression of EBNA2 into latent EBV-infected cells. Cells expressing EBNA2 are demonstrated to be more selectively sensitive to GCV. There is also a complete macroscopic regression of established B-cell lymphomas in SCID mice. However, gene therapy suffers from the common problem of accurate delivery to the appropriate disease sites.

5.3 SiRNA therapy

Therapies using drugs targeted at latent proteins mainly expressing in tumors such as LMP1, LMP2A, or EBNA1 are promising. These proteins are critical to the immortalisation and proliferation of cells and for evading immune responses. The efficacy of siRNA is manifested. Mei et al.[116] have constructed a plasmid stably encoding a 21-nt siRNA specifically and efficiently interfering with LMP1. The siRNA can induce apoptosis in EBV-positive lymphoma cells.

6. Conclusions

Asymptomatic EBV infection causes a few EBV-associated malignancies and autoimmune diseases. The prevention and treatment of these disorders are long-term and arduous. Chemotherapy based on chemical agents such as ACV and GCV can effectively inhibit the viral DNA polymerase used in the treatment of EBV infection and EBV-associated diseases. However, these agents are only effective in lyticly infected cells, but not in latently infected ones. Standard chemotherapy combined with chemical compounds that transform latent phase cells into lytic phase cells has apparently increased therapeutic efficiency. Employing immunotherapy after chemotherapy also has a prominent effect on prevention and treatment.

Gene and siRNA therapies effectively prevent or inhibit critical genes involved in EBV infection. However, they suffer from the same drawback of accurate delivery. Adaptive immunotherapy is a promising approach against EBV-associated neoplasias. Based on the reactivation and expansion of epitope-specific CTL clones in vitro, the epitope activates and increases the immune response against EBV-associated disorders. An effective vaccine that prevents primary EBV infection and produce long-lasting protective immunity may significantly lessen the occurrence of diseases caused by EBV. Abundant vaccines based on membrane glycoproteins or latent proteins against EBV have been developed, and have promisingresults. However, an effective vaccine should at least contain promising $CD4^+$ T cell and $CD8^+$ cell antigens for both the prevention of symptomatic EBV infection and immunotherapy against EBV-associated diseases.

7. References

[1] Bornkamm, G. W.; Hammerschmidt, W., Molecular virology of Epstein-Barr virus. *Philos Trans R Soc Lond B Biol Sci* 2001, *356* (1408), 437-59.

[2] Young, L. S.; Rickinson, A. B., Epstein-Barr virus: 40 years on. *Nat Rev Cancer* 2004, *4* (10), 757-68.

[3] Rowe, M.; Rowe, D. T.; Gregory, C. D.; Young, L. S.; Farrell, P. J.; Rupani, H.; Rickinson, A. B., Differences in B cell growth phenotype reflect novel patterns of Epstein-Barr virus latent gene expression in Burkitt's lymphoma cells. *EMBO J* 1987, *6* (9), 2743-51.

[4] Rickinson, A. B.; Moss, D. J., Human cytotoxic T lymphocyte responses to Epstein-Barr virus infection. *Annu Rev Immunol* 1997, *15*, 405-31.

[5] Khanna, R.; Burrows, S. R.; Moss, D. J., Immune regulation in Epstein-Barr virus-associated diseases. *Microbiol Rev* 1995, *59* (3), 387-405.

[6] Neri, A.; Barriga, F.; Inghirami, G.; Knowles, D. M.; Neequaye, J.; Magrath, I. T.; Dalla-Favera, R., Epstein-Barr virus infection precedes clonal expansion in Burkitt's and acquired immunodeficiency syndrome-associated lymphoma. *Blood* 1991, *77* (5), 1092-5.

[7] Bouffet, E.; Frappaz, D.; Pinkerton, R.; Favrot, M.; Philip, T., Burkitt's lymphoma: a model for clinical oncology. *European Journal of Cancer and Clinical Oncology* 1991, *27* (4), 504-509.

[8] Bonadonna, G., Historical Review of Hodgkin's Disease. *British Journal of Haematology* 2000, *110* (3), 504-511.

[9] Kasamon, Y. L.; Ambinder, R. F., Immunotherapies for Hodgkin's lymphoma. *Crit Rev Oncol Hematol* 2008, *66* (2), 135-44.

[10] Hjalgrim, H.; Smedby, K. E.; Rostgaard, K.; Molin, D.; Hamilton-Dutoit, S.; Chang, E. T.; Ralfkiaer, E.; Sundstrom, C.; Adami, H. O.; Glimelius, B.; Melbye, M., Infectious mononucleosis, childhood social environment, and risk of Hodgkin lymphoma. *Cancer Res* 2007, *67* (5), 2382-8.

[11] Beck, A.; Pazolt, D.; Grabenbauer, G. G.; Nicholls, J. M.; Herbst, H.; Young, L. S.; Niedobitek, G., Expression of cytokine and chemokine genes in Epstein-Barr virus-

associated nasopharyngeal carcinoma: comparison with Hodgkin's disease. *J Pathol* 2001, *194* (2), 145-51.

[12] Ohshima, K.; Muta, K.; Nakashima, M.; Haraoka, S.; Tutiya, T.; Suzumiya, J.; Kawasaki, C.; Watanabe, T.; Kikuchi, M., Expression of human tumor-associated antigen RCAS1 in Reed-Sternberg cells in association with Epstein-Barr virus infection: a potential mechanism of immune evasion. *Int J Cancer* 2001, *93* (1), 91-6.

[13] Zheng, H.; Li, L. L.; Hu, D. S.; Deng, X. Y.; Cao, Y., Role of Epstein-Barr virus encoded latent membrane protein 1 in the carcinogenesis of nasopharyngeal carcinoma. *Cell Mol Immunol* 2007, *4* (3), 185-96.

[14] Nonoyama, M.; Pagano, J. S., Homology between Epstein-Barr Virus DNA and Viral DNA from Burkitt's Lymphoma and Nasopharyngeal Carcinoma determined by DNA-DNA Reassociation Kinetics. *Nature* 1973, *242* (5392), 44-47.

[15] Nalesnik, M. A., Clinical and pathological features of post-transplant lymphoproliferative disorders (PTLD). *Springer Semin Immunopathol* 1998, *20* (3-4), 325-42.

[16] Crawford, D. H.; Thomas, J. A.; Janossy, G.; Sweny, P.; Fernando, O. N.; Moorhead, J. F.; Thompson, J. H., Epstein Barr virus nuclear antigen positive lymphoma after cyclosporin A treatment in patient with renal allograft. *Lancet* 1980, *1* (8182), 1355-6.

[17] Thomas, J. A.; Hotchin, N. A.; Allday, M. J.; Amlot, P.; Rose, M.; Yacoub, M.; Crawford, D. H., Immunohistology of Epstein-Barr virus-associated antigens in B cell disorders from immunocompromised individuals. *Transplantation* 1990, *49* (5), 944-53.

[18] Beral, V.; Peterman, T.; Berkelman, R.; Jaffe, H., AIDS-associated non-Hodgkin lymphoma. *The Lancet* 1991, *337* (8745), 805-809.

[19] Niller, H. H.; Wolf, H.; Minarovits, J., Regulation and dysregulation of Epstein-Barr virus latency: implications for the development of autoimmune diseases. *Autoimmunity* 2008, *41* (4), 298-328.

[20] Sumaya, C. V.; Myers, L. W.; Ellison, G. W., Epstein-Barr virus antibodies in multiple sclerosis. *Arch Neurol* 1980, *37* (2), 94-6.

[21] Ascherio, A.; Munch, M., Epstein-Barr virus and multiple sclerosis. *Epidemiology* 2000, *11* (2), 220-4.

[22] Levin, L. I.; Munger, K. L.; Rubertone, M. V.; Peck, C. A.; Lennette, E. T.; Spiegelman, D.; Ascherio, A., Multiple sclerosis and Epstein-Barr virus. *JAMA* 2003, *289* (12), 1533-6.

[23] Bray, P. F.; Culp, K. W.; McFarlin, D. E.; Panitch, H. S.; Torkelson, R. D.; Schlight, J. P., Demyelinating disease after neurologically complicated primary Epstein-Barr virus infection. *Neurology* 1992, *42* (2), 278-82.

[24] Craig, J. C.; Haire, M.; Merrett, J. D., T-cell-mediated suppression of Epstein-Barr virus-induced B lymphocyte activation in multiple sclerosis. *Clin Immunol Immunopathol* 1988, *48* (3), 253-60.

[25] James, J. A.; Harley, J. B.; Scofield, R. H., Epstein-Barr virus and systemic lupus erythematosus. *Current Opinion in Rheumatology* 2006, *18* (5), 462-467 10.1097/01.bor.0000240355.37927.94.

[26] James, J. A.; Kaufman, K. M.; Farris, A. D.; Taylor-Albert, E.; Lehman, T. J.; Harley, J. B., An increased prevalence of Epstein-Barr virus infection in young patients suggests a possible etiology for systemic lupus erythematosus. *J Clin Invest* 1997, *100* (12), 3019-26.

[27] Evans, A. S.; Rothfield, N. F.; Niederman, J. C., Raised antibody titres to E.B. virus in systemic lupus erythematosus. *Lancet* 1971, *1* (7691), 167-8.

[28] Verdolini, R.; Bugatti, L.; Giangiacomi, M.; Nicolini, M.; Filosa, G.; Cerio, R., Systemic lupus erythematosus induced by Epstein-Barr virus infection. *Br J Dermatol* 2002, *146* (5), 877-81.

[29] Tsokos, G. C.; Magrath, I. T.; Balow, J. E., Epstein-Barr virus induces normal B cell responses but defective suppressor T cell responses in patients with systemic lupus erythematosus. *J Immunol* 1983, *131* (4), 1797-801.

[30] Sawada, S.; Takei, M., Epstein-Barr virus etiology in rheumatoid synovitis. *Autoimmunity Reviews* 2005, *4* (2), 106-110.

[31] (a) Alspaugh, M. A.; Henle, G.; Lennette, E. T.; Henle, W., Elevated levels of antibodies to Epstein-Barr virus antigens in sera and synovial fluids of patients with rheumatoid arthritis. *J Clin Invest* 1981, *67* (4), 1134-40; (b) Tosato, G.; Steinberg, A. D.; Yarchoan, R.; Heilman, C. A.; Pike, S. E.; De Seau, V.; Blaese, R. M., Abnormally elevated frequency of Epstein-Barr virus-infected B cells in the blood of patients with rheumatoid arthritis. *J Clin Invest* 1984, *73* (6), 1789-95; (c) Balandraud, N.; Meynard, J. B.; Auger, I.; Sovran, H.; Mugnier, B.; Reviron, D.; Roudier, J.; Roudier, C., Epstein-Barr virus load in the peripheral blood of patients with rheumatoid arthritis: accurate quantification using real-time polymerase chain reaction. *Arthritis Rheum* 2003, *48* (5), 1223-8.

[32] Tosato, G.; Steinberg, A. D.; Blaese, R. M., Defective EBV-specific suppressor T-cell function in rheumatoid arthritis. *N Engl J Med* 1981, *305* (21), 1238-43.

[33] Origgi, L.; Hu, C.; Bertetti, E.; Asero, R.; D'Agostino, P.; Radelli, L.; Riboldi, P., Antibodies to Epstein-Barr virus and cytomegalovirus in primary Sjogren's syndrome. *Boll Ist Sieroter Milan* 1988, *67* (4), 265-74.

[34] Whittingham, S.; McNeilage, J.; Mackay, I. R., Primary Sjogren's syndrome after infectious mononucleosis. *Ann Intern Med* 1985, *102* (4), 490-3.

[35] Vrbikova, J.; Janatkova, I.; Zamrazil, V.; Tomiska, F.; Fucikova, T., Epstein-Barr virus serology in patients with autoimmune thyroiditis. *Exp Clin Endocrinol Diabetes* 1996, *104* (1), 89-92.

[36] Shore, A.; Klock, R.; Lee, P.; Snow, K. M.; Keystone, E. C., Impaired late suppression of Epstein-Barr virus (EBV)-induced immunoglobulin synthesis: a common feature of autoimmune disease. *J Clin Immunol* 1989, *9* (2), 103-10.

[37] (a) Morshed, S. A.; Nishioka, M.; Saito, I.; Komiyama, K.; Moro, I., Increased expression of Epstein-Barr virus in primary biliary cirrhosis patients. *Gastroenterol Jpn* 1992, *27* (6), 751-8; (b) Vento, S.; Guella, L.; Mirandola, F.; Cainelli, F.; Di Perri, G.; Solbiati, M.; Ferraro, T.; Concia, E., Epstein-Barr virus as a trigger for autoimmune hepatitis in susceptible individuals. *Lancet* 1995, *346* (8975), 608-9.

[38] Spieker, T.; Herbst, H., Distribution and phenotype of Epstein-Barr virus-infected cells in inflammatory bowel disease. *Am J Pathol* 2000, *157* (1), 51-7.

[39] Vergnon, J. M.; Vincent, M.; de The, G.; Mornex, J. F.; Weynants, P.; Brune, J., Cryptogenic fibrosing alveolitis and Epstein-Barr virus: an association? *Lancet* 1984, *2* (8406), 768-71.

[40] Heineman, T.; Gong, M.; Sample, J.; Kieff, E., Identification of the Epstein-Barr virus gp85 gene. *J Virol* 1988, *62* (4), 1101-7.

[41] Young, K. A.; Chen, X. S.; Holers, V. M.; Hannan, J. P., Isolating the Epstein-Barr virus gp350/220 binding site on complement receptor type 2 (CR2/CD21). *J Biol Chem* 2007, *282* (50), 36614-25.

[42] Spaete, R. R. C., CA, US), Jackman, Winthrop (Berkeley, CA, US) Non-splicing variants of gp350/220. 2007.

[43] Moutschen, M.; Leonard, P.; Sokal, E. M.; Smets, F.; Haumont, M.; Mazzu, P.; Bollen, A.; Denamur, F.; Peeters, P.; Dubin, G.; Denis, M., Phase I/II studies to evaluate safety and immunogenicity of a recombinant gp350 Epstein-Barr virus vaccine in healthy adults. *Vaccine* 2007, *25* (24), 4697-705.

[44] Sokal, E. M.; Hoppenbrouwers, K.; Vandermeulen, C.; Moutschen, M.; Leonard, P.; Moreels, A.; Haumont, M.; Bollen, A.; Smets, F.; Denis, M., Recombinant gp350 vaccine for infectious mononucleosis: a phase 2, randomized, double-blind, placebo-controlled trial to evaluate the safety, immunogenicity, and efficacy of an Epstein-Barr virus vaccine in healthy young adults. *J Infect Dis* 2007, *196* (12), 1749-53.

[45] Morgan, A. J.; Finerty, S.; Lovgren, K.; Scullion, F. T.; Morein, B., Prevention of Epstein-Barr (EB) virus-induced lymphoma in cottontop tamarins by vaccination with the EB virus envelope glycoprotein gp340 incorporated into immune-stimulating complexes. *J Gen Virol* 1988, *69* (Pt 8), 2093-6.

[46] Morgan, A. J.; Mackett, M.; Finerty, S.; Arrand, J. R.; Scullion, F. T.; Epstein, M. A., Recombinant vaccinia virus expressing Epstein-Barr virus glycoprotein gp340 protects cottontop tamarins against EB virus-induced malignant lymphomas. *J Med Virol* 1988, *25* (2), 189-95.

[47] Wilson, A. D.; Shooshstari, M.; Finerty, S.; Watkins, P.; Morgan, A. J., Virus-specific cytotoxic T cell responses are associated with immunity of the cottontop tamarin to Epstein-Barr virus (EBV). *Clin Exp Immunol* 1996, *103* (2), 199-205.

[48] Yao, Q. Y.; Rowe, M.; Morgan, A. J.; Sam, C. K.; Prasad, U.; Dang, H.; Zeng, Y.; Rickinson, A. B., Salivary and serum IgA antibodies to the Epstein-Barr virus glycoprotein gp340: incidence and potential for virus neutralization. *Int J Cancer* 1991, *48* (1), 45-50.

[49] Wolf, H. J. J. J. S., D-8130 Starnberg, DE) DNA sequences of the EBV genome, recombinant DNA molecules, processes for preparing EBV-related antigens, diagnostic compositions and pharmaceutical compositions containing said antigens. 1997.

[50] Mond, J. J. S. S., MD), Lees, Andrew (Silver Spring, MD) Enhancement of B cell activation and immunoglobulin secretion by co-stimulation of receptors for antigen and EBV Gp350/220. 2002.

[51] Khanna, R.; Sherritt, M.; Burrows, S. R., EBV Structural Antigens, gp350 and gp85, as Targets for Ex Vivo Virus-Specific CTL During Acute Infectious Mononucleosis: Potential Use of gp350/gp85 CTL Epitopes for Vaccine Design. *The Journal of Immunology* 1999, *162* (5), 3063-3069.

[52] Burrows, S. R. B. H., AU), Khanna, Rajiv (Herston, AU), Sherritt, Martina Alison (Kedron, AU) CTL epitopes from EBV. 2004.

[53] Khanna; Rajiv, K.; Beverley; Mavis, M.; Ihor; Spephan, M.; Denis; James, B.; Scott; Renton EBV CTL EPITOPES. 1997.

[54] Gu, S. Y.; Huang, T. M.; Ruan, L.; Miao, Y. H.; Lu, H.; Chu, C. M.; Motz, M.; Wolf, H., First EBV vaccine trial in humans using recombinant vaccinia virus expressing the major membrane antigen. *Dev Biol Stand* 1995, *84*, 171-7.

[55] Steinman, R. M. W., CT, US), Muenz, Christan (New York, NY, US) Protective antigen of epstein barr virus. 2008.

[56] Celis, E. R., MN, US) Epstein-barr-virus-specific immunization. 2008.

[57] Duraiswamy, J.; Sherritt, M.; Thomson, S.; Tellam, J.; Cooper, L.; Connolly, G.; Bharadwaj, M.; Khanna, R., Therapeutic LMP1 polyepitope vaccine for EBV-associated Hodgkin disease and nasopharyngeal carcinoma. *Blood* 2003, *101* (8), 3150-3156.

[58] Villegas, E.; Santiago, O.; Sorlozano, A.; Gutierrez, J., New strategies and patent therapeutics in EBV-associated diseases. *Mini Rev Med Chem* 2010, *10* (10), 914-27.

[59] Thomson, S., Anthony (AU); Duraiswamy, J., Kumar (AU); Moss, D., James (AU) Treatment of Epstein-Barr virus-associated diseases. 2007.

[60] Taylor, G. S.; Haigh, T. A.; Gudgeon, N. H.; Phelps, R. J.; Lee, S. P.; Steven, N. M.; Rickinson, A. B., Dual stimulation of Epstein-Barr Virus (EBV)-specific CD4+- and CD8+-T-cell responses by a chimeric antigen construct: potential therapeutic vaccine for EBV-positive nasopharyngeal carcinoma. *J Virol* 2004, *78* (2), 768-78.

[61] Elion, G. B.; Furman, P. A.; Fyfe, J. A.; de Miranda, P.; Beauchamp, L.; Schaeffer, H. J., Selectivity of action of an antiherpetic agent, 9-(2-hydroxyethoxymethyl) guanine. *Proc Natl Acad Sci U S A* 1977, *74* (12), 5716-20.

[62] Colby, B. M.; Shaw, J. E.; Elion, G. B.; Pagano, J. S., Effect of acyclovir [9-(2-hydroxyethoxymethyl)guanine] on Epstein-Barr virus DNA replication. *J Virol* 1980, *34* (2), 560-8.

[63] Martin, J. C.; Dvorak, C. A.; Smee, D. F.; Matthews, T. R.; Verheyden, J. P. H., 9-(1,3-Dihydroxy-2-propoxymethyl)guanine: a new potent and selective antiherpes agent. *Journal of Medicinal Chemistry* 1983, *26* (5), 759-761.

[64] Boyd, M. R.; Bacon, T. H.; Sutton, D.; Cole, M., Antiherpesvirus activity of 9-(4-hydroxy-3-hydroxy-methylbut-1-yl)guanine (BRL 39123) in cell culture. *Antimicrob Agents Chemother* 1987, *31* (8), 1238-42.

[65] Snoeck, R.; Sakuma, T.; De Clercq, E.; Rosenberg, I.; Holy, A., (S)-1-(3-hydroxy-2-phosphonylmethoxypropyl)cytosine, a potent and selective inhibitor of human cytomegalovirus replication. *Antimicrob Agents Chemother* 1988, *32* (12), 1839-44.

[66] De Clercq, E.; Holy, A.; Rosenberg, I., Efficacy of phosphonylmethoxyalkyl derivatives of adenine in experimental herpes simplex virus and vaccinia virus infections in vivo. *Antimicrob Agents Chemother* 1989, *33* (2), 185-91.

[67] Funch, D. P.; Walker, A. M.; Schneider, G.; Ziyadeh, N. J.; Pescovitz, M. D., Ganciclovir and acyclovir reduce the risk of post-transplant lymphoproliferative disorder in renal transplant recipients. *Am J Transplant* 2005, *5* (12), 2894-900.

[68] Bacon, T. H.; Boyd, M. R., Activity of penciclovir against Epstein-Barr virus. *Antimicrob Agents Chemother* 1995, *39* (7), 1599-602.

[69] Abdulkarim, B.; Sabri, S.; Zelenika, D.; Deutsch, E.; Frascogna, V.; Klijanienko, J.; Vainchenker, W.; Joab, I.; Bourhis, J., Antiviral agent Cidofovir decreases Epstein-Barr virus (EBV) oncoproteins and enhances the radiosensitivity in EBV-related malignancies. *Oncogene* 2003, *22* (15), 2260-2271.

[70] Kamp, W.; Schokker, J.; Cambridge, E.; De Jong, S.; Schuurman, R.; De Groot, T.; Boucher, C. A., Effect of weekly adefovir (PMEA) infusions on HIV-1 virus load: results of a phase I/II study. *Antivir Ther* 1999, *4* (2), 101-7.

[71] Purifoy, D. J. M.; Beauchamp, L. M.; De Miranda, P.; Ertl, P.; Lacey, S.; Roberts, G.; Rahim, S. G.; Darby, G.; Krenitsky, T. A.; Powell, K. L., Review of research leading to new anti-herpesvirus agents in clinical development: Valaciclovir hydrochloride (256u, the L-valyl ester of acyclovir) and 882c, a specific agent for varicella zoster virus. *Journal of Medical Virology* 1993, *41* (S1), 139-145.

[72] Jung, D.; Dorr, A., Single-dose pharmacokinetics of valganciclovir in HIV- and CMV-seropositive subjects. *J Clin Pharmacol* 1999, *39* (8), 800-4.

[73] Vere Hodge, R. A.; Sutton, D.; Boyd, M. R.; Harnden, M. R.; Jarvest, R. L., Selection of an oral prodrug (BRL 42810; famciclovir) for the antiherpesvirus agent BRL 39123 [9-(4-hydroxy-3-hydroxymethylbut-l-yl)guanine; penciclovir]. *Antimicrob Agents Chemother* 1989, *33* (10), 1765-73.

[74] Beutner, K. R.; Friedman, D. J.; Forszpaniak, C.; Andersen, P. L.; Wood, M. J., Valaciclovir compared with acyclovir for improved therapy for herpes zoster in immunocompetent adults. *Antimicrob Agents Chemother* 1995, *39* (7), 1546-53.

[75] Pescovitz, M. D.; Rabkin, J.; Merion, R. M.; Paya, C. V.; Pirsch, J.; Freeman, R. B.; O'Grady, J.; Robinson, C.; To, Z.; Wren, K.; Banken, L.; Buhles, W.; Brown, F., Valganciclovir results in improved oral absorption of ganciclovir in liver transplant recipients. *Antimicrob Agents Chemother* 2000, *44* (10), 2811-5.

[76] Clarke, S. E.; Harrell, A. W.; Chenery, R. J., Role of aldehyde oxidase in the in vitro conversion of famciclovir to penciclovir in human liver. *Drug Metab Dispos* 1995, *23* (2), 251-4.

[77] Coen, D. M.; Schaffer, P. A., Antiherpesvirus drugs: a promising spectrum of new drugs and drug targets. *Nat Rev Drug Discov* 2003, *2* (4), 278-288.

[78] Datta, A. K.; Hood, R. E., Mechanism of inhibition of Epstein-Barr virus replication by phosphonoformic acid. *Virology* 1981, *114* (1), 52-9.

[79] Hartline, C. B.; Harden, E. A.; Williams-Aziz, S. L.; Kushner, N. L.; Brideau, R. J.; Kern, E. R., Inhibition of herpesvirus replication by a series of 4-oxo-dihydroquinolines with viral polymerase activity. *Antiviral Research* 2005, *65* (2), 97-105.

[80] Masucci, M. G., Epstein-Barr virus oncogenesis and the ubiquitin-proteasome system. *Oncogene* 2004, *23* (11), 2107-15.

[81] Mei, Y. P.; Zhu, X. F.; Zhou, J. M.; Huang, H.; Deng, R.; Zeng, Y. X., siRNA targeting LMP1-induced apoptosis in EBV-positive lymphoma cells is associated with inhibition of telomerase activity and expression. *Cancer Lett* 2006, *232* (2), 189-98.

[82] Thorley-Lawson, D. A., Epstein-Barr virus: exploiting the immune system. *Nat Rev Immunol* 2001, *1* (1), 75-82.

[83] Uchida, J.; Yasui, T.; Takaoka-Shichijo, Y.; Muraoka, M.; Kulwichit, W.; Raab-Traub, N.; Kikutani, H., Mimicry of CD40 Signals by Epstein-Barr Virus LMP1 in B Lymphocyte Responses. *Science* 1999, *286* (5438), 300-303.

[84] Kulwichit, W.; Edwards, R. H.; Davenport, E. M.; Baskar, J. F.; Godfrey, V.; Raab-Traub, N., Expression of the Epstein-Barr virus latent membrane protein 1 induces B cell lymphoma in transgenic mice. *Proc Natl Acad Sci U S A* 1998, *95* (20), 11963-8.

[85] Stephenson, M. L.; Zamecnik, P. C., Inhibition of Rous sarcoma viral RNA translation by a specific oligodeoxyribonucleotide. *Proc Natl Acad Sci U S A* 1978, *75* (1), 285-8.

[86] Siakallis, G.; Spandidos, D. A.; Sourvinos, G., Herpesviridae and novel inhibitors. *Antivir Ther* 2009, *14* (8), 1051-64.

[87] Galletti, R.; Masciarelli, S.; Conti, C.; Matusali, G.; Di Renzo, L.; Meschini, S.; Arancia, G.; Mancini, C.; Mattia, E., Inhibition of Epstein Barr Virus LMP1 gene expression in B lymphocytes by antisense oligonucleotides: Uptake and efficacy of lipid-based and receptor-mediated delivery systems. *Antiviral Research* 2007, *74* (2), 102-110.

[88] Piche, A.; Kasono, K.; Johanning, F.; Curiel, T. J.; Curiel, D. T., Phenotypic knock-out of the latent membrane protein 1 of Epstein-Barr virus by an intracellular single-chain antibody. *Gene Ther* 1998, *5* (9), 1171-9.

[89] Miller, C. L.; Burkhardt, A. L.; Lee, J. H.; Stealey, B.; Longnecker, R.; Bolen, J. B.; Kieff, E., Integral membrane protein 2 of Epstein-Barr virus regulates reactivation from latency through dominant negative effects on protein-tyrosine kinases. *Immunity* 1995, 2 (2), 155-66.

[90] Caldwell, R. G.; Wilson, J. B.; Anderson, S. J.; Longnecker, R., Epstein-Barr virus LMP2A drives B cell development and survival in the absence of normal B cell receptor signals. *Immunity* 1998, 9 (3), 405-11.

[91] Monroe, J. G. P., PA, US), Katz, Elad (Manchester, GB), Murali, Ramachandran (Swarthmore, PA, US) Methods and compositions targeting viral and cellular ITAM motifs, and use of same in identifying compounds with therapeutic activity. 2007.

[92] Thorley-Lawson, D. A.; Gross, A., Persistence of the Epstein-Barr virus and the origins of associated lymphomas. *N Engl J Med* 2004, 350 (13), 1328-37.

[93] (a) Kennedy, G.; Komano, J.; Sugden, B., Epstein-Barr virus provides a survival factor to Burkitt's lymphomas. *Proc Natl Acad Sci U S A* 2003, 100 (24), 14269-74; (b) Yin, Q.; Flemington, E. K., siRNAs against the Epstein Barr virus latency replication factor, EBNA1, inhibit its function and growth of EBV-dependent tumor cells. *Virology* 2006, 346 (2), 385-93.

[94] Leight, E. R.; Sugden, B., EBNA-1: a protein pivotal to latent infection by Epstein-Barr virus. *Rev Med Virol* 2000, 10 (2), 83-100.

[95] Yates, J. L.; Warren, N.; Sugden, B., Stable replication of plasmids derived from Epstein-Barr virus in various mammalian cells. *Nature* 1985, 313 (6005), 812-5.

[96] (a) Bochkarev, A.; Barwell, J. A.; Pfuetzner, R. A.; Furey, W., Jr.; Edwards, A. M.; Frappier, L., Crystal structure of the DNA-binding domain of the Epstein-Barr virus origin-binding protein EBNA 1. *Cell* 1995, 83 (1), 39-46; (b) Bochkarev, A.; Barwell, J. A.; Pfuetzner, R. A.; Bochkareva, E.; Frappier, L.; Edwards, A. M., Crystal structure of the DNA-binding domain of the Epstein-Barr virus origin-binding protein, EBNA1, bound to DNA. *Cell* 1996, 84 (5), 791-800.

[97] Pagano, J. S.; Jimenez, G.; Sung, N. S.; Raab-Traub, N.; Lin, J. C., Epstein-Barr viral latency and cell immortalization as targets for antisense oligomers. *Ann N Y Acad Sci* 1992, 660, 107-16.

[98] Roth, G.; Curiel, T.; Lacy, J., Epstein-Barr viral nuclear antigen 1 antisense oligodeoxynucleotide inhibits proliferation of Epstein-Barr virus-immortalized B cells. *Blood* 1994, 84 (2), 582-7.

[99] Huang, S.; Stupack, D.; Mathias, P.; Wang, Y.; Nemerow, G., Growth arrest of Epstein-Barr virus immortalized B lymphocytes by adenovirus-delivered ribozymes. *Proc Natl Acad Sci U S A* 1997, 94 (15), 8156-61.

[100] Sun, X.; Barlow, E. A.; Ma, S.; Hagemeier, S. R.; Duellman, S. J.; Burgess, R. R.; Tellam, J.; Khanna, R.; Kenney, S. C., Hsp90 inhibitors block outgrowth of EBV-infected malignant cells in vitro and in vivo through an EBNA1-dependent mechanism. *Proc Natl Acad Sci U S A* 2010, 107 (7), 3146-51.

[101] Li, N.; Thompson, S.; Schultz, D. C.; Zhu, W.; Jiang, H.; Luo, C.; Lieberman, P. M., Discovery of selective inhibitors against EBNA1 via high throughput in silico virtual screening. *PLoS One* 2010, 5 (4), e10126.

[102] (a) Johannsen, E.; Koh, E.; Mosialos, G.; Tong, X.; Kieff, E.; Grossman, S. R., Epstein-Barr virus nuclear protein 2 transactivation of the latent membrane protein 1 promoter is mediated by J kappa and PU.1. *J Virol* 1995, 69 (1), 253-62; (b) Cordier, M.; Calender,

A.; Billaud, M.; Zimber, U.; Rousselet, G.; Pavlish, O.; Banchereau, J.; Tursz, T.; Bornkamm, G.; Lenoir, G. M., Stable transfection of Epstein-Barr virus (EBV) nuclear antigen 2 in lymphoma cells containing the EBV P3HR1 genome induces expression of B-cell activation molecules CD21 and CD23. *J Virol* 1990, *64* (3), 1002-13.

[103] Robertson, E. S.; Lin, J.; Kieff, E., The amino-terminal domains of Epstein-Barr virus nuclear proteins 3A, 3B, and 3C interact with RBPJ(kappa). *J Virol* 1996, *70* (5), 3068-74.

[104] Tomkinson, B.; Robertson, E.; Kieff, E., Epstein-Barr virus nuclear proteins EBNA-3A and EBNA-3C are essential for B-lymphocyte growth transformation. *J Virol* 1993, *67* (4), 2014-25.

[105] Farrell, C. J.; Lee, J. M.; Shin, E. C.; Cebrat, M.; Cole, P. A.; Hayward, S. D., Inhibition of Epstein-Barr virus-induced growth proliferation by a nuclear antigen EBNA2-TAT peptide. *Proc Natl Acad Sci U S A* 2004, *101* (13), 4625-30.

[106] Kempkes, B. D. Novel target for therapeutic intervention in EBV related diseases thereof. 2002.

[107] Knight, J. S.; Lan, K.; Bajaj, B.; Sharma, N.; Tsai, D. E.; Robertson, E. S., A peptide-based inhibitor for prevention of B cell hyperproliferation induced by Epstein-Barr virus. *Virology* 2006, *354* (1), 207-14.

[108] Hänel, M.; Fiedler, F.; Thorns, C., Anti-CD20 Monoclonal Antibody (Rituximab) and Cidofovir as Successful Treatment of an EBV-Associated Lymphoma with CNS Involvement. *Onkologie* 2001, *24* (5), 491-494.

[109] Heslop, H. E.; Ng, C. Y. C.; Li, C.; Smith, C. A.; Loftin, S. K.; Krance, R. A.; Brenner, M. K.; Rooney, C. M., Long-term restoration of immunity against Epstein-Barr virus infection by adoptive transfer of gene-modified virus-specific T lymphocytes. *Nat Med* 1996, *2* (5), 551-555.

[110] Heslop, H. E.; Slobod, K. S.; Pule, M. A.; Hale, G. A.; Rousseau, A.; Smith, C. A.; Bollard, C. M.; Liu, H.; Wu, M. F.; Rochester, R. J.; Amrolia, P. J.; Hurwitz, J. L.; Brenner, M. K.; Rooney, C. M., Long-term outcome of EBV-specific T-cell infusions to prevent or treat EBV-related lymphoproliferative disease in transplant recipients. *Blood* 2010, *115* (5), 925-35.

[111] Subklewe, M.; Sebelin, K.; Block, A.; Meier, A.; Roukens, A.; Paludan, C.; Fonteneau, J. F.; Steinman, R. M.; Munz, C., Dendritic cells expand Epstein Barr virus specific CD8+ T cell responses more efficiently than EBV transformed B cells. *Hum Immunol* 2005, *66* (9), 938-49.

[112] Steinman; Ralph; M, M.; Chriatian Protective Antigen of Epstein Barr Virus Therefor. 2001.

[113] Kuzushima, K. A., JP), Ito, Yoshinori (Aichi, JP), Okamura, Ayako (Aichi, JP), Akatsuka, Yoshiki (Aichi, JP), Morishima, Yasuo (Aichi, JP) Cytotoxic-cell epitope peptides that specifically attack Epstein Barr virus-infected cells and uses thereof. 2009.

[114] Liu; Feifei, K.; Henry; Joseph, C.; Marie, L.; Jianhua Nucleic acids and methods for treating EBV-positive cancers thereof. 2003.

[115] Franken, M.; Estabrooks, A.; Cavacini, L.; Sherburne, B.; Wang, F.; Scadden, D. T., Epstein-Barr virus-driven gene therapy for EBV-related lymphomas. *Nat Med* 1996, *2* (12), 1379-1382.

[116] Mei, Y.-P.; Zhu, X.-F.; Zhou, J.-M.; Huang, H.; Deng, R.; Zeng, Y.-X., siRNA targeting LMP1-induced apoptosis in EBV-positive lymphoma cells is associated with inhibition of telomerase activity and expression. *Cancer letters* 2006, *232* (2), 189-198.

12

Flavivirus Neurotropism, Neuroinvasion, Neurovirulence and Neurosusceptibility: Clues to Understanding Flavivirus and Dengue-Induced Encephalitis

Myriam Lucia Velandia[1] and Jaime E. Castellanos[1,2]
[1]Grupo de Virología, Universidad El Bosque
[2]Grupo Patogénesis Viral, Universidad Nacional de Colombia Bogotá,
Colombia

1. Introduction

Viral infections of the nervous system (NS) can be caused by many types of viruses, including rhabdoviruses, alpha and beta herpes viruses, retroviruses, picornaviruses, arenaviruses and flaviviruses (van den Pol, 2006) The replication of these viruses can occur both in neurons and in non-neuronal cells and each type of cell responds differently (Griffin, 2003). The final result of these infections is the alteration of function of the nervous system.

Flaviviruses are single-stranded positive sense RNA viruses of epidemiological and neurological importance because the majority of them infect the NS, causing severe damage to its function (Figure 1)(Lindenbach et al., 2007). The flaviviruses that most frequently infect nervous tissue are Japanese encephalitis virus (JEV), West Nile virus (WNV), Murray Valley encephalitis virus (MVEV) and tick-borne encephalitis virus (TBEV). However, other members of this family, such as yellow fever virus (YFV) and dengue virus (DENV), which preferentially infect hepatocytes and immune cells like monocytes and macrophages, can acquire the capacity to enter and infect nervous tissue (Misra et al., 2006).

Infection with flaviviruses occurs via an interaction between viral envelope (E) proteins and molecules on the cellular surface that act as receptors that promote endocytosis of the virus (Chambers et al., 1990; Lindenbach et al., 2007; van der Schaar et al., 2007, 2008). This initial interaction partially defines the virus tropism; however, the mechanisms that determine and promote infection of nervous tissue with neurotropic flaviviruses are not completely understood (Chambers & Diamond, 1999; McMinn, 1997). Furthermore, nervous tissue infected with DENV is of particular interest because although this virus is not neurotropic, it can induce alterations in nervous system function that are being reported with increasing frequency. In some cases, virus-specific IgMs have been isolated from the cerebrospinal fluid, which suggests the presence of the virus in the NS (Domingues et al., 2008; Lum et al., 1996).

In severe cases of dengue fever, neurological alterations including encephalitis, encephalomyelitis, transverse myelitis, flaccid paralysis, Guillain-Barre Syndrome, cerebrovascular accident and behaviour disorders have been reported (Domingues et al.,

2008; Mathew & Pandian, 2010; Misra et al., 2006; Solomon, 2003, 2004). Frequently, neurological signs manifest as a consequence of viral infection in organs such the liver (encephalopathies) (Gulati & Maheshwari, 2007; Row et al., 1996). Despite the fact that little is known about the mechanisms that favour DENV infection of nervous tissue (Chien et al., 2008; Kumar et al., 2008; Malavige et al., 2007), it has been postulated that the individuals' age, genetic background and immune status, in addition to the viral serotype and genotype, may explain both the ability of the virus to infect the NS and the appearance of neurological manifestations as a result of this virus infection. This chapter will review the interactions between nervous tissue and certain flaviviruses, including DENV, such as neuroinvasion, neurotropism, neurovirulence and neurosusceptibility.

A.   Genomic RNA of Flavivirus

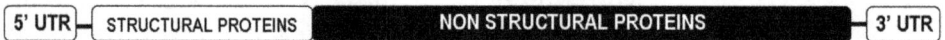

B.   Polypeptide Translation

C.   Organization of structural proteins in flavivirus

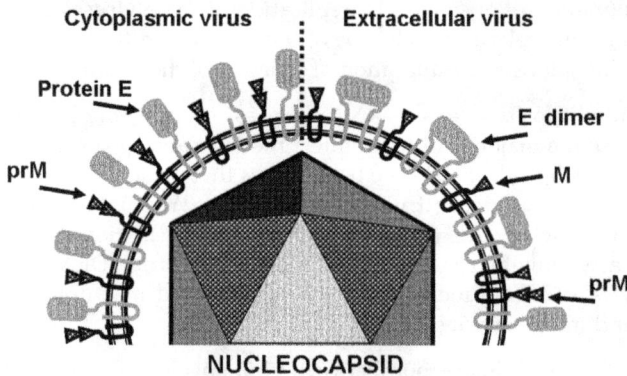

Fig. 1. Schematic organization of genome and polyprotein of a flavivirus. A. The flaviviruses are enveloped single strand RNA viruses with a unique open reading frame coding for both structural and non structural genes. B. Diagram of entire translated polyprotein which must be cleaved by viral and cellular proteases to release mature and active proteins. C. Schematic representation of final disposition of structural proteins in the virions. After proteolytic cleavage or prM, glycoprotein E exposes its homodimerization domains to activate the cell-binding site. Core protein specifically encapsidates the recently synthesized genomic RNA to form the nucleocapsid.

2. Neuroinvasion

Neuroinvasion is the ability of viruses to enter nervous tissue and cause neurological alterations. The majority of viruses in the Flavivirus genus are transmitted via the bite of an arthropod vector (mosquito or tick), and once inoculated in the dermis, these viruses spread to infect target cells such as dendritic cells or monocytes/macrophages or enter directly into the lymph nodes, muscles, liver, spleen or nervous system via nerve endings (Chambers & Diamond, 2003; McMinn, 1997). In some cases during infection with these viruses, the blood-brain barrier (BBB) is disturbed as a result of cytokines and chemokines that favour the entry of WNV and JEV into nervous tissue (Chambers & Diamond 2003; Chaturvedi et al., 1991) (Figure 2).

2.1 The blood-brain barrier disruption and axonal transport

The BBB is formed by specialised endothelial cells, surrounded by a basal lamina, pericytes, astrocytes and neurons that together form the neurovascular unit (NVU). This structure acts as a physical and metabolic barrier that restricts the type of nutrients and molecules that can enter the cerebral parenchyma (Banerjee & Bhat, 2007; Calabria & Shusta, 2006; Cardoso et al., 2010). Inter-endothelial junctions formed by membrane proteins present at tight junctions (claudins, occludins and Juctional Adhesion Molecules (JAM) and adherens junctions (cadherins and catenins) filter nutrients and metabolites and regulate the passage of immune cells into nervous tissue (Cardoso et al., 2010).

However, during infection with neurotropic flaviviruses such as WNV, JEV and MVEV, the over-expression of cytokines, such as tumour necrosis factor-alpha (TNF-alpha), or enzymes, such as matrix metalloproteinase (MMP), affects the permeability of the endothelium and permits the entry of viruses into the cerebral parenchyma (Chambers & Diamond, 2003). Wang et al. (2004) reported that during WNV infection, the over-expression of TNF-alpha and interleukin 6 (IL-6) affects the integrity of the BBB because they alter expression of the proteins responsible for inter-endothelial junctions (Wang et al., 2004). Additionally, MMP enzymes digest the basal lamina, weakening the interactions between endothelial cells and other elements forming the NVU (Cardoso et al., 2010; Petty & Lo, 2002; Wang et al., 2004), favouring the entry of viral particles or infected leukocytes into the cerebral parenchyma and facilitating the spreading and replication of the virus in nervous tissue (Wang et al., 2004).

Additionally during infection, endothelial cells are activated and overexpress cellular adhesion molecules that favour the transmigration of immune cells into the cerebral parenchyma, such as E-selectin, VCAM-1 and ICAM-1 (Shen et al., 1997; Verna et al., 2009). For example, the overexpression of ICAM-1 promotes the adhesion and diapedesis of infected and activated leukocytes that can enter and alter the brain. These cells can also amplify the infection, acting as *Trojan horses* that introduce viral particles into nervous tissue (Ben-Nathan et al., 1996; Cardosa et al., 1986; King et al., 2007). Infection with flaviviruses and the signalling induced by some cytokines expressed extraneurally during such an infection can activate macrophages and microglial cells. These cells then acquire an antigen-presenting phenotype and produce and spread pro- and anti-inflammatory molecules such as IL-6, IL-1β, IL-10, TNF-alpha, type I and II interferon (IFN) and the monocytes chemotactic protein (MCP-1) in the brain microenvironment (Ghoshal et al., 2007), which promotes the disturbance of the endothelium and increases and sustains the activation of glial cells, promoting the infiltration of leukocytes (Muñoz-Fernández & Fresno, 1998) Another aspect that contributes to the disturbance of the BBB and the transport of flaviviruses into the cerebral parenchyma is the infection of endothelial cells (Avirutnan et al., 1998), which allows replication of the virus and its subsequent movement toward the cerebral parenchyma (Liu et al., 2008; Lopes et al., 2007; Mathur et al., 1992; Mishra et al., 2009) (Figure 3).

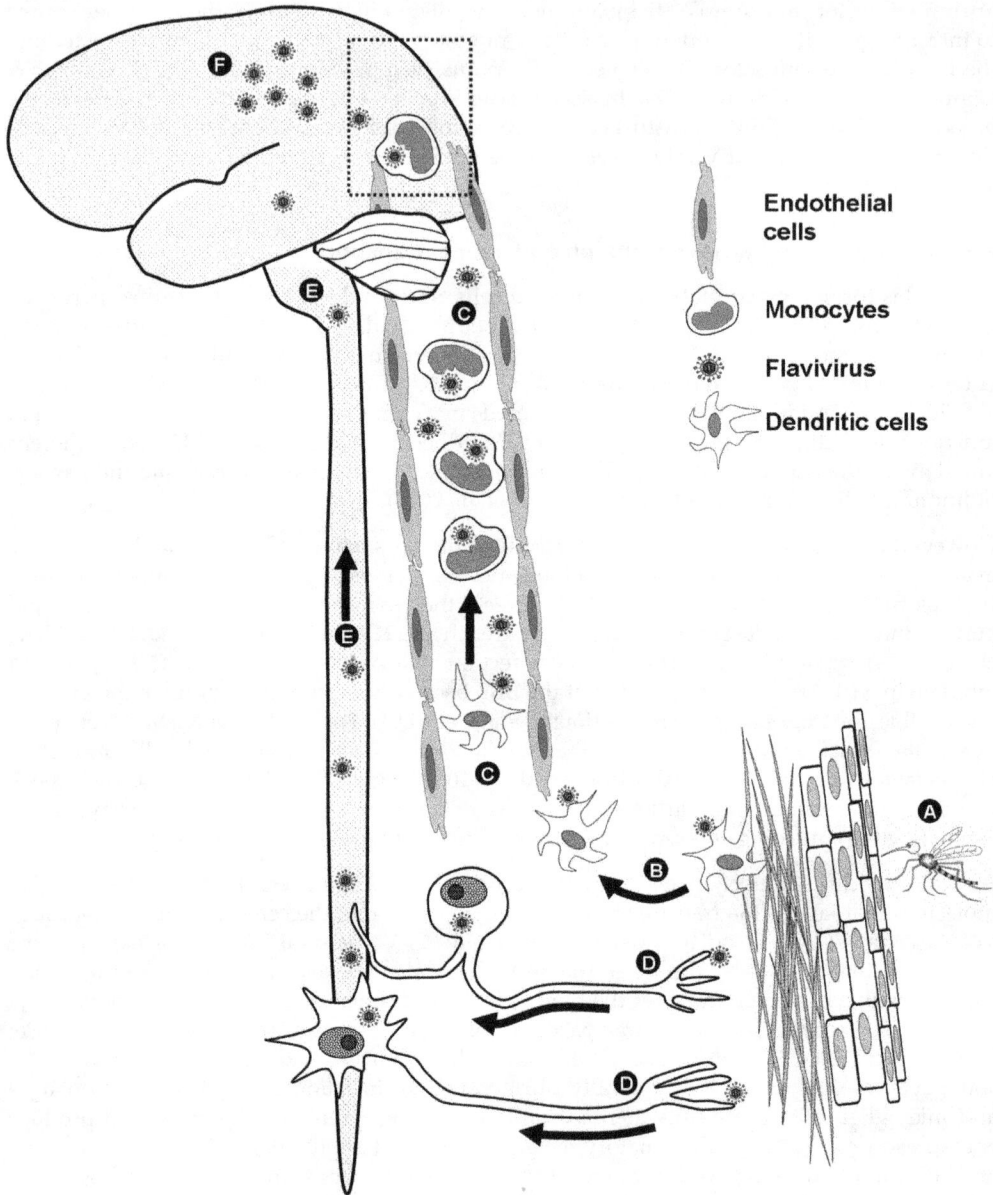

Fig. 2. Mechanisms for flavivirus entry into nervous systmem during neurologic disease. Neurotropism could be explained by direct binding of virus with neurons or ability of neurons to replicate the virus, while neuroinvasión and neurovirulence depends on ability of virus to enter to CNS and disrupt the brain architecture or function. A. - Flavivirus inoculation by an arthropod bite in the dermis. B. - Dendritic cells or Langerhans cells take the inoculated virus and migrates to lymph nodes to infect other immune cells. C. - Virus-

free or cell-associated virus spreading using hematogenous pathway to enter CNS and infect neurons and glia. D. - Virus capture and spreading by sensory and motor peripheral fibers. Virus is transported by retrograde axonal transport to CNS, and E. - spread following connecting neurons. F. - Then; virus can infect neighborhood neurons affecting its metabolism and function. During extraneural infection, immune cells produce cytokines and chemokines, which induce adhesion molecules expression in brain endothelial cells, favouring rolling of monocytes and macrophages. In addition to cells, the over expression of pro-inflammatory mediators and proteolytic enzymes result in an increase in blood-brain barrier permeability (see explanation of inset box in Figure 3).

However, infection and the damage to endothelial cells *in vivo* is not always evident, and Liou and Hsu (1998) demonstrated that JEV passes through brain endothelial cells via transcytosis (Liou & Hsu, 1998), suggesting that flaviviruses can exploit a diversity of mechanisms to penetrate nervous tissue. Other routes used by flaviviruses to enter the NS include the hematogenous route and axonal transport. The haematogenous route is the most likely dispersion route of various viruses, including flaviviruses, because these viruses, after being inoculated by mosquitoes or ticks, infect monocytes/macrophages, dendritic cells and Langerhans cells, which then transport viruses to a second cell type, such as epithelial, endothelial, fibroblast or muscle cells (Chambers & Diamond 2003; Lindenbach et al., 2007).

WNV, JEV and TBEV also utilise axonal retrograde or anterograde transport in olfactory epithelial neurons and motor neurons to penetrate and spread within the central nervous system (CNS) (Charles et al., 1995; Monath et al., 1983; Ramos et al., 1994; Sriurairatna et al., 1973), and peripheral nervous system (PNS) (An et al., 2003; McMinn et al., 1996; Samuel et al., 2007a; Silvia et al., 2003). Samuel et al. (2007b) using *in vitro* and *in vivo* infection models demonstrated that WNV enters to CNS and is transported efficiently using axonal transport in spinal medullary neurons and superior cervical ganglion neurons. Additionally, they demonstrated that in a hamster model, after viral inoculation and sciatic nerve transection, animals exhibited neurological alterations such as paralysis and prostration, suggesting that WNV utilises both the nervous and hematogenous routes to penetrate and replicate in nervous tissues (Samuel et al., 2007b).

2.2 DENV neuroinvasion

With regard to infection of the NS by DENV, it has been reported that this virus can infect NS cells *in vitro* and *in vivo* and could use axonal transport to spread inside the brain. Moreover, it has been demonstrated that *in vivo* infection with DENV can alter the integrity of the BBB (Chaturvedi et al., 1991), which has been associated with high levels of MMP-9 in plasma. This enzyme can degrade the basal lamina of the NVUs and facilitate the free passage of the virus and infected leukocytes into the cerebral parenchyma (Luplertlop et al., 2006). In animal models, NS infection by DENV has been reported after the virus was tissue-adapted, as reported by Cole and Wisseman (1969) and Sriurairatna et al. (1973). These authors achieved infection and virus production in tissue, accompanied by neurological signs associated with infection, such as paralysis of the posterior limbs. This infection was achieved after adapting a DENV strain via numerous passages in mice brains (Cole & Wisseman, 1969; Sriurairatna et al., 1973).

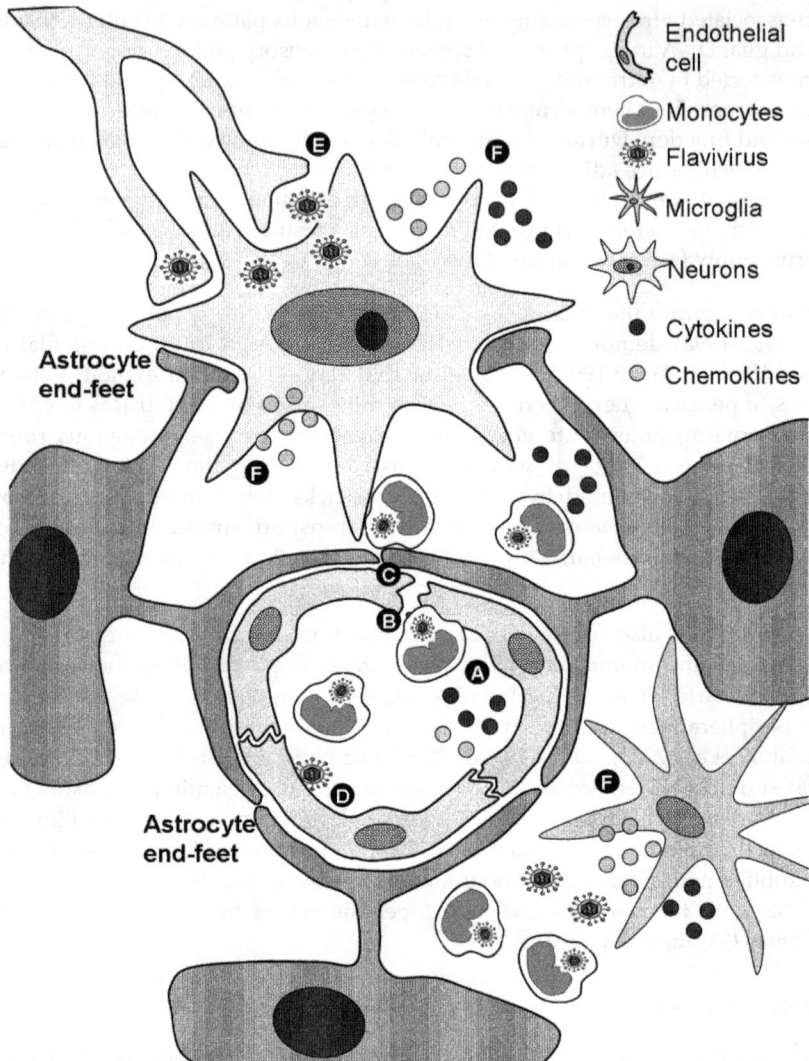

Fig. 3. Diagram of a neurovascular unit, a capillary enclosed in astrocyte end-feet during a nervous system infection by a flavivirus. A. - Infected and non-infected immune cells secretion of cytokines and chemokines. B. - Disruption of thigh junctions that seal the pathway between the capillary endothelial cells, caused by cells and inflammatory mediators. C. - Transcytosis of infected cells to brain parenchyma. D. - Direct infection of brain endothelial cells by flavivirus. E. - Virus neuronal infection by neighbor infected neurons. F. - Secretion of cytokines and chemokines by infected neurons and glial cells, causing change of structure and function of nervous cells.

Desprès et al. (1996 and 1998) also reported nervous tissue infection in 2-day-old mice after viral adaptation of a DENV-1 strain in nervous tissue and mosquito cells. This model of neuroinfection was the first to suggest that neurological alterations exhibited by infected animals, such as paralysis of posterior limbs, are principally associated with the death of infected neurons (Desprès et al., 1996, 1998). Another model of nervous tissue infection was developed by An et al. (2003), who infected young adult SCID mice (severe combined immunodeficiency) with an non adapted DENV-2 strain; these animals developed high viral titres in their nervous tissue, and viral particles were observed in spinal medullary motor neurons, axons and ependymal cells (An et al., 2003), using transmission electron microscopy, which suggests that DENV can penetrate and infect CNS and PNS neurons using axonal transport. Finally, it has been reported that DENV, regardless of whether it is adapted to nervous tissue, can infect a low proportion of primary neurons or cell lines *in vitro* (Desprès et al., 1996; Imbert et al., 1994; Ramos et al., 1994, 1998), thus demonstrating DENV weak neurotropism.

This evidence demonstrates that neuroadapted DENV, can infect the brain and can utilise axonal transport to enter and spread throughout the nervous system (An et al., 2003; Desprès et al., 1998; Liou & Hsu, 1998; Lum et al., 1996), Nevertheless, this virus continues to be classified as a non-neurotropic virus, and the elements that confer the capacity to enter, infect and spread in nervous tissue in this virus are completely unknown.

3. Neurotropism

The ability of some viruses to infect and replicate in neurons is called neurotropism and is determined by viral and cellular factors. Mostly virus determinants are associated with envelope glycoprotein gene mutations that favour interactions between the virus and molecules on the neuron surface. These interactions promote the fusion of the virus with the plasma membrane and can also trigger endocytosis or transcytosis of the virus. A well-known example of viral neurotropism is that mediated by glycoprotein G of the rabies virus (a highly neurotropic virus), which interacts with the neurotrophin low-affinity receptor, the neural cell adhesion molecule or the nicotinic receptor present in muscular and neuron cells to infect cells. The interaction of rabies virus protein G with some of these molecules promotes virus entry and replication in the nervous system (Lafon, 2005).

3.1 Viral and cellular proteins

The virus and cell interactions occur first with molecules that act as low-affinity receptors to bring virus closer to membrane, and then viral co-receptors bind to proteins on the viral surface and promote infection, which can be mediated by endocytosis or through the fusion of membranes such that the nucleocapsid is released into the cytoplasm. The use of co-receptors is common in many different viruses. The human immunodeficiency virus uses co-receptors to infect T lymphocytes and macrophages via the CD4 receptor and the CCR5 and CXCR4 molecules (Moiser, 2009).

In flaviviruses, the envelope protein (E) is the principal component of the virion surface. It participates in the recognition and subsequent binding to the receptor and the fusion of the virus with the cell membranes (Lindenbach et al., 2007). This protein is formed by three beta-barrel domains known as domains I, II and III, and these last two are responsible for interacting with putative receptor molecules (Pastorino et al., 2010). Variations in the amino acid sequence in domains II and III, which are associated with the lack of proofreading activity of the flavivirus RNA polymerase, bear directly upon the changes of viral tropism and can promote the neurotropism of flaviviruses including DENV (Lee et al., 2006a).

The molecules that have been reported as possible receptors for DENV, JEV, WNV and TBEV in different cell populations include ICAM-3 (Jindadamrongwech & Smith, 2004), CD209 (DC-SIGN) (Tassaneetrithep et al., 2003), DC-SIGNR (Davis et al., 2006), integrins (Chu & Ng, 2004), the mannose receptor (Miller et al., 2008), HSP70 and HSP90 (Das et al., 2009; Reyes del Valle et al., 2005), the laminin receptor (Tio et al., 2005) and heparan sulphate (HS) (Germi et al., 2002) among others (Barba-Spaeth et al., 2005; Upanan et al., 2008). It seems that HS favours the attraction and recruitment of viral particles to the cellular surface, thus favouring the direct entry of the virus or interaction with a second receptor molecule (Germi et al., 2002; Lee et al., 2002, 2004, 2006b). Additionally, in human (SK-N-SH) and murine (N1E-115) neuroblastoma cell lysates, the presence of a 65 kDa protein that binds specifically to DENV-2 has been reported (Ramos et al., 1997), and these findings suggest that neuronal cells express a receptor that permits the binding of the virus with these cell membranes; however, the characterisation of this protein has not been reported.

As was previously stated, the E flavivirus protein partially determines cellular tropism. However, the molecular determinants that promote the entry into susceptible cells are not well known, and the mechanisms that define their neurotropism are even less clear (Lobigs et al., 1990). So far, the RGD Motif (Arg-Gly-Asp), present in the E protein of JEV, YFV, TBEV and WNV, has been identified as being responsible for promoting the interaction between these viruses and integrins present on the surface of susceptible cells. This was demonstrated by modification of position 390 of the E protein of MVEV, which changes the tropism and virulence of this virus. Thus, this amino acid motif is proposed as the main site of interaction between flaviviruses and their viral receptors (Becker, 1990; Lee & Lobigs, 2000; Lobigs et al., 1990). However, the RGD motif has not been identified in the E protein of DENV, which suggests that this virus possesses different domains or mechanisms for interacting with receptor molecules.

3.2 DENV neurotropism

Various authors have reported that during *in vitro* or *in vivo* passages associated with the adaptation of DENV, mutations occur throughout the genome, primarily in glycoprotein E, which seemingly confers neurotropism and the ability to enter nervous tissue and cause neurological alterations. For example, the $Glu_{126}Lys$ change in the DENV-2 E protein changed the virus tropism and conferred the capacity to infect nervous tissue (Gualano et al., 1998). Similarly, the mutations $Asp_{390}His$ and $Phe_{402}Leu$ in DENV-4 E protein conferred a neurotropic and neurovirulent phenotype on the virus (Bray et al., 1998; Kawano et al., 1993; Sanchez & Ruiz, 1996). These findings suggest that for DENV to acquire a neurotropic phenotype, certain variations must occur in the sequence located in specific regions of the E protein (Desprès et al., 1996; Lee et al., 2006a). However, mutations in non-structural viral proteins could also determine the success of the infection, particularly the replication of the virus in neurons (Duarte dos Santos et al., 2000). Consequently, it should be determined whether the neurotropism of adapted DENV depends on the viral serotype used and the type of cell to which the virus is adapted. Additionally, the mechanisms of DENV transport and dispersion throughout the nervous tissue and whether these mechanisms depend on changes in the viral genome acquired during the process of neuroadaptation should be evaluated.

4. Neurovirulence

Neurovirulence is the capacity of viruses to cause disease and alterations in the nervous system and can be affected by both viral and host-related factors. The viral factors that affect neurovirulence are viral serotype and genotype. Beasley et al. (2002) reported that genotypes I and II of WNV, which are very similar in sequence, cause different neurological alterations in mice and hamsters (Beasley et al., 2002).

4.1 Serotypes and genotypes

The dengue serocomplex is formed by 4 viral serotypes that possess a high genome homology and all cause dengue symptoms. However, it has been demonstrated that the genomic differences among DENV serotypes and genotypes induce clinical manifestations of the disease that vary in intensity, as has been shown for DENV-2 and DENV-3, which have been associated mainly with haemorrhagic symptoms and cases of severe dengue in some patients (Clyde et al., 2006; Tsia et al., 2009).

For example, the best-studied genotypes of DENV-2 are the Asian and American genotypes. When the Asian strain started circulation in the American continent, it was caused serious dengue outbreaks with haemorrhagic symptoms in patients with primary infections who were from Central and South American countries (Clayde et al., 2006). The Asian genotype is frequently associated with severe dengue and haemorrhagic symptoms in Asian patients, and experimentally, this genotype is more virulent and replicates with higher efficiency in macrophages, while the American genotype is associated with signs of dengue fever and its replication is slower in cultured macrophages (Barreto dos Santos et al., 2002; Guzmán et al., 2002a; Rico-Hesse et al., 1997).

These differences could partially explain the changes in symptoms exhibited by patients infected with DENV. When the genotype sequences were compared, significant differences were found in the 5′ and 3′ untranslated regions (UTR) of the genomic RNA, and it was observed that the 3′ UTR of the Asian genotype generates secondary structures that permit better interaction with the viral RNA polymerase and enhance its processivity. This difference could explain the efficiency of virus replication and virus production in infected cells with this genotype, which in turn could be related to the inefficiency of the immune system to control and eliminate this virus (Cologma & Rico-Hesse, 2003; Leitmeyer et al., 1999).

Additionally, it has been reported that genotype I of DENV-3 can induce different symptoms in infected mice, when was intracerebrally inoculated. An effective infection was observed with high viral titer detected in tissue. Infiltration of monocyte cells into the cerebral parenchyma was also detected, as was the appearance of neurological symptoms such as meningo-encephalitis and paralysis associated with neuronal degeneration. In contrast, intracerebral inoculation of mice with genotype III induced a less intense immune response, with less tissue damage and low viral production (Ferreira et al., 2010), confirming that differences among genotypes and serotypes can be related to flavivirus virulence.

Beyond viral factors, the neurovirulence caused by flaviviruses can be related to the type of immune response that the individual generates against infection at the local and systemic levels. This response is similar among flaviviruses, although DENV and YFV mainly induce alterations in vascular permeability and in coagulation (Avirutnan et al., 2010). The immune response that occurs in nervous tissue during flavivirus infection varies in intensity and can support the control and clearence of the virus and establish a neuroprotective state that stimulates the repair of tissue damaged by the infection (Griffin, 2003). To promote virus clearance, monocytes and lymphocytes enter the cerebral parenchyma, attack infected cells, and release soluble mediators, which stimulate and maintain the local immune response activating astrocytes, microglia and the cerebrovascular endothelium. Additionally, the infected or damaged neurons themselves can express and release some of these mediators (Chakraborty et al., 2010), amplifying the local immune response.

4.2 The host factor: Immune and nervous system

As was mentioned above, the activated cerebrovascular endothelium can facilitate the passage of T and B lymphocytes and macrophages and allow the leakage of soluble factors and toxins that increase inflammation and damage the cerebral parenchyma, causing neuron death (Less et al., 2006; Lin et al., 2002; Wrona, 2006). Inflammation of nervous tissue is associated with the activation of astrocytes and microglia, as indicated by morphological changes and changes in the expression profile of adhesion molecules, cytokines and interleukins (TNF-alpha, IL-1β, IL-6, IL-10, IFNs, MCP-1 and TGF-β), which combined with factors secreted by infiltrating immune cells, can increase the nervous system damage (Muñoz-Fernández & Fresno, 1998). The activation of glial cells is partially due to their infection by flaviviruses. For example, astrocytes infected with WNV express the chemokine CXCL10 and other neuroinflammatory and neurotoxic molecules that can increase nervous system injury and induce the death of both infected and uninfected neurons. These data suggest that the activation of glial cells depends on viral replication and that the signalling induced by some inflammatory mediators within the nervous tissue can increase the neuropathogenesis caused by WNV (Van Marle et al., 2007) and other flaviviruses.

Immune cells that infiltrate nervous tissue are mainly CD4+ and CD8+ T lymphocytes and macrophages. CD4+ lymphocytes producing IL-12 stimulate the cytotoxic activity of CD8+ lymphocytes that arrive to the nervous tissue. These lymphocytes secrete proinflammatory mediators such as IFN-gamma, TNF-alpha and IL-6 that alter tissue homeostasis when expressed consistently during the infection (Chaturvedi et al., 2000; Sánchez-Burgos et al., 2004; Swarup et al., 2007). Additionally, CD8+ lymphocytes embedded in the tissue promote the death of both infected and uninfected cells via the release of perforin and granzymes and the expression of Fas ligand (FasL) (Courageot et al., 2003; Marques-Deak et al., 2005; Mellor & Munn, 2006; Rempel et al., 2004, 2005). Lastly, infiltrating macrophages modulate the type of immune response that occurs in the tissue during infection and can clear free viral particles and infected and damaged cells present in the tissue, although these cells can also seemingly promote the entering of some flaviviruses, acting as *Trojan horses* by releasing viral particles within the nervous tissue (Chaturvedi, 2006; Chaturvedi et al., 2006).

Finally, neurotropic flaviviruses as well as non-neurotropic flaviviruses preferably infect neurons *in vitro* and *in vivo* (Chambers & Diamond, 2003; Johnson & Roehring, 1999; Samuel & Diamond, 2006; Shrestha et al., 2003). Nevertheless, it has been reported that other nervous tissue cells such as oligodendrocytes, astrocytes and microglial cells (Chen et al., 2000; Jordan et al., 2000) may be susceptible to infection. The cellular, metabolic and molecular factors that increase the susceptibility of neurons to flaviviruses are unknown. Additionally, it has been reported that infection with some flaviviruses, including DENV, induces death in infected neurons. This response may be mediated by TNF-alpha, the Fas/FasL complex or the release of cytochrome c and the presence of free radicals; during this process, caspases 3, 8, and 9 have been found to be activated (Courageot et al., 2003; Marianneau et al., 1998; Samuel et al., 2007a). Thus, cell death in nervous tissue, such as neurons, has been associated with the development of neurological alterations resulting from infection.

5. Neurosusceptibility

Neurosusceptibility refers to the vulnerability of a host to neurological alterations during an infection with neurotropic viruses. This vulnerability can be affected by the age, species, immune status and genetic background of the individual.

5.1 Age

Using animal models, it has been demonstrated that physiological immaturity increases the susceptibility of the nervous system to WNV and JEV infection (Ogata et al., 1991; Weiner et al., 1970). For example, JEV infects neurons in the cortex, hippocampus and brainstem of 2-day-old mice, but the areas susceptible to infection and the numbers of infected neurons diminish as age increases. This resistance to infection is maintained, even if previously-infected neonatal neurons are implanted in animals greater than 14 days old. Similarly, the mortality of rats infected with JEV is 100% when they are inoculated between 2 and 12 days after birth but diminishes to 50%, 8.3% and 0% when they are inoculated at 13, 14 and 17 postnatal days. These results demonstrate that the neuronal and physiological maturity of nervous tissues is determining factors in favouring infection and neuronal alteration (Ogata et al., 1991; Weiner et al., 1970).

With respect to DENV-2 infection, Guzmán et al. (2002b) reported that children between the age of 3 and 4 years old were more vulnerable to developing symptoms of dengue compared to older children and adults (Guzmán et al., 2002b). This vulnerability is due principally to the type of response that neonates generate against viral, bacterial, fungal and parasitic infections (Maródi, 2006). Some clinical reports demonstrate that neonatal immunity is predominantly of the Th2 type, which specifically stimulates immune tolerance and inhibits the Th1 type response, which in turn activates immune cells to control and eliminate pathogens.

This tolerance and ineffectiveness in young individuals is related to the type of cytokines that are released and circulate before and after infection, such as the immunomodulatory molecule IL-10 that negatively regulates the activation of cells such as macrophages, NK cells and T and B lymphocytes. This hypo-reactivity of antigen-presenting cells causes them to inefficiently recognise and present viral or bacterial antigens. Additionally, in neonates, the absence of specific antibodies against microorganisms and the low level of production of molecules like IFN-gamma and TNF-alpha further reduce the activation of Th1 lymphocytes (Kemp & Campbell, 1996; Maródi, 2006; Wilson et al., 1999), and thus the cytotoxic activity and pathogen control exhibited by lymphocytes is not established.

5.2 Genetic background

The species and genetic background of vectors and hosts are other determinants that can favour the dispersion of flaviviruses and neuroinfection. The enzootic life cycle of flaviviruses includes vectors and reservoirs such as birds, monkeys or other wildlife, as well as humans. Flavivirus vectors can include mosquitoes of the *Aedes* and *Culex* genera and ticks of the *Ixodes* genus, which transmit TBEV (Lindenbach et al., 2007). The known reservoirs for flaviviruses are birds and small mammals, which suggests that there are some species-specific characteristics that restrict the transmission of these viruses. These ecological restrictions are evident during DENV infection in some experimental infection models in mice and monkeys, which reproduce some signs of disease that manifest in infected humans. Additionally, in these models, certain symptoms associated with infection are exhibited that are uncommon in infected humans, such as neurological alterations (Tan et al., 2010), which render interpretation of the data more difficult.

Murine models commonly used to reproduce certain symptoms associated with DENV infection are mice models such as SCID (Lin et al., 1998), AG129 (lacking functional IFN-α/β and-γ receptors) (Johnson & Roehring, 1999; Williams et al., 2009; Tan et al., 2010) and NOD/SCID (non-obese diabetic/severe combined immunodeficient) (Bente et al., 2005; Huang et al., 2000; Mota & Rico-Hesse, 2009), which upon being infected by DENV develop some signs of disease such as haemorrhage, thrombocytopenia and plasma leakage (Shresta et al., 2006). Nevertheless, while these models have increased our understanding of some of the cellular and molecular mechanisms involved in the development of the haemorrhagic signs observed during infection, their interpretation should be tentative given that these animals present an incomplete immune response to the virus due to their modified genomes.

With these differences in mind, other models have been established using immunocompetent animals such as C57BJ/C, ICR, A/J (Shresta et al., 2004) and Balb/C (Barreto et al., 2007) mice, which present robust immune responses to the virus, and possibly are less susceptible to infection as a result. However, these animals contract the virus and develop symptoms when infected with mouse cell- or tissue-adapted DENV or following intravenous or intracerebral inoculation with high viral titres in suckling or young adult mice (Yauch & Sheresta, 2008; Wu-Hsieh et al., 2009). These models allow the acquisition of other data that allow a different understanding of the molecular mechanisms associated with DENV immunopathogenesis. Nevertheless, independent of the strain of animals employed, one must keep in mind that these models are experimental tools that so far only allow the *in vivo* reproduction of some symptoms of the very complicated disease induced by DENV.

6. Conclusion

The infection and pathogenesis caused by neurotropic flaviviruses is a product of a series of complex interactions between the virus and nervous tissues and is affected by viral diversity and the host's immune response and susceptibility. Therefore, it will be necessary to perform new studies with new experimental strategies to expand our knowledge and understand the interactions between flaviviruses and nervous tissue. It will be necessary to identify those factors affecting DENV and the nervous system that favour neuroinfection and the increasingly frequent appearance of neurological symptoms stemming from this virus. The study of this phenomenon will provide information that permits an understanding of viral pathogenesis that is of great importance for public health in tropical countries. In addition, this proposal will uncover new strategies for antiviral and vaccine research that will be useful for fighting DENV.

7. Acknowledgment

We are grateful to Dr. Jacqueline Chaparro-Olaya who spared her time to go through the manuscript at various stages and offered valuable suggestions. This work was funded by División de Investigaciones – Universidad El Bosque, Colciencias-Colombia (Project 130 848925267) and Universidad Nacional de Colombia.

8. References

An, J.; Zhou D. Kawasaki, K. & Yasui, K. (2003). The pathogenesis of spinal cord involvement in dengue virus infection. *Virchows Arch*, Vol.442, No.5, (May 2003), pp. 472-81, ISSN 0945-6317

Avirutnan, P. Malasit, P. Seliger, B. Bhakdi, S. & Husmann, M. (1998). Dengue virus infection of human endothelial cells leads to chemokine productions, complement activation, and apoptosis. *J Immunol*, Vol.161, No.11, (December 1998), pp. 6338-46, ISSN 0022-1767

Avirutnan ,P; Fuchs, A. Hauhart, R. Somnuke, P. Youn, S. Diamond, M. & Atkinson, J. (2010). Antagonism of the complement component C4 by flavivirus nonstructural protein NS1. *J Exp Med*, Vol.207, No.4, (April 2010), pp. 793-806, ISSN 0022-1007

Banerjee, S. & Bhat, M. (2007). Neuron-Glial interactions in blood brain barrier formation. *Ann Rev Neurosci*, Vol.30, (July 2007), pp. 235-58, ISSN 1545-4126

Barba-Spaeth, G.; Longman, R. Albert, M. & Rice, C. (2005). Live attenuated yellow fever 17D infects human DCs and allows for presentation of endogenous and recombinant T cell epitopes. *J Exp Med*, Vol.202, No.9, (October 2005), pp. 1179-84, ISSN 0022-1007

Barreto dos Santos, FB.; Miagostovich, MP. Nogueira, RM. Edgil, D. Schatzmayr, HG. Riley, LW. & Harris E. (2002). Complete nucleotide sequence analysis of a Brazilian Dengue Virus Type 2 Strain. *Mem Inst Oswaldo Cruz*. Vol.97, No.7, (October 2002), pp. 991-995, ISSN 1678-8060

Barreto, D.; Takiya, C. Schatzmayr, H. Ribeiro, R. Farias, J. & Barth, O. (2007) Histological and ultrastructural aspects of mice lungs experimentally infected with dengue virus serotipe 2. *Mem Inst Oswaldo Cruz*, Vol.102, No.2 (May 2007), pp. 175-82, ISSN 1678-8060

Beasley, D.; Li, L. Suderman, & M. Barrett, A. (2002). Mouse neuroinvase phenotype of West Nile virus strain varies depending upon virus genotype. *Virology*, Vol.296, No.1, (April 2002), pp. 17-23, ISSN 0042-6822

Becker, Y. (1990). Computer analysis antigenic domains and RGD-like sequences (RGWG) in the glycoprotein of flavivirus: a approach to vaccine development. *Virus Genes*, Vol.4, No.3, (September 1990), pp. 267-82, ISSN 0920-8569

Ben-Nathan, D.; Huitinga, I. Lustig, S. van Rooijen, N. & Kobiler, D. (1996). West Nile virus neuroinvasion and encephalitis induced by macrophage depletion in mice. Arch Virol, Vol.141, No.3-4, pp. 459-69, ISSN 0304-8608

Bente, D.; Melkus, M. Garcia, J. & Rico-Hesse, R. (2005). Dengue fever in humanized NOD/SCID mice. *J Virol*, Vol.79, No.21, (November 2005), pp. 13797–13799, ISSN 1098-5514

Bray, M.; Men, R. Tokimatsu, I. & Lai, C. (1998). Genetic determinant responsible for acquisition of dengue type 2-virus mouse neurovirulence. *J Virol*, Vol.72, No.2, (February 1998), pp. 1647-51, ISSN 1098-5514

Calabria, A.; & Shusta, E. (2006). Blood–brain barrier genomics and proteomics: elucidating phenotype, identifying disease targets and enabling brain drug. *Drug Discov Today*, Vol.11, No.17-18, (September 2006), pp. 792-99, ISSN 1878-5832

Cardosa, M.; Gordon, S. Hirsch, S. Springer, T. & Porterfield, J. (1986). Interaction of West Nile Virus with Primary Murine Macrophages: Role of Cell Activation and Receptors for Antibody and Complement. *J Virol*, Vol.57, No.3, (March 1986), pp. 952-959, ISSN 1098-5514

Cardoso, F.; Brites, D. & Brito, M. (2010). Looking at the blood-brain barrier: Molecular anatomy and possible investigation approaches. *Brain Res Rev*, Vol.64, No.2, (May 2010), pp. 328-63, ISSN 1872-6321

Chakraborty, S.; Nazmi, A. Dutta, K. & Basu, A. (2010). Neurons under viral attack: Victins or Warriors?. *Neurochem Int*. Vol.56, No.6-7, (March 2010), pp. 727-35, ISSN 1872-9754

Chambers, T.; & Diamond, M. (1990). Pathogenesis of flavivirus encephalitis. *Adv Virus Res*, Vol.60, pp. 273-42, ISSN 0065-3527

Chambers, T.; Hahn, C. Galler, R. & Rice, C. (1990). Flavivirus genome organization, expression and replication. *Ann Rev Microbiol*, Vol.44, pp. 649-88, ISSN 0066-4227

Charles, P.; Walters, E. Margolis, F. & Johnston, R. (1995). Mechanisms of neuroinvasion of Venezuelan Equine Encephalitis Virus in the mouse. *Virology*, Vol.208, No.2, (April 1995), pp. 662-671, ISSN:0042-6822

Chaturvedi, U.; Dhawan, R. Khanna, M. & Mathur, A. (1991). Breakdown of the blood-brain barrier during dengue virus infection of mice. *J Gen Virol*, Vol.72, No.4, (April 1991), pp. 859-66, ISSN 1465-2099

Chaturvedi, U.; Agarwal, R. Elbishbishi, E. & Mustafa, A. (2000). Cytokine cascade in dengue hemorrhagic fever: implications for pathogenesis. *FEMS Immunol Med Micro*, Vol.28, No.3, (July 2000), pp. 183-88, ISSN 1574-695X

Chaturvedi, U. (2006). Tumor necrosis factor & dengue. *Indian J Med Res*, Vol.123, No.1, (January 2006), pp. 11-14, ISSN 0971-5916

Chaturvedi, U.; Nagar, R. & Shrivastava, R. (2006). Macrophage & dengue virus: friend or foe?. *Indian J Med Res*, Vol.124, No.1, (July 2006), pp. 23-40, ISSN 0971-5916

Chen, C.; Liao, S. Kuo, M. & Wang, Y. (2000). Astrocytic alteration induced by Japanese encephalitis virus infection. *NeuroReport*, Vol.11, No.9, (June 2000), pp. 1933-1937, ISSN 1473-558X

Chien, J.; Ong, A. & Low, S. (2008). An unusual complication of dengue infection. *Singapore Med J*, Vol.49, No.12, (December 2008), pp. e340-41, ISSN 0037-5675

Chu, J.; & Ng, M. (2004). Interaction of West Nile Virus with alpha v beta 3 integrin mediates
    virus entry into cells. *J Biol Chem*, Vol. 279, No.24, (October 2004), pp. 54533-41,
    ISSN 1083-351X

Clyde, K.; Kyle, J. & Harris, E. (2006). Recent advances in deciphering viral and host
    determinants of dengue virus replications and pathogenesis. *J Virol*, Vol.80, No.23,
    (August 2006), pp. 11418-31, ISSN 0022-538X

Cole, G.; & Wisseman, C. (1969). Pathogenesis of type1 dengue virus infection in sckling,
    weanling and adult mice. 1. The relation of virus replication to interferon and
    antibody formation. *Am J Epidemiol*, Vol.89, No.6, (June 1969), pp. 669-80, ISSN
    0002-9262

Cologma, R.; & Rico-Hesse, R. (2003). American genotype structure decrease dengue virus
    output from human monocytes and dendritic cells. *J Virol*, Vol.77, No.7, (April
    2003), pp.3929-38, ISSN 0022-538X

Courageot, M.; Catteau, A. & Després, P. (2003). Mechanisms of dengue virus induced cell
    death. *Adv Virus Res*, Vol.60, pp. 157-86, ISSN 1557-8399

Das, S.; Laxminarayana, S. Chandra, N. Ravi, V. & Desai, A. (2009). Heat shock protein 70 in
    Neuro2a cell is a putative receptor for Japanese Encephalitis virus. *Virology*,
    Vol.385, No.1, (December 2008), pp. 47-57, ISSN 1096-0341

Davis, C.; Nguyen, H. Hanna, S. Sánchez, M. Doms, R. & Pierson T. (2006). West Nile virus
    discriminates between DC-SIGN and DC-SIGNR for cellular attachment and
    infection. *J Virol*, Vol. 80, No.3, (February 2006), pp. 1290-301, ISSN:0022-538X

Desprès, P.; Flamand, P. Ceccaldi, P. & Deubel V. (1996). Human isolates of dengue type I
    virus induce apoptosis in mouse neuroblastoma cells. *J Virol*, Vol.70, No,6, (June
    1996), pp. 4090-5, ISSN:0022-538X

Desprès, P.; Frenkiel, M. Caccaldi, P. Duarte dos Santos, C. & Deubel, V. (1998). Apoptosis in
    the mouse central nervous system in response to infection with mouse
    neurovirulent dengue virus. *J Virol*, Vol.72, No.1, (January 1998), pp. 823-29,
    ISSN:0022-538X

Domingues, R.; Kuster, G. Onuki-Castro, F. Souza, V. Levi, J. & Pannuti, C. (2008).
    Involvement of the central nervous system in patients with dengue virus infection.
    J Neurol Sci, Vol.267, No.1-2, (October 2007), pp. 36-40, ISSN 0392-0461

Duarte dos Santos, C.; Frenkiel, M. Courageot, M. Rocha, C. Vazeille, M. Wien, M. Rey, F.
    Deubel, V. & Desprès, P. (2000). Determinants in the envelope E protein and viral
    RNA helicase NS3 that influence the induction of apoptosis in response to infection
    with dengue type 1 virus. *Virology*, Vol.274, No.2, (September 2000), pp. 292-08,
    ISSN 1096-0341

Ferreira, G. Figueiredo, L. Coelho, L. Junior, P. Cecilio, A. Ferreira, P. Borjardim, C. Arantes,
    R. Campos, M. & Kroon, E. (2010). Dengue virus, clinical isolates show different
    patterns of virulence in experimental mice infection. *Microbes Infection*, Vol.12,
    No.7, (April 2010), pp. 546-54, ISSN 1769-714X

Germi, R.; Crance, J. Garin, D. Guimet, J. Lortat-Jacob, H. Ruigrok, R. Zarski, J. & Drouet, E.
    (2002). Heparan sulfate-mediated binding of infectious dengue virus type 2 and
    yellow fever virus. *Virology* Vol.292, No.1, (January 2002), pp. 162-68, ISSN 1096-
    0341

Ghoshal, A.; Das, S. Ghosh, S. Mishra, M. Sharma, V. Koli, P. Sen, E. & Basu, A. (2007).
    Proinflammatory mediators released by activated microglía induces neuronal death

in Japanese encephalitis. *Glia,* Vol.55, No.5, (April 2007), pp. 483:496, ISSN 1098-1136

Griffin, D. (2003). Immune responses to RNA-Virus infections of the CNS. *Nat Rev Immunol.* Vol.3, No.6, (June 2003), pp. 493-502, ISSN 1474-1741

Gualano, R.; Pryor, M. Cauchi, M. Wright, P. & Davidson, A. (1998). Identification of a major determinant of Mouse neurovirulence of dengue virus type 2 using stably cloned genomic-length cDNA. *J Gen Virol,* Vol.79, No.3,   (March 1998), pp. 437-46, ISSN 0022-1317

Gulati, S.; & Maheshwari, A. (2007). Atypical manifestations of dengue. *Tropl Med Int Health,* Vol.12, No.9, (September 2007), pp. 1087-95, ISSN 1360-2276

Guzmán, M.; Kouri, G. Valdés, L. Bravo, J. Vázquez, S. & Halstead, S. (2002a). Enhanced severity of secondary dengue-2 infections: death rates in 1981 and 1997 Cuban outbreaks. *Rev Panam Salud Publica,* Vol.11, No.4, (April 2002), pp. 223-227, ISSN 1680-5348

Guzmán, M.; Kouri, G. Valdes, L. Bravo, J. Vazquez, S. & Halstead, S. (2002b). Effect of age on outcome of secondary dengue 2 infections. *Int J Infect Dis,* Vol.6, No.2, (June 2002), pp. 118–24, ISSN 1878-3511

Huang, K.; Li, S. Chen, S. Liu, H. Lin, Y. Yeh, T. Liu, C. & Lei, H. (2000). Manifestation of thrombocytopenia in dengue 2 virus infected mice. *J Gen Virol,* Vol.81, No.9, (September 2000), pp. 2177-82, ISSN 0022-1317

Imbert, J.; Guevara, P. Ramos-Castañeda, J. Ramos, C. & Sotelo, J. (1994). Dengue virus infects mouse cultured neurons but not astrocytes. *J Med Virol,* Vol.42, No.3, (March 1994), pp. 228-33, ISSN 0146-6615

Jindadamrongwech, S.; & Smith, D. (2004). Virus Overlay Protein Binding Assay (VOPBA) reveals serotype specific heterogeneity of dengue virus binding proteins on HepG2 human liver cells. *Intervirology,* Vol.47, No.6, pp. 370-3, ISSN 1423-0100

Johnson, A.; & Roehring, J. (1999). New model for dengue virus vaccine testing. *J Virol,* Vol.73, No.1, (January 1999), pp. 783-86, ISSN 0022-538X

Jordan, I.; Briese, T. Fischer, N. Lau, J. & Lipkin, W. (2000). Ribavirin inhibits West Nile virus replication and cytopathic effect in neural cells. *J Infect Dis,* Vol.182, No.4, (August 2000), pp. 1214-7, ISSN 1537-6613

Kawano, H.; Rostapshov, V. Rosen, L. & Lai, C. (1993). Genetic determinants of dengue type 4 virus neurovirulence for mice. *J Virol,* Vol.67, No.11, (November 1993), pp. 6567-75, ISSN 0022-538X

Kemp, A.; & Campbell, D. (1996). The neonatal immune system. *Semin Neonatal, Vol.*1, No.2, (May 1996), pp. 67-75, ISSN 1744-165X

King, N.; Getts, D. Getts, M. Rana, S. Shrestha, B. & Kesson, A. (2007). Immunopathology of flavivirus infections. *Immunol Cell Biol,* Vol.85, No.1, (December 2006), pp. 33-42, ISSN 1440-1711

Kumar, R.; Tripathi, S. Tambe, J. Arora, V. Srivastava, A. & Nag, V. (2008). Dengue encephalopathy in children in Northern India: clinical factures and comparison with non dengue. *J Neurol Sci,* Vol.269, No. 1-2, (January 2008), pp. 41-48, ISSN 1878-5883

Lafon, M. (2005). Rabies virus receptors. *J NeuroVirol,* Vol.11, No.1, (February 2005), pp. 82-7, ISSN 1355-0284

Lee, E.; & Lobigs, M. (2000). Substitutions at the putative receptor-binding site of an encephalitic flavivirus alter virulence and host cell tropism and reveal a role for glycosaminoglycans in entry. *J Virol*, Vol.74, No.19, (October 2000), pp. 8867-8875, ISSN 0022-538X

Lee, E.; & Lobigs, M. (2002). Mechanism of virulence attenuation of glycosaminoglycan-binding variants of Japanese encephalitis virus and Murray Valley encephalitis virus. *J Virol*, Vol.76, No.1o, (May 2002), pp. 4901-4911, ISSN 0022-538X

Lee, E.; Hall, R. & Lobigs, M. (2004). Common E protein determinants for attenuation of glycosaminoglycan-binding variants of Japanese encephalitis and West Nile viruses. *J Virol*, Vol.78, No.15, (August 2004), pp. 8271-8280, ISSN 0022-538X

Lee, E.; Wright, P. Davidson, A. & Lobigs, M. (2006a). Virulence attenuation of Dengue virus due to augmented glycosaminoglycan-binding affinity and restriction in extraneural dissemination. *J Gen Virol*, Vol.87, No.10, (October 2006), pp. 2791-2801, ISSN 1465-2099

Lee, E.; Pavy, M. Young, N. Freeman, C. & Lobigs, M. (2006b). Antiviral effect of the heparan sulfate mimetic, PI-88, against dengue and encephalitic flaviviruses. *Antiviral Res*, Vol.69, No.1, (November 2005), pp. 31–38, ISSN 1872-9096.

Lees, J.; Archambault, A. & Russell, J. (2006). T-cell trafficking competence is required for CNS invasion. *J Neuroimmunol*, Vol.177, No.1-2, (July 2006), pp. 1-10, ISSN 1872-8421

Leitmeyer, K.; Vaughn, D. Watts, D. Salas, R. Villalobos, I. Ramos, C. & Rico-Hesse, R. (1999). Dengue virus structural differences that correlate with pathogenesis. *J Virol*, Vol.73, No.6, (June 1999), pp. 4738-47, ISSN 0022-538X.

Lin, C.; Lei, H. Shiau, A. Liu, H. Yeh, T. Chen, S. Liu. C, Chiu S, & Lin Y. (2002). Endothelial cell apoptosis induced by antibodies against dengue virus nonstructural protein 1 via production of nitric oxide. *J Immunol*, Vol.169, No.2, (July 2002), pp. 657-64, ISSN 0022-1767

Lin, Y.; Liao, C. Chen, L. Yeh, C. Liu, C. Ma, S. Huang, Y. Kao, C. & King, C. (1998). Study of dengue virus infection in SCID mice engrafted with human K562 cells. *J Virol*, Vol. 72, No. 12, (December 1998), pp. 9729–9737, ISSN 1098-5514

Lindenbach, B.; Thiel, H. & Rice, C. (2007). Flavivirus: the virus and their replication, In: *Fields Virology*, David Knipe, Peter Howley, Diane Griffin, Robert A Lamb, Malcolm Martin, Bernard Roizman, Stephen Straus, pp. 1101-52, Lippincott Williams & Wilkins, ISBN 0781760607, Philadelphia, USA

Liou, M.; & Hsu, C. (1998). Japanese encephalitis virus is transported across the cerebral blood vessels by endocytosis in mouse brain. *Cell Tissue Res*, Vol.293, No.3, (September 1998), pp. 389-94, ISSN 0302-766X

Liu, T.; Llang, L. Wang, C. Liu, H. & Chen, W. (2008). The blood-brain barrier in the cerebrum is the initial site for the Japanese Encephalitis Virus entering the central nervous system. *J NeuroVirol*, Vol.14, No.6, (November 2008), pp. 514-21, ISSN 1538-2443

Lobigs, M.; Usha, R. Nestorowicz, A. Marshall, I. Weir, R. & Dalgarno, L. (1990). Host cell selection of Murray Valley encephalitis virus protein and in mouse virulence. *Virology*, Vol.176, No.2, (June 1990), pp. 587-95, ISSN 1096-0341

Lopes, H.; Redig, P. Glaser, A. Armien, A. & Wuschmann, A. (2007). Clinical findings, lesions, and viral antigen distribution in great owls (Strix nebulosa) and barred owls (Strix varia) with spontaneous West Nile Virus infections. *Avian Dis*, Vol.51, No.1, (March 2007), pp. 140-5, ISSN 0005-2086

Lum, L.; Lam, S. Choy, Y. George, R. & Harun, F. (1996). Dengue encephalitis: A true entity? *Am J Trop Med Hyg*, vol 54, No.3, (March 1996), pp. 256-59, ISSN 1476-1645

Luplertlop, N.; Missé, D. Bray, D. Deleuze, V. Gonzalez, J. Leardkamolkarn, V. Yssel, H. & Veas F. (2006). Dengue-virus-infected dendritic cells trigger vascular leakage through metalloproteinase overproduction. *EMBO Rep*, Vol. 11, No.11, (October 2006), pp. 1176-81, ISSN 1469-3178

Malavige, G.; Ranatunga, P. Jayaratne, S. Wijesiriwardana, B. Seneviratne, S. & Karunatilaka D. (2007). Dengue viral infections as a cause of encephalopathy. *Indian J Med Microbiol*, vol.25, No.2, (April 2007), pp. 143-5, ISSN 1998-3646

Marianneau, P.; Flamand, M. Deubel, V. & Desprès, P. (1998). Apoptotic cell death in response to dengue virus infection: the pathogenesis of dengue hemorrhagic fever revisited. *Clin Diagn Virol*, Vol.10, No.2-3, (July 1998), pp. 113-9, ISSN 0928-0197

Maródi, L. (200065). Innate cellular immune responses in newborns. *Clin Immunol*, Vol.118, No.2-3, (December 2005), pp. 137-144, ISSN 1521-7035

Marques-Deak, A.; Cizza, G. & Sternberg, E. (2005). Brain immune interactions and disease susceptibility. *Mol Psychiatry*, vol.10, No.3, (March 2005), pp. 239-50, ISSN 1476-5578

Mathew, S.; & Pandian, J. (2010). Stroke in patients with dengue. *J Stroke Cerebrovasc Dis*, Vol.19, No.3, (May 2010), pp. 253-6, ISSN 1532-8511

Mathur, A.; Khanna. N, & Chaturvedi, U. (1992). Breakdown of blood-brain barrier by virus induced cytokine during Japanese Encephalitis virus infection. *Int J Exp Pathol*, Vol.73, No.5, (October 1992), pp. 603-11, ISSN 1365-2613

McMinn, P.; Dalgarno, L. & Weir, R. (1996). A comparison of the spread of Murray Valley Encephalitis Viruses of high or low neuroinvasiveness in the tissues of swiss mice after peripheral inoculation. *Virology*, Vol.220, No.2, (June 1996), pp. 414–423, ISSN 1096-0341

McMinn, P. (1997). The molecular basis of virulence of the encefalitogenic flaviviruses. *J Gen Virol*, Vol.78, No.11, (November 1997), pp. 2711-22, ISSN 1465-2099

Mellor, A.; & Munn, D. (2006). Immune privilege: a recurrent theme in immunoregulation?. *Immunol Rev*, Vol.213, No.1, (October 2006), pp. 5-11, ISSN: 1600-065X

Miller, J.; de Wet, B. Martínez-Pomares, L. Radcliffe, C. Dwek, R. Rudd, P. & Gordon, S. (2008). The Mannose Receptor Mediates Dengue Virus Infection of Macrophages. *Plos Pathog*, Vol.4, No.2, (February 2008), pp. e17, ISSN 1553-7374

Mishra, M.; Dutta, K. Saheb, S. Basu, A. (2009). Understanding the molecular mechanism of blood-brain barrier damage in a experimental model f Japanese encephalitis: Correlation with minocucline administration as a therapeutic agent. *Neurochem Int*, Vol.55, No.8, (July 2009), pp. 717-23, ISSN 1872-9754

Misra, U.; Kalita, J. Syam, U. & Dhole, T. (2006). Neurological manifestations of dengue virus infection. *J Neurol Sci*, Vol.244, No.1-2, (March 2006), pp. 117-22, ISSN 0392-0461

Monath, T.; Cropp, C. & Harrison, A. (1983). Mode of entry of a neurotropic arbovirus into the central nervous system: reinvestigation of an old controversy. *Lab Invest*, Vol.48, No.4, (April 1983), pp. 399-410, ISSN 1530- 0307

Mosier, D. (2009). How HIV changes its tropism: evolution and adaptation? *Curr Opin HIV AIDS*, Vol.4, No.2, (March 2009), pp. 125-30, ISSN 1746-6318

Mota, J.; & Rico-Hesse, R. (2009). Humanized mice show clinical signs of dengue fever according to infecting virus genotype. *J Virol*, Vol.83, No.17, (June 2009), pp. 8638–8645, ISSN 1098-5514

Muñoz-Fernández M.; & Fresno, M. (1998). The role of tumor necrosis factor, interleukin 6, interferon gamma, and Inducible nitric oxide synthase in the development and pathology of the nervous system. *Prog Neurobiol*, Vol.56, No.3, (October 1998), pp. 307:40, ISSN 1873-5118

Ogata, A.; Nagashima, K. Hall, W. Ichikawa, M. Kimura-Kuroda, J. & Yasui, K. (1991). Japanese Encephalitis Virus neurotropism is dependent on the degree of neuronal maturity. *J Virol*, Vol.65, No.2, (February 1991), pp. 880-86, ISSN 0022-538X

Pastorino, B.; Nougairède, A. Wurtz, N. Gould, E. & de Lamballerie X. (2010). Role of host cell factors in flaviviurs infection: implications for pathogenesis and development of antiviral drugs. *Antiviral Res*, Vol.87, No3, (May 2010), pp. 281-94, ISSN 1872-9096

Petty, M.; & Lo, E. (2002). Juctional complexes of the blood-brain barrier permeability changes in neuroinflammation. *Prog Neurobiol*, Vol.68, No.5, (December 2002), pp. 311-23, ISSN 1873-5118

Ramos, J.; Imbert, J. Ortega, A. & Ramos, C. (1994). Synaptophysin and neurofilament expression in neurons infected with dengue virus. *Arch Med Res*, Vol.25, No.2, (Summer 1994), pp. 215-7, ISSN 0188-4409

Ramos, J.; Imbert, J. Barrón, B. & Ramos, C. (1997). A 65-KDa trypsin-sensible membrane cell protein as a possible receptor for dengue virus in cultured neuroblastoma cells. *J Neurovirol*, Vol.3, No.6, (December 1997), pp. 435-40, ISSN 1538-2443

Ramos, J.; Sánchez, G. Hernández, R. Baquera, J. Hernández, D. Mota, J. Ramos, J. Flores, A. & Llausás, E. (1998). Dengue virus in the brain of a fatal case of hemorrhagic dengue fever. *J Neurovirol*, Vol.4, No.4, (August 1998), pp. 465-68, ISSN 1538-2443

Rempel, J.; Murray, S. Meisner, J. & Buchmeier, M. (2004). Differential regulation of innate and adaptative immune responses in viral encephalitis. *Virology*, Vol.318, No.1, (January 2004), pp. 381-92, ISSN 1096-0341

Rempel, J.; Quina, A. Blakely-Gonzales, P. Buchmeier, M. & Gruol, D. (2005). Viral induction of central nervous system innate immune responses. *J Virol*, Vol.79, No.7, (April 2005), pp. 4369-81, ISSN 1098-5514

Reyes del Valle, J.; Chávez-Salinas, S. Medina, F. & Del AngelÁngel RM. (2005). Heat shock protein 90 and heat shock protein 70 are components of dengue virus receptor complex in human cells. *J Virol*, Vol.79, No.8, (April 2005), pp. 4557-67, ISSN 1098-5514

Rico-Hesse, R.; Harrison, L. Salas, R. Tovar, D. Nisalak, A. Ramos, C. Boshell, J. de Mesa, M. Nogueira, R. & da Rosa A. (1997). Origins of dengue type 2 viruses associated with increased pathogenicity in the Americas. *Virology*, Vol.230, No.2, (April 1997), pp. 244-251, ISSN 1096-0341

Row, D.; Weinstein, P. & Murray-Smith, S. (1996). Dengue fever with encephalopathy in Australia. *Am J. Trop Med Hyg*, Vol.54, No.3, (March 1996), pp. 253-255, ISSN 1476-1645

Samuel, M. & Diamond, M. (2006). Pathogenesis of West Nile Infection: a Balance between virulence, innate and adaptive immunity, and viral evasion. *J Virol*, Vol.80, No.19, (October 2006), pp. 9349-60, ISSN 1098-5514

Samuel, M.; Morrey, J. & Diamond M. (2007a). Caspase 3 dependent cell death of neurons contributes to the pathogenesis of West Nile virus encephalitis. *J Virol*, Vol.81, No.6, (December 2006), pp. 2614-23, ISSN 1098-5514

Samuel, M.; Wang, H. Siddharthan, V. Morrey, J. & Diamond, M. (2007b). Axonal transport mediates west nile virus entry into the central nervous system and induces acute flaccid paralysis. *Proc Natl Acad Sci U S A*, Vol.104, No.43, (October 2007), pp. 17140-45, ISSN 1091-6490

Sánchez, I.; & Ruiz, B. (1996). A single nucleotide change in the E protein gene of dengue virus 2 Mexican strain affects neurovirulence in mice. *J Gen Virol*, Vol.77, No.10, (October 1996), pp. 2541-45, ISSN 1465-2099

Sánchez-Burgos, G.; Hernández-Pando, R. Campbell, I. Ramos-Castañeda, J. & Ramos C. (2004). Cytokine production in brain of mice experimentally infected with dengue virus. *Neuroreport*, Vol.15, No.1, (Janaury 2004), pp. 37-42, ISSN 1473-558X

Shen, J.; T-To. S. Schrieber, L. & King, N. (1997). Early E-Selecting, VCAM-1, ICAM-1 and late mayor histocompatibility complex antigen induction on human endothelial cells by flavivirus and comodulation of adhesion molecule expression by immune cytokines. *J Virol*, Vol.71, No.12, (December 1997), pp. 9323-32, ISSN 1098-5514

Shresta, S.; Kyle, J. Beatty, P. & Harris, E. (2004). Early activation of natural killer and B cells in responses to primary dengue virus infection in A/J mice. *Virology*, Vol.319, No.2, (February 2004), pp. 262-73, ISSN 1096-0341

Shresta, S.; Sharar, K. Prigozhin, D. Beatty, P. & Harris, E. (2006). Murin model for dengue virus induced lethal diseases with increased vascular permeability. *J Virol*, Vol.80, No.20, (October 2006), pp. 10208-17, ISSN 1098-5514

Shrestha, B.; Gottlieb, D. & Diamond, M. (2003). Infection and injury of neurons by West Nile encephalitis virus. *J Virol*, Vol.77, No.24, (December 2003), pp. 13203-13, ISSN 1098-5514

Silvia, O.; Pantelic, L. Mackenzie, J. Shellam, G. Papadimitriou, J. & Urosevic, N. (2004). Virus spread, tissue inflammation and antiviral response in brains of flavivirus susceptible and resistant mice acutely infected with Murray Valley Encephalitis Virus. *Arch Virol*, Vol.149, No.3, (November 2003), pp. 447-64, ISSN 1432-8798

Solomon, T. (2003). Exotic and emerging viral encephalitides. *Curr opin Neurol*, Vol.16, No.3, (June 2003), pp. 411-18, ISSN 1473-6551

Solomon, T. (2004). Flavivirus encephalitis. *N Engl J Med*, Vol.351, No.4, (July 2004), pp. 370-8, ISSN 1533-4406

Sriurairatna, S.; Bhamarapravati, N. & Phalavadhtana, O. (1973). Dengue virus infection of mice: morphology and morphogenesis of dengue type-2 in suckling mouse neurons. *Infect Immun*, Vol.8, No.6, (December 1973), pp. 1017-28, ISSN 1098-5522

Swarup, V.; Ghosh, J. Duseja, R. Ghosh, S. & Basu, A. (2007). Japanese Encephalitis Virus infection decrease endogenous IL-10 production: Correlation with microglial activation and neuronal death. *Neurosci Letter*, Vol.420, No.2, (May 2007), pp. 144-49, ISSN 1872-7972

Tan, G.; Ng, J. Trasti, S. Schul, W. Yip, G. & Alonso, S. (2010). A Non Mouse adapted dengue virus strain as a new model of severe dengue infection in AG129 mice. *Plos Negl Trop Dis*, Vol.4, No.4, (April 2010), pp. e672, ISSN 1935-2735

Tassaneetrithep, B.; Burgess, T. Granelli-Piperno, A. Trumpfheller, C. Finke, J. Sun, W. Eller, M. Pattanapanyasat, K. Sarasombath, S. Birx, D. Steinman, R. Schlesinger, S. & Marovich, M. (2003). DC-SING (CD209) mediates dengue virus infection of human dendritic cells. *J Exp Med*, Vol.197, No.7, (April 2003), pp. 823-29, ISSN 1540-9538

Tio, P.; Jong, W. & Cardosa, M. (2005). Two-dimensional VOPBA reveals laminin receptor (LAMR1) interaction with dengue virus serotypes 1, 2 and 3. *Virol J*, Vol.2, No.2, (March 2005), pp. 25-36, ISSN 1743-422X

Tsia, J.; Chan, K. Chang, J. Chang K, Lin, C. Huang, J. Lin, W. Chen, J. Hsien, H. Lin, S. Lin, J. Lu, P. Chen, Y. & Lin, C. (2009). Effect of serotype on clinical manifestation of dengue fever in adults. *J Microbiol Immunol Infect* Vol.42, No.6 (December 2009), pp. 471-78, ISSN 1684-1182

Upanan, S.; Kuadkitkan, A. & Smith, D. (2008). Identification of dengue virus binding proteins using affinity chromatography. *J Virol Method*, Vol.151, No.2, (June 2008), pp. 325-8, ISSN 1879-0984

van den Pol, A. (2006). Viral infections in the developing and mature brain. *Trens Neurosci*, Vol.29, No.7, (June 2006), pp. 398-06, ISSN 1878-108X

van der Schaar, H.; Rust, M. Waarts, B. van de Ende-Metselaar, H. Kuhn, R. Wilschut, J. Zhuang, X. & Smit, J. (2007). Characterization of the early events in Dengue Virus cell entry by biochemical assays and single-virus tracking. *J Virol*, Vol.81, No.21, (August 2007), pp. 12019-28, ISSN 1098-5514

van der Schaar, H.; Rust, M. Chen, C. van der Ende-Metselaar, H. Wilschut, J. Zhuang, X. & Smit, J. (2008). Dissecting the cell entry pathway of Dengue Virus by single-particle tracking in living cells. *PLOS Pathogens*, Vol.4, No.12, (December 2008), pp. e1000244, ISSN 1553-7374

Van Marle, G.; Antony, J. Ostermann, H. Dunhanm, C. Hunt, T. Holliday, W. Maingat, F. Urbanoswski, M. Hobman, T. Peeling, J. & Power, C. (2007). West Nile virus induced neuroinflammation: glial infection and capsid protein-mediated neurovirulence. *J Virol*, Vol.420, No.20, (August 2007), pp. 144-149, ISSN 1098-5514

Verna, S.; Lo, Y. Chapagain, M. Lum, S. Kumar, M. Gurjav, U. Luo, H. Nakatsuka, A. & Nerurkar, V. (2009). West Nile Virus infection modulates human brain microvascular endothelial cells tight junctions proteins and cell adhesion molecules transmigration across the in vitro blood brain barrier. *Virology*, Vol.385, No.2, (January 2009), pp. 425-33, ISSN 1096-0341

Wang, T.; Town, T. Alexopoulou, L. Anderson, J. Fikrig, E. & Flavell, R. (2004). Toll-Like receptor 3 mediates west nile virus entry into the brain causing lethal encephalitis. *Nat Med*, Vol.10, No.12, (November 2004), pp, 1366-73, ISSN 1546-170X

Weiner, L.; Cole, G. & Nathanson N. (1970). Experimental encephalitis following Peripherals inoculation of West Nile virus in mice of different ages. *J Hyg (Lond)*, Vol.68, No. 3, (September 1970), pp. 435-46, ISSN 0022-1724

Williams, K.; Zompi, S. Beatty, P. & Harris, E. (2009). A mouse model for studying dengue virus pathogenesis and immune response. Immunology and pathogenesis of viral hemorrhagic fevers. *Ann NY Acad Sci*, Vol.1171, Suppl.1, (September 2009), pp. E12–E23, ISSN 1749-6632

Wilson, C.; Lewis, D. & English, K. (1999). Immunity in the neonate. *Semin Pediatr Infect Dis*, Vol.10, No.2, (April 1999), pp. 83-90, ISSN 1045-1870

Wrona, D. (2006). Neural-immune interactions: an integrative view of the bidirectional relationship between the brain and immune systems. *J Neuroimmunol*, Vol.172, No.1-2, (Janaury 2006), pp. 38-58, ISSN 1872-8421

Wu-Hsieh, B.; Yen, Y. & Chen, H. (2009). Dengue hemorrhage in a Mouse model: Immunology and pathogenesis of viral hemorrhagic fevers. *Ann NY Acad Sci*, Vol.1171, Suppl.1, (September 2009), pp. E42–E47, ISSN 1749- 6632

Yauch, L.; & Sheresta S. Mouse models of dengue virus infection and disease. *Antiviral Res*, Vol.80, No.2, (July 2008), pp. 87–93, ISSN 1872-9096

13

Identification of Aquatic Birnavirus VP3 Death Domain and Its Dynamic Interaction Profiles in Early and Middle Replication Stages in Fish Cells

Jiann-Ruey Hong[1],* and Jen-Leih Wu[2]
*¹Laboratory of Molecular Virology and Biotechnology,
Institute of Biotechnology, National Cheng Kung University*
*²Laboratory of Marine Molecular Biology and Biotechnology,
Institute of Cellular and Organismic Biology, Academia Sinica, Nankang*
Taiwan

1. Introduction

Infectious pancreatic necrosis virus (IPNV) is a fish pathogen and the prototype of the *Birnaviridae* virus family (Dobos et al., 1979). IPNV causes infectious pancreatic necrosis, an acute and serious disease in juvenile salmonid fish worldwide (Hill and Way, 1995) Birnaviruses possess a bi-segmented, double-stranded RNA genome contained within a medium-sized, unenveloped, icosahedral capsid. The protein products of four unrelated major genes undergo various post-translational cleavage processes to generate three to five different structural proteins (Dobos, 1995). The largest of these proteins (VP1; 90–110 kDa) is encoded by the smaller segment B RNA (Wu et al., 1998). The larger genome segment A contains a large open reading frame (ORF; which encodes VP3 [32-kDa; a minor capsid protein] [Wu et al., 1998], VP4 [28 kDa], and VP2 [the major capsid protein; 46 kDa] [Wu et al., 1998]) and a small ORF, which encodes VP5 (17 kDa; a non-structural protein with anti-apoptotic activity) (Hong et al., 2002a; Hong and Wu, 2002c).

Previously, it was found that IPNV-induced apoptosis (Hong et al., 1998; Hong, et al., 1999a; Hong et al., 1999b; Hong and Wu, 2002) may be mediated through activation of caspase-8 and -3 (Hong et al., 2005) and requires new protein synthesis (Hong et al., 1999). This pathway may be triggered through NF-κB transcription factor transactivation of downstream effector genes such as Bad (Hong et al., 2008). Recently, it was found that IPNV-induced loss of $\Delta\Psi m$ can be blocked by the adenine nucleotide translocase (ANT) inhibitor bongkrekic acid (BKA) (Chen et al., 2009), and that IPNV-induced expression of annexin 1 can have an anti-death function (Hwang et al., 2007). IPNV was also found to induce apoptotic cell death and necrotic cell death in the same cells through TNFα up-regulation (Wang et al., 2011).

The structures associated with IPNV replication and genome and particle assembly in infected cells have only started to be elucidated. Recently, viral particles of different sizes

---

*Corresponding Author

were identified during the IPNV infective cycle (Villanueva et al., 2004). Immediately after synthesis, non-infectious, immature particles 66 nm in diameter appeared. These provirion particles were detected simultaneously with viral double-stranded (ds) RNA in infected cells, suggesting that viral assembly occurs as soon as dsRNA replication has begun. Subsequent maturation into smaller virions (60 nm in diameter) was found to proceed through proteolytic cleavage of the viral precursors within the capsid. An early study of IPNV also suggested an association of VP3 with viral RNA. VP3-containing ribonucleoprotein core structures were identified by electron microscopy studies, and an association of the basic C-terminal end of the protein with the viral genome was proposed (Hudson et al., 1986). Furthermore, Pedersen found that VP3 can interact with itself, co-precipitate dsRNA, and interact with VP1 (Pedersen et al., 2007). The dynamics of VP3 interaction with other proteins or RNA are still unknown.

In the present study, we located the death domain of VP3 protein and examined its interactions with other viral molecules culminating in the formation of the VP3-VP1-RNA-VP2 complex during viral replication.

2. Materials and methods

2.1 Cell line and virus

Chinook salmon embryo cells (CHSE-214 obtained from the American Type Culture Collection, ATCC, Manassas, VA, USA) were grown at 18°C in plastic tissue-culture flasks (Nalge Nunc International, Rochester, NY, USA) containing Eagle's minimum essential medium (MEM) supplemented with 10% (v/v) fetal bovine serum (FBS) and gentamycin (25 µg/ml). ZLE cells were grown at 28°C in Leibovitz's L-15 medium (GibcoBRL, Grand Island, NY) supplemented with 5% v/v FBS and 25 µg/ml of gentamycin. An isolate of the Ab strain of IPNV, designated E1-S, was obtained from Japanese eels in Taiwan (Wu et al., 1987). The virus was propagated in CHSE-214 cell monolayers at a multiplicity of infection (MOI) of 0.01 per cell. Infected cultures were monitored as described previously (Dobos, 1977) and Tissue Culture Infectious Dose 50 (TCID$_{50}$) assay was performed on confluent monolayers (Nicholson and Dunn, 1974).

2.2 VP3-enhanced green fluorescent protein (EGFP) gene fusion and different mutant form constructions

A VP3 coding sequence from IPNV E1S VP3 (VP3 cDNA in pGEMT-easy plasmid) was amplified using the sense primer (1-P1; in Table 1) 5'-CCgCTCgAgCCATggACgAggAgCTg CAA-3' (the *Xho* I site is underlined and the start codon of VP3 is in boldface) and an anti-sense primer (1-P2) 5'-CgggATCCATTCACCTCCgCATCTT-3' (the *Bam*HI site is underlined). This allowed for the construction of VP3-EGFP fusion genes to monitor VP3-induced cellular morphological changes. The other VP3 mutant forms (pEGFP-VP3: 1–158, pEGFP-VP3: 80–237, pEGFP-VP3: 80–158, pEGFP-VP3: 1–79, and pEGFP-VP3: 159–237) were amplified using the sense primers (2-P1, 2-P1, 4-P1, 5-P1, and 6-P1; in Table 1) and an anti-sense primer (2-P2, 3-P2, 4-P2, 5-P2, or 6-P2; in Table 1). The *Xho* I and *Bam*HI sites are underlined in the constructions shown in Table 1. PCR products were ligated with the identically predigested pEGFP (enhanced yellow fluorescent protein)-C1 vector (BD Biosciences ClonTech, Palo Alto, CA, USA) after restriction digestion with *Xho* I and *Bam*HI to create pEGFP-VP3. Cell transfection was implemented by seeding 3 x 10^5 ZLE cells in 60-

mm culture dishes one day prior to the transfection procedure. pEGFP, pEGFP-VP3, and each of the VP3 deletion mutant forms (2 μg) were separately transfected into cells using Lipofectamine-Plus (Life Technologies, Inc., Gaithersburg, MD). Expression of the EYFP fusion proteins was visualized using a fluorescence microscope with an FITC filter set, as previously described (Hong et al., 1999b). The number of apoptotic cells containing EGFP or EGFP-VP3 (200 cells per sample) was assessed for three individual experiments. Each point represents the mean number of apoptotic cells ± the standard error of the mean (SEM). Data were analyzed using either the paired or unpaired Student's t-test as appropriate. Statistical significance of between-group differences in mean values was defined at $P < 0.05$.

2.3 Western blot analysis

CHSE-214 cells (4.0 ml) were seeded at 10^5 cells per ml per 60 mm Petri dish for at least 20 h prior to cultivation. The resulting monolayers were rinsed twice with PBS. Cells were infected with virus at an MOI of 1.0 and were incubated for 0, 4, 6, 8, 12, or 24 h. Culture media were aspirated at the end of each time point. Cells were washed with PBS and lysed in 0.3 ml of lysis buffer (10 mM Tris base, 20% glycerol, 10 mM sodium dodecyl sulfate, and 2% β-mercaptoethanol; pH = 6.8). ZLE cells (4.0 ml) were seeded at 1×10^5 cells per ml and cultivated as previously described for VP3 to allow mutant gene overexpression. The cells after 20 h were transfected with pEGFP, pVP3-EGFP, pEGFP-VP3: 1–158, pEGFP-VP3: 80–237, pEGFP-VP3: 80–158, pEGFP-VP3: 1–79, and pEGFP-VP3: 159–237 plasmids (2 μg added to each dish) using Lipofectamine-Plus (Life Technologies). Transfection was allowed to proceed for 4 h. ZLE cells were incubated at 28°C for 0, 24, and 48 h. Culture media were aspirated at the end of each incubation period. The cells were washed with PBS and finally lysed in 0.3 ml of lysis buffer.

Proteins were separated by sodium dodecyl sulfate-polyacrylamide gel electrophoresis (Laemmli, 1970), electroblotted, and visualized using a previously described method (Kain et al., 1994). Blots were incubated first with anti-IPNV E1-S polyclonal antibodies (PolyAb, 1:1500) (Hong et al., 1999a), anti-mouse β-actin MAb (1:5000; Chemicon, Temecula, CA, USA), or anti-mouse EGFP MAb (1:3000, Clontech Laboratories, Mountain View, CA, USA), then with peroxidase-labeled goat anti-mouse conjugate (1:7500 to 1:10,000; Amersham, Piscataway, NJ, USA) or peroxidase-labeled goat anti-rabbit conjugate (1:7500 to 1:10,000; Amersham).

Chemiluminescence detection was performed using the Western Exposure Chemiluminescence Kit (Amersham) according to the manufacturer's instructions. Chemiluminescence was visualized on Kodak XAR-5 film (Eastman Kodak, Rochester, NY, USA), and protein expression level amounts were quantified using a Personal Densitometer (Molecular Dynamic, Sunnyvale, CA).

2.4 Radioactive labeling of infected cells for determination of VP3 interactions

Radioactive labeling of infected cells: A confluent CHSE-214 cell monolayer in a plastic tissue culture plate (60 mm diameter, Nunc) was infected with IPN virus at 18°C at an MOI of 1. After a 1-h adsorption period, the medium was removed, the monolayer was rinsed three times with phosphate-buffered saline (PBS), and starved for 2 h at 18°C. Then, growth medium containing 50 μCi ml^{-1} of [^{35}S]methionine was added. At different times in the labelling period (3–20 h) or continuously labelled for 20 h as a positive control, the cell layer was washed with PBS, and the monolayer was lysed in 1 ml of lysis buffer (0.5% [v/v] Nonidet P-40 in phosphate-buffered saline) (Wu et al., 1998).

VP3 interaction assay: Cell extracts were incubated for 18 h at 4°C in the presence of 0.1% (w/v) *N*-laurolysarcosine, 1% bovine serum albumin, 10 M phenylmethanesulfonyl fluoride (PMSF), and anti-mouse E1S VP3 B9 and E7 monoclonal antibodies (home-made). The mixture was incubated with anti-rabbit IgG serving as a negative control) 1 µg/ ml each antibody; Transduction Laboratories, Lexington, KY) at 4°C for 12 h, and then with 160 µl of 10% (v/v) protein A-Sepharose (Zyme) in 150 mM NaCl/10 mM Tris-HCl, pH 7.8/1% *N*-lauroylsarcosine at 4°C for 1 h. The protein A-Sepharose-antibody complexes so formed were washed three times in the same buffer, boiled in 200 µl of 10 mM Tris-HCl, pH 7.8/1 mM EDTA/0.1% SDS (Hong and Wu. 2002), and boiled in 60 µl of sodium dodecyl sulfate (SDS)-polyacrylamide gel electrophoresis (PAGE) sample buffer containing 5% SDS and 4% β-mercaptoethanol to release bound proteins. Proteins present in the cell lysate were separated by SDS-polyacrylamide gel electrophoresis (Laemmli, 1970), and the dried gel (dried under vacuum) (Studier, 1973) was autoradiographed.

3. Results

3.1 VP3 expressed early in the replication cycle

The present study confirmed that a minor viral capsid protein VP3 is expressed early in the IPNV replication cycle (Wu et al., 1987; Wu et al., 1998). VP3 expression levels increased 5-fold between 4 and 6 h post-infection (p.i.) (Fig. 1A, lanes 2 and 3) compared to levels in mock-infected normal control cells (Fig. 1A, lane 1). The elevated expression was maintained for 8–24 h p.i. (Fig. 1A, lanes 4–7).

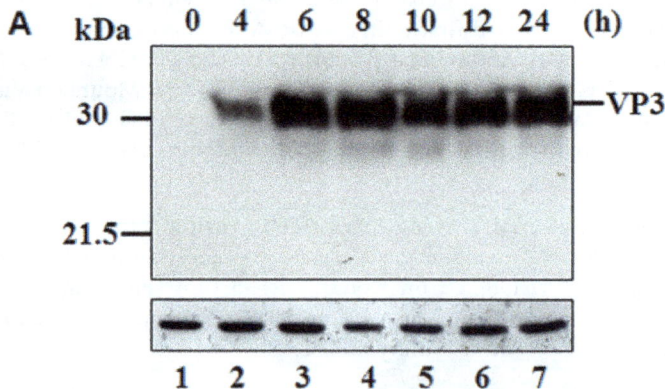

Fig. 1. IPNV VP3 protein expression in CHSE-214 cells.
The pattern of VP3 expression in CHSE-214 cells was revealed using a anti-IPNV E1-S particle polyclonal antibody. The cells were infected with IPNV (MOI=1), lysed at different times p.i., and the proteins in lysates were separated electrophoretically. Lane 1 corresponds to IPNV-uninfected cells (0 h; the control). Lanes 2–7 correspond to IPNV-infected cells lysed 4, 6, 8, 10, 12, and 24 h p.i., respectively.

3.2 VP3 localization

Green fluorescent protein (GFP) from the jellyfish *Aequorea victoria* has become an invaluable
tool for monitoring gene expression and fusion protein localization *in vivo* or *in situ*, as well
as for determination of cellular morphological changes (Chalfie et al., 1994; Hong et al.,
1999b; Maniak et al., 1995; Oparka et al., 1997). A variant type of GFP (EGFP), a marker for
visualization of apoptotic cell morphological changes (Hong et al., 1999b), was used in the
present study to trace VP3 localization and monitor interaction with other proteins.

The tracking of VP3 with EGFP in ZLE cells has revealed their presence in plasma (Espinoza
et al., 2000). However, VP3 was absent from endoplasmic reticulum (ER) when tracked with
an ER marker (data not shown).

3.3 Identification of the VP3 death-inducing domain

Recently, we demonstrated that IPNV VP3 induces apoptotic cell death through the up-
regulation of the pro-apoptotic effector Bad, which triggers the mitochondria-mediated cell
death pathway (Chiu et al., 2010). A deletion series was used to determine which VP3 region
might be associated with death induction. Different VP3 constructs were designed (Fig. 2A;
Table 1) and inserted into pEGFP to produce pEGFP-VP3 (VP3 full length, 1–237 aa),
pEGFP-VP3: 1–158, pEGFP-VP3: 80–237, pEGFP-VP3: 80–158, pEGFP-VP3: 1–79, and
pEGFP-VP3: 159–237. All were amplified for further identification. Western blotting with
anti-EGFP monoclonal antibody confirmed the different sizes of EGFP (lane 1), EGFP-VP3
(lane 2), EGFP-VP3: 1–158 (lane 3), EGFP-VP3: 80–237 (lane 4), EGFP-VP3: 80–158 (lane 5),
EGFP-VP3: 1–79 (lane 6), and EGFP-VP3: 159–237 (lane 7) transiently expressed 48 h p.i.
(Fig. 2B).

Construction of vectors	Primer sequences
pEGFP-VP3	1-P1: 5'-CCG**CTCGAG**CCATGGACGAGGAGCTGCAA-3' 1-P2: 5'-CG**GGATCC**ATTCACCTCCGCATCTT-3'
pEGFP-VP3:1-158	2-P1: 5'-CCG**CTCGAG**CCATGGACGAGGAGCTGCAA-3' 2-P2: 5'-CG**GGATCC**GCCGTACACACTGTTGG-3'
pEGFP-VP3:80-237	3-P1: 5'-CCG**CTCGAG**CCGAGGCGAGAGCCGTTCGCATCT-3' 3-P2: 5'-CG**GGATCC**ATTCACCTCCGCATCTT-3'
pEGFP-VP3:80-158	4-P1: 5'-CCG**CTCGAG**CCGAGGCGAGAGCCGTTCGCATCT-3' 4-P2: 5'-CG**GGATCC**GCCGTACACACTGTTGG-3'
pEGFP-VP3:1-79	5-P1: 5'-CCG**CTCGAG**CCATGGACGAGGAGCTGCAA-3' 5-P2: 5'-CG**GGATCC**TTGGGCGGCATGCTGGTT-3'
pEGFP-VP3:159-237	6-P1: 5'-CCG**CTCGAG**CCCTCCCGCACCAGGAACCAGCC-3' 6-P2: 5'-CG**GGATCC**ATTCACCTCCGCATCTT-3'

Notes: P1, designed as for forward primer and P2 as for reverse primer.

Table 1. The primers were used to VP3 and different mutant constructions of EGFP-VP3
fusion genes in plasmids.

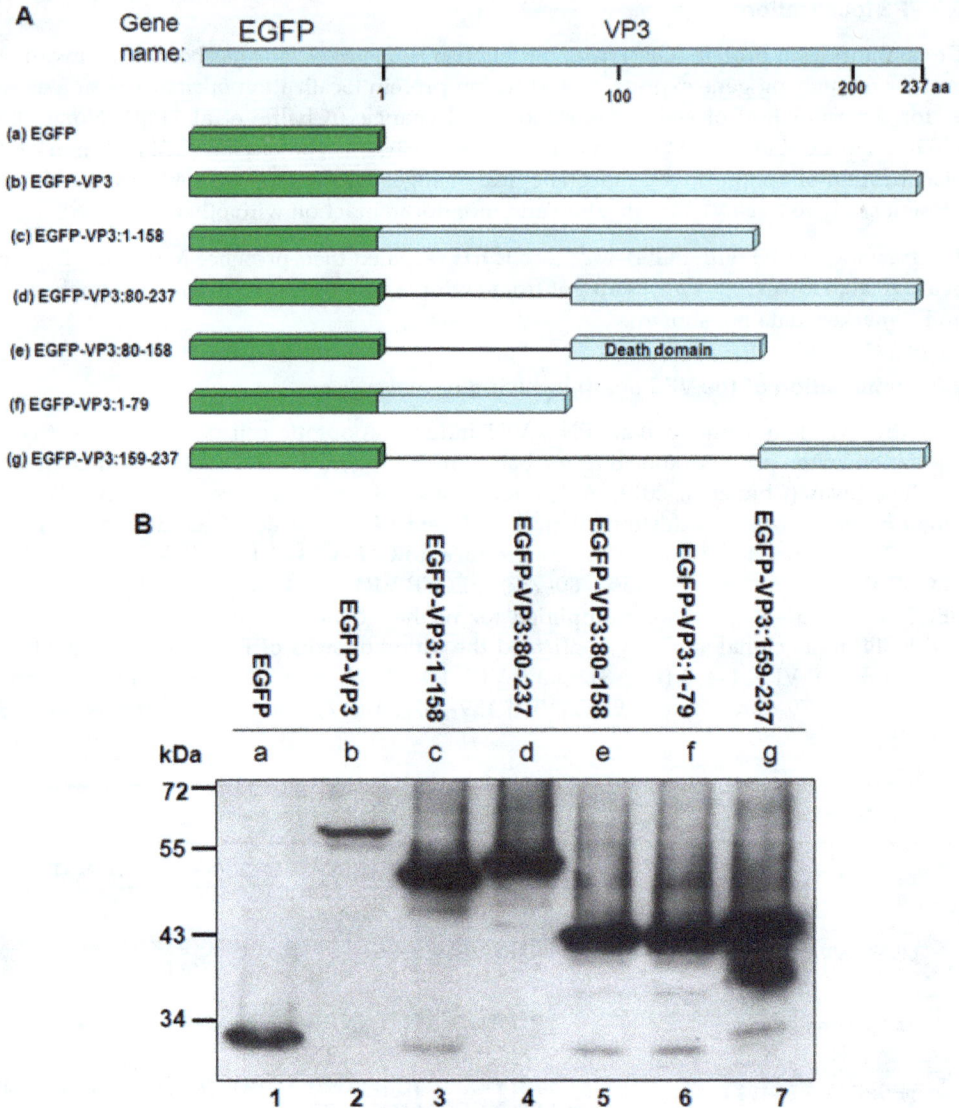

Fig. 2. Mapping and identification of the IPNV VP3 death-inducing domain (80–158 aa) in fish cells. (A) Schematic representation of various IPNV VP3 deletion mutant constructs N-terminally fused with EGFP and transfected into ZLE cells. (B) Identification of EGFP, EGFP-VP3, and EGFP-VP3 deletion mutants. The pEGFP, pEGFP-VP3, and different pEGFP-VP3 mutants were transfected into ZLE cells with Lipofectamine-Plus. The cells were incubated for 48 h post-transfection (lane 1, EGFP; lane 2, EGFP-VP3; lane 3, EGFP-VP3: 1–158; lane 4, EGFP-VP3: 80–237; lane 5, EGFP-VP3: 80–158; lane 6, EGFP-VP3: 1–79, and lane 7, EGFP-VP3: 159–237) and lysed. The lysates were analyzed by Western blotting using a monoclonal antibody to EGFP.

EGFP served as a tracer to allow direct visualization of apoptotic changes in ZLE cells overexpressing VP3 and VP3 peptides at 48 h p.t. The site of the death domain was found to be in a 78-aa region between aa 80 and aa 157 (Fig. 3A). Compared to EGFP-VP3 fusion proteins without this region, those containing this domain induced a higher rate of apoptosis (Fig. 3B; see phase-contrast micrographs in panels b [EGFP-VP3], c [EGFP-VP3: 1–158], d [EGFP-VP3: 80–237], and i [EGFP-VP3: 80–158] and green fluorescence micrographs in panels f [EGFP-VP3], g [EGFP-VP3: 1–158], h [EGFP-VP3: 80–237], and l [EGFP-VP3: 80–158; arrows indicate EGFP-positive cells]). By contrast, the rate of apoptosis was much less in cells expressing EGFP, EGFP-VP3: 1–79, and EGFP-VP3: 159–237 (panels c [EGFP], m [EGFP-VP3: 1–79], and f [EGFP-VP3: 159–237]).

As shown in Figure 3B, these rates at 48 h p.i. were higher in the presence of the death domain (69% [EGFP-VP3], 36% [EGFP-VP3: 1–158], 49% [EGFP-VP3: 80–237], and 52% [EGFP-VP3: 80–158]) than in its absence (3% [EGFP], 18% [EGFP-VP3: 1–79], and 12% [EGFP-VP3: 159–237]).

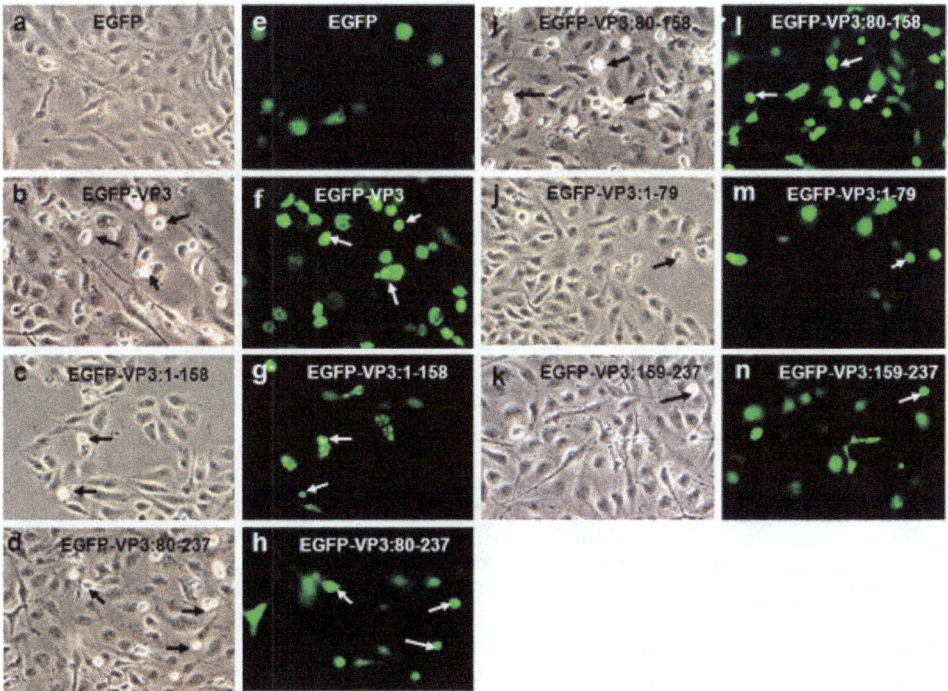

A

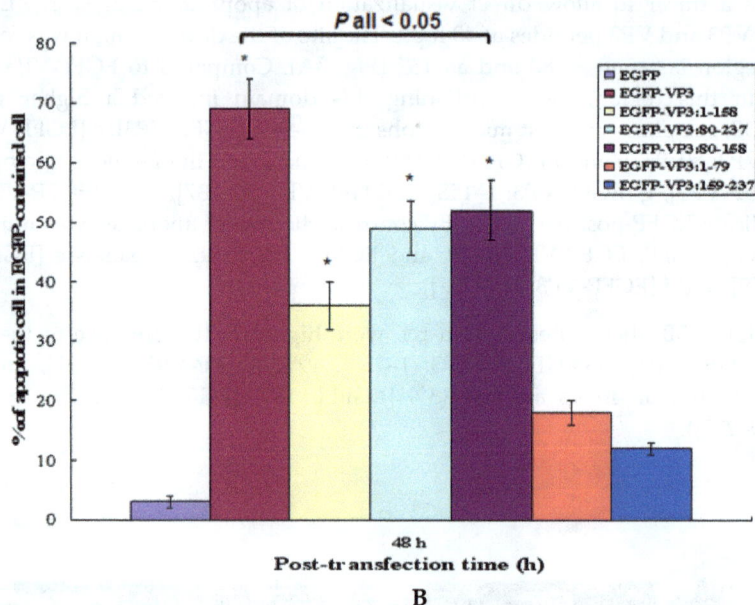

P all < 0.05

Legend:
- EGFP
- EGFP-VP3
- EGFP-VP3:1-158
- EGFP-VP3:80-237
- EGFP-VP3:80-158
- EGFP-VP3:1-79
- EGFP-VP3:159-237

y-axis: % of apoptotic cell in EGFP-contained cell

x-axis: 48 h
Post-transfection time (h)

B

Fig. 3. Identification of the death domain of VP3 in ZLE cells.
Expression of EGFP, EGFP-VP3, and EGFP-VP3 deletion mutants was assessed in ZLE cells. The pEGFP, pEGFP-VP3, and different pEGFP-VP3 mutants were transfected into ZLE cells with Lipofectamine-Plus, and the cells were incubated for 48 h post-transfection (p.t.). (A) Phase-contrast and fluorescence micrographs of transfected and untransfected apoptotic ZLE cells at 48 h p.t. Rounded up cells and plasma membrane blebbing are indicated by arrows (Bar = 10 μm). (B) The percentage of apoptotic cells containing EGFP, EGFP-VP3, or the different EGFP-VP3 deletion mutant constructs at 48 h p.t. The number of apoptotic cells per 200 cells per sample was assessed. Each point represents the mean of three independent experiments and the vertical bars indicate ± the standard error of the mean (SEM). Data were analyzed using either paired or unpaired Student's t-test as appropriate. Statistical significance was defined at $P < 0.05$.

3.4 Kinetics of VP3 interaction during IPNV infection

Although VP3 participated in many viral processes (apoptosis induction, binding to viral RNA, binding to RdRp, and self interaction), it did not associate directly with capsid protein VP2. So, we were interested in the mechanism of interaction between VP3 protein and other proteins. The binding of native VP3 to two VP3 monoclonal antibodies was assessed at 0, 2.5, 4, 6, 8, and 24 h p.i. (Fig. 4). VP3 protein was found to bind to both anti-VP3 B9 (lanes, 15–20) and anti-VP3 E7 (lanes, 21–26) Mabs at 6, 8, and 24 h p.i. but not to the negative controls protein A (lanes, 3–8) and secondary anti-rabbit Ig (lanes, 9–14). The positive control ([35S]methionine continuously labelled for 20 h) is shown in lane 2 and protein markers labelled with [35S]methionine are shown in lane 1.   Furthermore, we found just only anti-VP3 E7 that at 6 h p.i., VP3 associated with VP1, p85 protein, and RNA to form a complex capable of binding VP2 protein during the period 8–24 h p.i (Fig. 5), indicating that viral particle assembly begins as early as 6 h p.i.

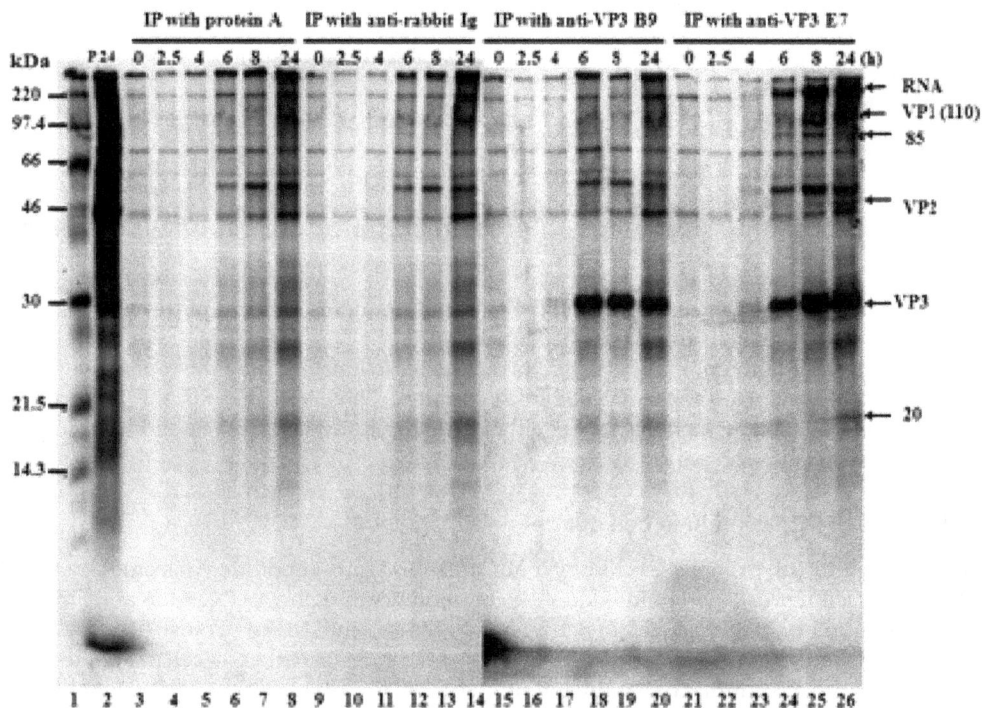

Fig. 4. Identification of profiles of IPNV-VP3 protein interaction with anti-VP3 monoclonal
antibody at different replication stages in CHSE-214 cells.
The non-infected and IPNV-infected cells were labelled with [^{35}S]methionine at different
times after infection and lysed. The lysates were immunoprecipitated with protein A
Sepharose (Zyme) coupled with anti-rabbit Ig, and anti-IPNV-VP3 B9 and anti-IPNV-VP3 E7
monoclonal antibodies, and analyzed on 12% SDS-PAGE. Lane 1, [^{35}S]methionine protein
markers; lane 2, lysate of IPNV-infected CHSE-214 cells labelled with [^{35}S]methionine for 20
h; lanes 3–8, [^{35}S]methionine-labelled CHSE-214 cells lysed and reacted with protein A
Sepharose (Zyme) at 0, 2.5, 4, 6, 8, and 24 h p.i., respectively; lanes 9–14, [^{35}S]methionine-
labelled CHSE-214 cells lysed and reacted with protein A Sepharose (Zyme) coupled to anti-
rabbit Ig (Zyme) at 0, 2.5, 4, 6, 8, and 24 h p.i., respectively; lanes 15–20, [^{35}S]methionine-
labelled CHSE-214 cells lysed and reacted with protein A Sepharose (Zyme) coupled to anti-
IPNV-VP3 B9 (home-made) at 0, 2.5, 4, 6, 8 and 24 h p.i., respectively; lanes 21–26,
[^{35}S]methionine-labelled CHSE-214 cells lysed and reacted with protein A Sepharose (Zyme)
coupled to anti-IPNV-VP3 E7 (home-made) at 0, 2.5, 4, 6, 8 and 24 h p.i., respectively.

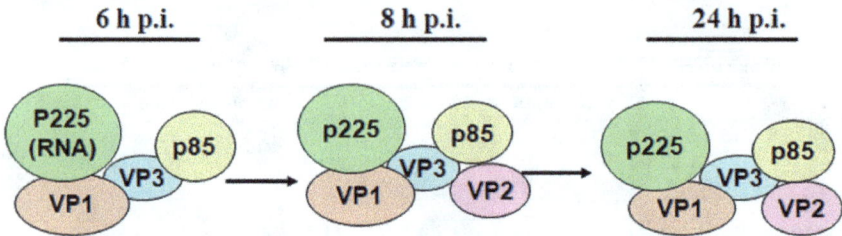

The formation of VP3 interaction protein complex during IPNV infection

Fig. 5. The interaction between VP3 and other viral components and complexes in IPNV-infected CHSE-214 cells during the replication cycle.

Schematic representation of the IPNV VP3 protein interactions. Early in the replicative cycle (at 6 h p.i.), VP3 interacted with VP1-p85-RNA complex. Then, VP1-p85-RNA-VP3 recruited VP2 at 8 h and 24 h p.i., which may affect either viral replication or viral assembly during the mid to late replication period.

3.5 The role of VP3 protein in cell death

The Bcl-2 family of proteins, including both anti- and pro-apoptotic molecules, act at a critical, intracellular decision point along the common death pathway (Newton and Strasser, 1998). The ratio of antagonist (Bcl-2, Bcl-x_L, Mcl-1, Bcl-W, and A1) to agonist (Bax, Bak, Bcl-x_s, Bid, Bik, Bad, PUMA, and NOXA) molecules dictates whether a cell responds to a proximal apoptotic stimulus (Newton and Strasser, 1998; Galluzzi et al., 2008). Homologues of Bcl-2 (Bcl-xL, Bcl-W, Mcl-1, and A1) reside in mitochondria and stabilize the barrier function of mitochondrial membranes. In contrast, pro-apoptotic proteins can shuttle between non-mitochondrial locations (the cytosol for Bax, Bad, and Bid) and mitochondrial membranes where they can insert and permeabilize the mitochondrial membrane (Zamzami and Kroemer, 2001; Galluzzi et al., 2008). In our system, the Bcl-2 family member, zebrafish Bcl-xL, was found to play a role in blocking both apoptotic (Yang et al., 1995) and necrotic cell death. Recently, VP3 overexpression was found to up-regulate Bad protein (a pro-apoptotic Bcl-2 family member), but how VP3 overexpression was induced remained unknown. Moreover, the interactions of the death domain (80–158 aa) of VP3 protein with other molecules also remained unknown. From the recently published literature, we found that VP3 protein contains a potential phosphokinase C phosphorylation site in Domain 2 corresponding to residues 122–124 (Domain A2; TGR) (Chiu et al., 2010), which could have an important role in cell death induction.

3.6 Novel role for VP3 in the replication cycle

Viruses infect specific target cells, replicate in them to produce large numbers of progeny virions, which then spread to other susceptible cells to initiate new rounds of infection. They also encode proteins that are highly efficient for the optimization of such replication.

However, target organisms use both systemic and cell-based defence mechanisms to limit the extent of viral infection, including immune and inflammatory processes and the execution or suicide of infected cells (Benedict et al., 2002). When we investigated the involvement of viral protein (VP3) in host cell death later in the viral replication cycle (VP3, apoptosis), we found that VP3 interacts with VP1, p85, and RNA initially (Fig. 4, lane 24; Fig. 5), then the VP1-p85-RNA-VP3 complex recruits VP2 at 8 h (Fig. 4, lane 25; Fig. 5) and 24 h (Fig. 4, lane 26; Fig. 5), suggesting that VP3 protein performs many functions. For example, VP3 may stimulate VP1 expression and prime viral assembly. In the present study, VP2 interacted with VP3 protein and a new protein p85 (which may have a role in viral replication), but the role of p85 remains to be elucidated.

4. Conclusion

IPNV E1-S is a fish pathogen of the IPNV Ab strain. It induces apoptotic cell death in CHSE-214 cells (Hong et al., 1998; Hong et al., 1999a; Hong et al., 1999b) and zebrafish ZLE cells (Hong et al., 2005). In summary, we provided evidence that IPNV minor capsid proteins were specifically involved in the induction of necrotic cell death. The death domain of VP3 was found to be located within the stretch 80–158 aa, which contains a protein kinase C phosphorylation site. We also found that the VP3 protein can interact with VP1-p85-RNA complex early in the replicative cycle (at 6 h p.i.), and that VP3-VP1-p85-RNA binds VP2 at 8 h and 24 h p.i. thereby affecting either viral replication or viral assembly. Our study adds important new information concerning the IPNV VP3-host cell interaction and provides the basis for study of the viral pathogenesis of VP3-mediated necrotic cell death and viral assembly.

5. Summary

Aquatic birnavirus induces secondary necrotic cell death through the synthesis of new protein. Very recently we found that the viral genome-encoded minor capsid protein VP3 can induce cell death in fish and mouse cells. In the present study, we identified the death domain of VP3 and the role of VP3 in processes such as interaction with VP1 and viral assembly during late stages of replication. Aquatic birnavirus-encoded *VP3* was mildly expressed in CHSE-214 cells at 4 h post-infection (p.i.), but its expression increased up to 3.5- to 4-fold by 6 h p.i. Furthermore, using a deletion series, the *VP3* death domain was localized to a 78-amino acid (aa) segment (80–158 aa), which was separated from the VP3 self-binding domain (1–101 aa) and VP1 binding domain (the so-called RNA-dependent RNA polymerase, RdRp binding domain; 171–236 aa). Using two anti-VP3 monoclonal antibodies, VP3 was also found to interact with VP1, VP2, viral RNA, and host protein-85. Our results suggested that aquatic birnavirus VP3 not only triggers Bad-mediated cell death, but also stabilizes viral RNA, and promotes viral particle assembly. Thus, VP3 may be a good target for antiviral drug-therapy.

6. Acknowledgments

This work was supported by grants from the National Science Council, Taiwan, Republic of China awarded to Dr. Jainn-Ruey Hong (NSC 96-2313-B-004-MY3 and NSC 99-2321-B-006-010-MY3).

7. References

Benedict, C. A., Norris P. S. and Ware C. F. (2002) To kill or be killed: viral evasion of apoptosis. Nature Immunol. 3:1013-1018.

Chalfie, M., Tu Y., Euskirchen G., Ward W. W. and Prasher D. C. (1994) Green fluorescent protein as a marker for gene expression. Science 263:802-805.

Chen P. C., Wu J. L., Her G. M. and Hong J. R. (2009) Aquatic birnavirus induces necrotic cell death via mitochondria-mediated caspases pathway that inhibited by bongkrekic acid. Fish Shellfish Immunol. 28:344-353.

Chiu C. L., Wu J. L., Her G. M., Chou Y. L. and Hong J. R. (2010) Aquatic birnavirus capsid protein, VP3, induces apoptosis via the Bad-mediated mitochondria pathway in fish and mouse cells. Apoptosis 15(6):653-668.

Dobos, P. (1977). Virus-specific protein synthesis in cells infected by infectious pancreatic necrosis virus. J. Virol. 21:242-258.

Dobos, P., Hill B. J., Hallett R., Kells D. T. C., Becht H. and Tenings D. (1979) Biophysical and biochemical characterization of five animal viruses with bisemented double-stranded RNA genomes. J. Virol. 32:593-605.

Dobos, P. (1995) The molecular biology of infectious pancreatic necrosis virus (IPNV). Annu. Rev. Fish Dis. 5:25-54.

Espinoza, J. C., Hjalmarsson A., Everitt E. and Kuznar J. (2000) Temporal and subcellular localization of infectious pancreatic necrosis virus structural proteins. Arch. Virol. 145:739-748.

Galluzzi L., Brenner C., Morselli E., Touat Z. & Kroemer G. (2008) Viral control of mitochondria apoptosis. PLoS Path 4:1-15.

Hill, B. J. and Way, K. (1995) Serological classification of infectious pancreatic necrosis (IPN) virus and other aquatic birnaviruses. Annu. Rev. Fish Dis. 5:55-77.

Hong, J. R., Lin T. L., Hsu Y. L. and Wu J. L. (1998) Apoptosis precedes necrosis of fish cell line with infectious pancreatic necrosis virus infection. Virology 250:76-84.

Hong, J. R., Hsu Y. L. and Wu J. L. (1999a) Infectious pancreatic necrosis virus induces apoptosis due to down-regulation of survival factor MCL-1 protein expression in a fish cell Virus Res. 63:75-83.

Hong, J. R., Lin T. L., Yang J. Y., Hsu Y. L. and Wu J. L. (1999b) Dynamics of nontypical apoptotic morphological changes visualized by green fluorescent protein in living cells with infectious pancreatic necrosis virus infection. J. Virol. 73:5056-5063.

Hong, J. R. and Wu J. L. (2002a) Molecular regulation of cellular apoptosis by fish infectious pancreatic necrosis virus (IPNV) infection. Curr. Top. Virol. 2:151-160.

Hong, J. R. and Wu J. L. (2002b) Induction of apoptotic death in cells via bad gene expression by infectious pancreatic necrosis virus infection. Cell Death Differ. 9:113-124.

Hong, J. R., Gong H. Y. and Wu J. L. 2002c. IPNV VP5, a novel anti-apoptosis gene of the Bcl-2 family, regulates Mcl-1 and viral protein expression. Virology 295:217-229.

Hong, J. R., Huang L. J. andWu J. L. (2005) Aquatic birnavirus induces apoptosis through activated caspase-8 and -3 in a zebrafish cell line. J. Fish Dis. 28:133-140.

Hong, J. R., Guan B. J., Her G. M., Evensen O., Santi N. and Wu L. L. (2008) Aquatic birnavirus infectious activates the transcription factor NF-kB via tyrosine kinase signaling leading to cell death. J. Fish Dis. 31:451-460.

Hudson P. J, McKern N. M, Power B. E, and Azad A. A. (1986) Genomic structure of the large RNA segment of infectious bursal disease virus. Nucleic Acids Res. 14:5001-5012.

Hwang H. J., Moon C. H., Kim J. Y., Lee J. M., Park J. W. and Chung D. K. (2007) Identification and functional analysis of salmon annexin 1 induced by a virus infection in a fish cell line. J. Virol. 81:13816-24.

Kain, S. R., Mai K. and Sinai P. (1994) Human multiple tissue Western blots: a new immunological tool for the analysis of tissue-specific protein expression. BioTechniques 17:982-987.

Laemmli, U. K. (1970) Cleavage of structural proteins during the assembly of the head of bacteriophage T4. Nature 227:680-685.

Maniak, M., Rauchenberger R., Albrecht R., Murphy J. and Gerisch G. (1995) Coronin involved in phagocytosis: dynamics of particle-induced relocalization visualized by a green fluorescent protein tag. Cell 83:915-924.

Newton, K. and Strasser A. (1998) The Bcl-2 family and cell death regulation. Curr. Opin. Genet. Dev. 8:68-75.

Nicholson, B. and Dunn J. (1974) Homologous viral interference in trout and Atlantic salmon cell cultures infected with infectious pancreatic necrosis. J. Virol. 14:180-182.

Oparka, K. J., Roberts A. G., Santa-Cruz S., Boevink P., M. Prior D. A. and Smallcobe A. (1997) Using GFP to study virus invasion and spread in plant tissues. Nature 388:401-402.

Pedersen T., Skjesol A. and Jorgensen J. A. (2007) VP3, a structural protein of infectious pancreatic necrosis virus, interacts with RNA-dependent polymerase VP1 and with double-stranded RNA. J Virol. 81:6652-6663.

Studier F. W. (1973) Analysis of bacteriophage T7 early RNA's and proteins in slab gels. Journal of Molecular Biology 79:237-248

Wang W. L., Hong J. R., Lin G. H., Liu W. T., Gong H. Y., Lu M. W., Lin C. C. and Wu J. L. (2011) Stage-Specific Expression of TNFα Regulates Bad/Bid-Mediated Apoptosis and RIP1/ROS-Mediated Secondary Necrosis in Birnavirus-Infected Fish Cells. PloS ONE 6: e16740.

Wu, J. L., Chang C. Y. and Hsu Y. L. (1987) Characteristics of an infectious pancreatic necrosis like virus isolated from Japanese eel (Anguilla japonina). Bull. Inst. Zool. Acad. Sinica 26:201-214.

Wu, J. L., Hong J. R., Chang C. Y., Hui C. F., Liao C. F. and Hsu Y. L. (1998) Involvement of serine proteinase in infectious pancreatic necrosis virus capsid protein maturation and NS proteinase cleavage in CHSE-214 cells. J. Fish Dis. 21:215-220.

Villanueva R. A, Galaz J. L, Valdes J. A, Jashes M. M. and Sandina A. M. (2004) Genome assembly and maturation of the birnavirus infectious pancreatic necrosis virus. J. Virol. 78:13829-13838.

Yang, E., Zha J., Jokel J., Boise L. H., Thompson C. B. and Korsmeyer S. J. (1995) Bad, a heterodimeric partner for Bcl-XL and Bcl-2, displaces Bax and promotes cell death. Cell 80:285-291.

Zamzami, N. and Kroemer G. (2001) The mitochondria in apoptosis: how Pandora's box opens. Nat. Rev. Mol. Cell Biol. 2:67-71.

14

Molecular Virology and Pathogenicity of *Citrus tristeza virus*

Maria R. Albiach-Marti

Instituto Valenciano de Investigaciones Agrarias, Valencia

Spain

1. Introduction

Citrus tristeza virus (CTV) (genus *Closterovirus*, family *Closteroviridae*) is the largest identified RNA virus infecting plants and the second largest worldwide after the animal *Coronaviruses*. Unlike from most elongated viruses, CTV particles are bipolar flexuous helicoidal filaments of 2000 x 11 nm, having two different capsid proteins that coat the opposite ends of the virions (Febres et al., 1996; Kitajima et al., 1964; Satyanarayana; et al., 2004). The viral genome consists of a long single-stranded positive-sense RNA molecule (gRNA) of around 19.3 kb (Karasev et al., 1995), a size that defies the theoretical predictions on the upper limit of RNA size found in nature, based on the high error frequency of RNA polymerases (Domingo & Holland, 1997). The viral genome encodes twelve open reading frames (ORFs), which potentially express at least nineteen protein products, and contains two non-translated regions (NTR) at the 5´ and 3´ terminus (Figure 1) (Karasev et al., 1995). CTV infects most species, cultivars and hybrids of *Citrus* and related genera, and it is transmitted vegetatively (via infected budwood) and by aphids (*Hemiptera: Aphididae*) in a semipersistent manner. While aphid transmission is responsible of local spread, CTV dispersal to new areas or countries occurs by graft propagation of virus-infected plant tissues (Bar-Joseph et al., 1989).

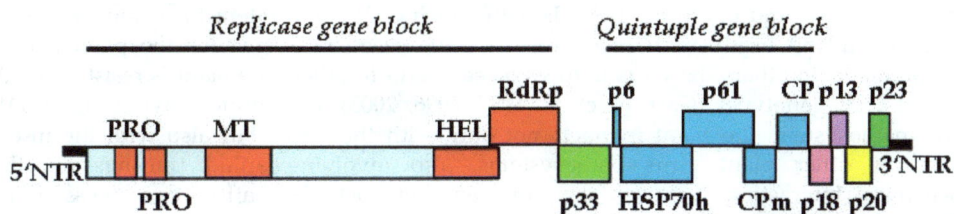

Fig. 1. Organization of CTV genome. PRO, MT, HEL and RdRp indicate protein domains of papain-like protease, methyltranferase, helicase and RNA-dependent RNA polymerase, respectively. HSP70h, CPm and CP indicate ORFs encoding a homologue of heat shock protein 70, the minor and the major capsid proteins, respectively. NTR indicates non-translated region.

Citrus and related genera were originated 20 million years ago in South Eastern Asia and the Malayan archipelago (Scora, 1988). CTV probably co-evolved with its host plant during centuries. There are at least four *Citrus* progenitors reported to be the origin of all of the current citrus varieties in agriculture: citron (*C. medica* L.), pummelo [*C. grandis* (L.) Osb.], mandarin (*C. reticulata* Blanco) and *C. micrantha* L. (Nicolosi et al., 2000). The fact that many CTV isolates are symptomless in most of these citrus species supports this hypothesis (Moreno et al., 2008). The CTV-citrus pathosystem was created by mankind with the advent of the commercial citrus industries during the last two hundred years (Bar-Joseph et al., 1989). Since transport of citrus plants was difficult for centuries, they were transported and grown from seeds, which were free of CTV. Improvements in maritime transport during the XIX and XX centuries enabled the movement of intact citrus plants, which often contained different CTV genotypes, from Asia to other regions, where the virus interacted and evolved in different environmental conditions. The establishment of the modern commercial citrus industries also brought the use of new varieties and cultivars, along with new combinations citrus variety/rootstocks (which is the way as the commercial citrus trees are generally grown in the field), to increase the fruit productivity. After two hundred years, both mankind and aphid transmission have made this virus endemic in most of the citrus-growing areas, with only a few places in the Mediterranean basin and Western USA free of CTV infections. The spreading of the virus to new citrus areas resulted in the death of one hundred million trees. The actual economic damages to the citrus industries worldwide depend on the environmental conditions, the resident CTV pathotypes and the sensitivity to CTV infection of the local citrus varieties or scion-rootstock combinations. The endemic CTV population of some citrus areas is composed by mild or severe, but controllable, strains. These commercial citrus industries are continuously threatened by the possible introduction of exotic and almost uncontrollable CTV isolates of higher virulence (Bar-Joseph & Dawson, 2008). In order to avoid risks derived from introduction and dispersal of highly virulent isolates, methods are needed to discriminate the indigenous from any newly introduced severe isolates. Consequently, there is a considerable interest in mapping CTV disease determinants as well as studying their pathogenicity mechanisms.

In order to infect a plant, the virus first needs to enter in the cell and to overcome the constitutive and/or inducible plant defences to program the plant cellular machinery for its viral multiplication, followed by the systemic invasion of the plant, thus interacting with the host and inducing disease (Culver & Padmanabhan, 2007). *Citrus* genus contains many species, varieties, and intergenic hybrids with which CTV could interact causing a range of physiological and biochemical responses. In some cases, CTV invades the plant and is asymptomatic, in others the virus induces disease, and in others the plant is resistant to all or some CTV genotypes (Garnsey et al., 1987, 1996, 2005). Furthermore, within the CTV-*Citrus* pathosystem, the plant interacts not only with the virus, but also with the insect vector. In other plant virus pathosystems, also involving aphid transmission, the biochemical interactions between the virus and its host plant affect the fitness of its arthropod vectors, therefore modulating their own spread and connecting pathogenicity with effective viral transmission (Fereres & Moreno, 2009). In order to develop methods to discriminate between pathotypes (severe, disease-causing isolates and mild or symptomless isolates) and to engineer reliable and enduring biotechnological strategies to control the diseases induced by CTV, it is essential to understand the processes that occur during CTV-citrus interactions and that connect CTV-infected plants and insect vector interactions, which lead to symptoms development and viral spread.

The genetic research on CTV was hindered for a long time due to the difficulties for experimenting with a virus with a large RNA genome, encapsidated in fragile particles and present in reduced amounts in a tree host, thus difficult to isolate and characterize. Afterwards, the elevated complexity of CTV genetics, aphid transmission and viral populations of highly divergent genotypes, and the myriad of phenotypes induced depending on the *Citrus* host and the CTV strain, challenged the study of the viral genetics and the virus-host interactions (Bar-Joseph & Dawson, 2008; Moreno et al., 2008). In the last decades, efforts were made to develop molecular techniques to improve CTV detection and genotype differentiation. However, a remarkable achievement was the development of reverse genetics to overcome the challenges of mapping the CTV genes involved in CTV-citrus interactions, particularly the pathogenic determinants (Albiach-Marti et al., 2010; Satyanarayana et al., 1999, 2001). Therefore, the generation of a cDNA clone (*T36-CTV9*, Figure 2, left panel) of the Florida (USA) isolate T36 and the development of a *in vitro* genetic system to analyze CTV genotypes, mutants and self replicating constructs in *Nicotiana benthamiana* protoplasts or indexing citrus plants (Satyanarayana et al., 1999; 2001), allowed examining CTV viral replication, gene expression and assembly. Furthermore, the T36 genetic system was modified by Gowda et al. (2005) to allow CTV agroinoculation and replication in *N. benthamiana* plants. Afterwards, the *T36-CTV9* construct was adapted to be employed as a virus-based vector (Folimonov et al., 2007) (Figure 2, center and right panels) for the study of movement and virus-host interactions (Folimonova el al., 2008; Tatineni et al., 2008). The complexity of the CTV genetics and the pathosystem established by the interaction of *Citrus*, CTV and the insect vector, plus the molecular virology advances in identification of the genetic determinants of the diseases induced by CTV, and the different technology approaches used in these studies, are discussed.

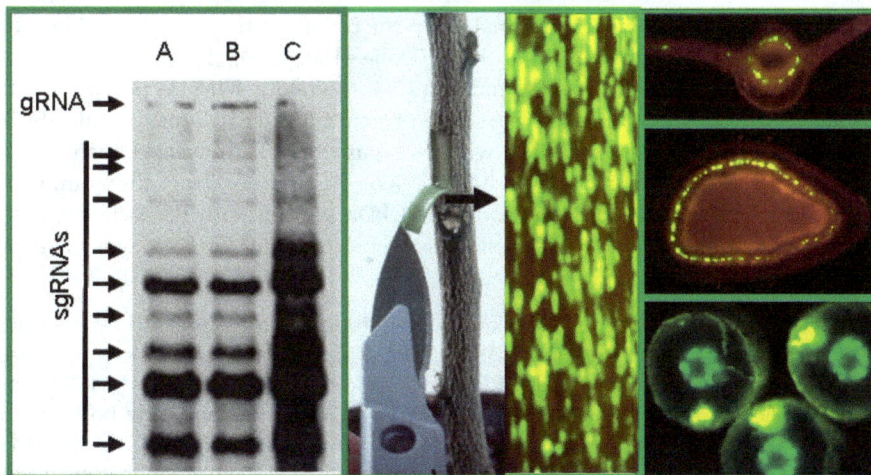

Fig. 2. *T36-CTV9* infectious clone and the *in vitro* genetic system to manage CTV. (Left panel): Northern-Blot hybridization using minus-stranded riboprobes specific to the 3´ proximal genomic region. Accumulation of total RNAs from (line A) construct *T36-CTV9* in *N. benthamiana* protoplast; (line B) T36 isolate in *C. macrophylla* plants; (line C) dsRNA accumulation of contruct *T36-CTV9* in *C. macrophylla* plants. (Center and right panels): Sections of *C. macrophylla* infected with construct *BCN5-GFP* visualized in a confocal microscope under UV light, indicating location of CTV in plants. (Center panel): bark flap; (Right panel): (Top) leaf, (Center) shoot, (Bottom) roots. Photos from Folimonov et al. (2007).

2. The complexity of the *Citrus tristeza virus* molecular genetics

The virus needs to program the plant cellular machinery for the viral multiplication and for the viral movement through the plasmodesmata and the plant vascular system to colonize the plant. Although each viral gene product seems to have primary functions that are required for the survival of the virus, there are secondary interactions that cause disease or trigger resistance response in the plant host (Culver & Padmanabhan, 2007). Additionally, the genetic variability of CTV viral populations is extremely important in order to design strategies to study pathogenic determinants and to develop reliable and perdurable biotechnological strategies of viral control (Albiach-Marti et al., 2010; Folimonova et al., 2010). In this section, the primary functions of CTV genes and the genetic variability of CTV genotypes, composing a wild isolate, will be reviewed.

2.1 Genome organization and functions of viral proteins

The expression of the twelve CTV ORFs is a remarkable process that includes at least three different RNA expression mechanisms widely used by positive-strand RNA viruses: proteolytic processing of the polyprotein precursor, translational frameshifting and the generation of a nested set of ten 3'-coterminal sub-genomic RNAs (sgRNA) (Karasev et al., 1997). The organization and expression of the 19.3 Kbs of the CTV genome resembles that of *Coronaviruses*, but phylogenetically the CTV polymerase, like in other *Closteroviruses*, belongs to the Sindbis virus-like lineage (Karasev et al., 1997). The *replication gene block*, which is conserved in the family *Closteroviridae* and in the supergroup of sindbis-like viruses, comprises ORF 1a and 1b and makes up the 5´ half of the genome and encodes, as indicates the replication machinery (Figure 1) (Dolja et al, 2006). The ORFs 1a and 1b are directly translated from the positive stranded gRNA to yield a 400 kDa polyprotein that is later proteolytically processed into at least nine protein products (Karasev et al., 1995). The ORF1a encodes a 349 kDa polyprotein with two papain-like protease domains, a type I methyltransferase-like domain, and a helicase-like domain bearing the motifs of the superfamily I helicases. The ORF1b encodes a 54 kDa protein with RNA-dependent RNA polymerase (RdRp) domains that is occasionally translated after ORF 1a by a +1 ribosomal frameshifting (Karasev et al., 1995).

The other 10 ORFs, located at the 3' half of the CTV genome, are expressed by the synthesis of a set of 3' co-terminal subgenomic RNAs (sgRNAs). Each 3' sgRNA serves as a messenger for the translation of its 5' proximal ORF (Hilf et al., 1995; Karasev et al., 1997) and the expression of each of the ten 3' proximal ORFs is regulated independently both in amount and timing (Hilf et al., 1995; Navas-Castillo et al., 1997). Part of the CTV 3´ ORFs are enclosed in the conserved *quintuple gene block* (Figure 1), another hallmark of the *Closteroviridae* family that is related with virion assembly and trafficking (Dolja et al., 2006). This consists of the major coat protein (CP of 25kDa), the minor capsid protein (CPm of 27kDa) (Febres et al., 1996), p61, HSP70h (a p65 kDa protein homologue of the HSP70 plant heat-shock proteins) and p6 (a small hydrophobic protein that belongs to the single-span transmembrane proteins) (Karasev et al., 1995). About the 97% of the CTV genome is coated by CP, while the remainder 3% is encapsidated by CPm resulting in viral particles with the emblematic tail of the members of the *Closteroviridae* family (Dojla et al., 2006; Febres et al., 1996; Satyanarayana et al., 2004). The coordinate action of HSP70h and p61, in addition to the CP and CPm coat proteins, are required for proper assembly of CTV virions (Satyanarayana et al., 2000). During CTV assembly, HSP70h or p61 bind to the transition zone between CP and CPm (around 630 nt) and restrict CPm to the virion tail (Satyanarayana et al., 2004; Tatineni et al., 2010). The protein homologous to HSP70h and p61 in *Beet yellows virus* (BYV) (genus *Closterovirus*) are coordinately assembled with CPm in the virion structure and remain attached to the viral particles (Dolja et al., 2006). However, the assembly of HSP70h and p61 has not been directly confirmed for CTV (Satyanarayana et al., 2004). Although unnecessary for virus assembly or replication, p6 is required for systemic invasion of host plant (Tatineni et al., 2008) and probably functions as a movement protein, similarly to its homologue in BYV (Dolja et al., 2006).

The additional five CTV ORFs located at the 3´ half of the genome (Figure 1) are the p20 ORF, an homologue of p21 of BYV, and four genes encoding proteins with no homologue in other closteroviruses (p33, p18, p13 and p23) (Dolja et al., 2006). The p20 protein is the main component of the CTV-induced amorphous inclusion bodies (Gowda et al., 2000) and, as well as p6, is needed for CTV systemic infection, thus suggesting a possible role in CTV translocation in the citrus plant (Satianarayana et al., 2001; Tatineni et al., 2008). Unexpectedly, ORFs that encode proteins p33, p18 and p13 are not required either for replication or assembly (Satyanarayana et al., 1999, 2000) or for systemic infection of Mexican lime [*C. aurantifolia* (Christm.) Swing.] and *C. macrophylla* Wester plants (Tatineni et al., 2008). These three genes are CTV host range determinants (Tatineni et al., 2011). The p33 gene is essential for complete infection of sour orange (*C. aurantium* L.) and lemon [*C. limon* (L.) Burm. f.] trees. The p33 plus p18 and the p33 plus p13 are required for systemic infection of grapefruit (*C. paradisi* Macf.) and of calamondin (*C. madurensis* Lour.) trees, respectively (Tatineni et al., 2011). Part of the plant antiviral defense consists of the post-transcriptional gene silencing (PTGS) mechanism. Viruses have evolved developing genes to suppress this plant mechanism (Qu & Morris, 2005; Voinnet, 2005). Unusually, CTV evolved ending up with three proteins that act as RNA silencing suppressors in *N. benthamiana* and *N. tabacum* plants. The p23 inhibits intercellular RNA silencing, while CP impedes intracellular RNA silencing and p20 obstructs both inter and intracellular RNA silencing (Lu et al., 2000). Additionally, the multifunction protein p23 contains a Zn finger domain that binds cooperatively both ssRNA and dsRNA molecules in a non-sequence specific manner (López et al., 2000) and it controls asymmetrical accumulation of positive and negative RNA strands during viral replication, ensuring the presence of enough quantity of positive gRNA ready for virion assembly (Satyanarayana et al., 2002b). The CTV genes or sequences related with aphid transmission are unknown. However, for viral transmission, unknown *helper component* or CTV virions have to interact with the mouthparts and the foregut of the aphids (Ng & Falk, 2006). Therefore, the structural proteins, CP, HSP70h, p61 (Satyanarayana et al., 2004) and especially the CPm, which composes the CTV particle tail structure, are suspected to affect aphid transmission (Barzegar et al., 2009; Febres et al., 1996). In fact, the CPm of *Lettuce infectious yellows virus* (LIYV), a close relative to CTV (genus *Crinivirus*, family *Closteroviridae*), is involved in viral transmission (Steward et al., 2010).

The 3´and 5´ non-translated regions (Figure 1) contain the cis-acting elements indispensable for CTV replication (Satyanarayana et al., 1999). The 5´ termini of the CTV genome is protected with a cap structure (Karasev et al., 1995). CTV 5´ NTR is the most variable genomic region with nucleotide identities as low as 42% among some CTV isolates (Figure

3) (Albiach-Marti et al., 2000b; López et al., 1998). Remarkably, the CTV 5´NTR secondary structure is similar even for divergent genotypes and folded in two stem-loops separated by a short spacer region (Gowda et al, 2003b, 2009; López et al., 1998). This secondary structure contains the sequences necessary for both replication and particle assembly (Gowda et al, 2003b; Satyanarayana et al., 2004; Tatineni et al., 2010). Opposite to the sequence divergence of the CTV 5´NTRs, the 3´ NTR sequences are almost identical (Figure 3) (Harper et al, 2010; López et al., 1998). The 3´NTR lacks a poly-A tract and does not appear to fold in a tRNA-like structure (Karasev et al., 1995) but instead consist in a secondary structure of 10 stem-loop structures, which contain the sequences necessary for minus-strand initiation for the CTV gRNA and the sgRNAs (Satyanarayana et al., 2002a)

2.2 Viral RNA species generated during *Citrus tristeza virus* replication

CTV replication is an extraordinary process that generates at least 35 different species of viral RNA in CTV-infected cells (Gowda et al., 2001) plus a myriad of defective RNAs (D-RNAs) (Albiach-Marti et al., 2000a; Ayllon et al., 1999a; Mawassi et al., 1995). The positive to negative-stranded total RNA (gRNAs plus sgRNAs) ratio, approximately 40 to 50:1, falls within the range of the genomic RNAs of most positive-stranded RNA viruses, particularly the more similar *Alphavirus* supergroup and large complex viruses of the *Nidovirales* (Satanarayana et al., 2002a). Viral replication starts when the CTV replicase generates a negative-stranded gRNA using as template the positive stranded CTV genome. Apart from the gRNA molecules, CTV accumulates high quantities of single- and double- stranded sgRNAs generated during the expression of the ten ORFs situated at the 3´half of the CTV genome (Hilf et al., 1995). Unlike the large animal viruses of the *Nidovirales*, the 3´ sgRNAs of CTV do not share a common 5´ terminus. The synthesis of each sgRNA is controlled by its corresponding cis-acting element [controller elements (CEs)] (Ayllon et al., 2005; Gowda et al., 2001; Karasev et al., 1997). In addition of the plus and minus stranded 3´ coterminal sgRNAs, the CEs corresponding to each of the ten 3´ ORFs produce a reduced amount of a set of 5´ coterminal positive-stranded sgRNAs, probably due to premature termination during the synthesis of the gRNA (Gowda et al., 2001). In addition, CTV generates significant amounts of *low molecular-weight tristeza*, LMT1 and LMT2, a two positive-stranded 5´ coterminal sgRNAs population with heterogeneous 3´ termini at nt 842-854 and 744-746, respectively (Che at al., 2001), which are produced and accumulated differently (Gowda et al., 2003a; 2009). LMT1 is likely created by premature termination during genomic RNA synthesis at a 5´ CE sited in the PRO I domain of replicase (Gowda et al., 2003a). Nevertheless, the LMT2 production is correlated to virion assembly although its exact viral function is unknown (Gowda et al., 2009).

In addition to the 35 different species of RNA created during replication, CTV could accumulate considerable amounts of D-RNAs in infected cells, probably originated during the generation of the positive-stranded sgRNA or gRNA by a template-switching mechanism (Ayllón et al., 1999a; G. Yang et al., 1997). Generally, D-RNAs bear a genome from about 2.0 to 12.0 kb and are composed by variable portions of the 3' and 5' termini of CTV genomic RNA with large internal delections (Mawassi et al., 1995a; Ayllón et al., 1999a; Che et al., 2003). Moreover, some of them resembled the RNAs 1 and 2 distinctive of the bipartite *Criniviruses*, also included in the *Closteroviridae* family (Che et al., 2003). Apart from of their size, they vary also in sequence, and could be encapsidated into particles and be transmitted by aphids (Albiach-Martí et al., 2000a; Mawassi et al, 1995a, 1995b). D-RNAs need to use the viral machinery for their survival (Pathak & Nagy, 2009). CTV D-RNAs replication *in trans* required a minimal 5´ proximal region of 1kb and a 3´ termini limited to the 3´ NTR. In addition, efficient replication involves some spacing between these terminal cis-acting signals and a continuous ORF through most of the 5´ proximal regions of the D-RNA sequence (Mawassi et al., 2000). In other viral pathosystems, D-RNAs have the capacity of interfering in the viral replication of their helper virus (named defective interfering (DI) RNAs) (Pathak & Nagy, 2009), but this function was not confirmed for CTV (Mawassi et al., 2000) and thus, their biological role is presently unknown.

2.3 *Citrus tristeza virus* viral diversity: From the extreme genomic divergence to the genetic stasis

CTV isolates from different host and areas exhibit great variability either biologically or genetically (Moreno et al., 2008). There are CTV isolates that consist basically of a main genotype and its quasispecies (Albiach-Marti et al., 2000b; Satyanarayana et al., 2001), but others are composed by two or more different CTV genotype-related groups which are recognized as strains. These could vary in the pathogenicity induced depending on the citrus host and in transmission efficiency by aphids (Bar-Joseph & Dawson, 2008; Moreno et al., 2008). The strains composing a field population are unequally distributed within a CTV infected tree (d'Urso et al., 2000) and could be somehow separated from the former mixture by aphid or graft transmission or by host passage creating a new mixture of strains, thus a new isolate, which could generate a completely distinct symptomatology in citrus plants (Albiach et al., 2000a; Ayllon et al., 1999b; Moreno et al., 1993; Roy & Brlansky, 2009; Velazquez-Monreal et al., 2009; Weng et al., 2007). Additionally, aphid or graft transmission and host passage could modify the composition of the D-RNAs population in the CTV isolates (Albiach-Martí et al., 2000a; Mawassi et al., 1995b)

The sequencing of the genomes of nineteen CTV isolates from distant places in the planet, which represented a subset of its local CTV population, helped the understanding of the evolution of CTV and the complex structure of the actual CTV isolates. These genomic sequences are T36 and T30 from Florida, USA (Albiach-Martí et al., 2000b; Karasev et al., 1995); VT from Israel (Mawassi et al., 1996); SY568R from California (Z.N. Yang et al., 1999; Vives et al., 2005); T385 and T318A from Spain (Ruiz-Ruiz et al., 2006; Vives et al., 1999); NuagA from Japan (Suastika et al., 2001); Qaha (AY340974) from Egypt; Mexican isolate (DQ272579); B165 from India (Roy & Brlansky, 2010); NZ-M16, NZ-B18, NZRB-TH28, NZRB-TH30, NZRB-M12, NZRB-M17 and NZRB-G9 from New Zealand (Harper et al., 2009, 2010); and HA16-5 and HA18-9 from Hawaii (Melzer et al., 2010). Comparison of CTV genomes yielded nucleotide identities from 79.9% (between Qaha and VT) to 99.3% (between T30 and T385) (Melzer et al 2010). Phylogenetic analysis grouped the CTV genome diversity in seven clades reflecting six main severe sequence-related groups [(1) T36-like (T36, Qaha, Mexican); (2) the NZRB strains plus HA18-9; (3) the VT-like (VT, NUagA, T318A, SY568); (4) HA16-5; (5) B165 and NZ-B18; (6) NZ-M16] and one asymptomatic genotype [the T30- like (T30 ,T385)] (Harper et al., 2009; Melzer et al. 2010). Moreover, the T36-like genotypes and close relatives [clades (1) and (2)] show an unusually high genetic distance to the other CTV genotypes in spite of belonging to the same taxonomic entity. The divergence between these two groups of genotypes is mostly concentrated at the 5´ half of the genome and increases towards the 5´NTR (Figure 3) (Bar-Joseph & Dawson, 2008; Hilf et al., 1999; Mawassi et al., 1996). In this way, the comparison of the genomic sequences of T30 and T36 diverged from 5% in the 3´NTR to as high as 58% in the 5´NTR (Figure 3) (Albiach-Marti et al., 2000b). Conversely, genomic sequence divergence between the CTV genotypes included in clades (3) to (7) increase slightly in the 5´NTR region but is relatively constant in proportion and distribution along the genome (Figure 3) (Hilf et al., 1999; Melzer et al. 2010). Based in this two paths of sequence divergence (Hilf et al., 1999) it was speculated that the T36 genotype and relatives evolved from a recombinant of a CTV genome and an unknown virus millions of years ago in Asia (Mawassi et al., 1996). This high genomic divergence between distinct genotypes confirmed the genetic variability found between strains composing the CTV isolates.

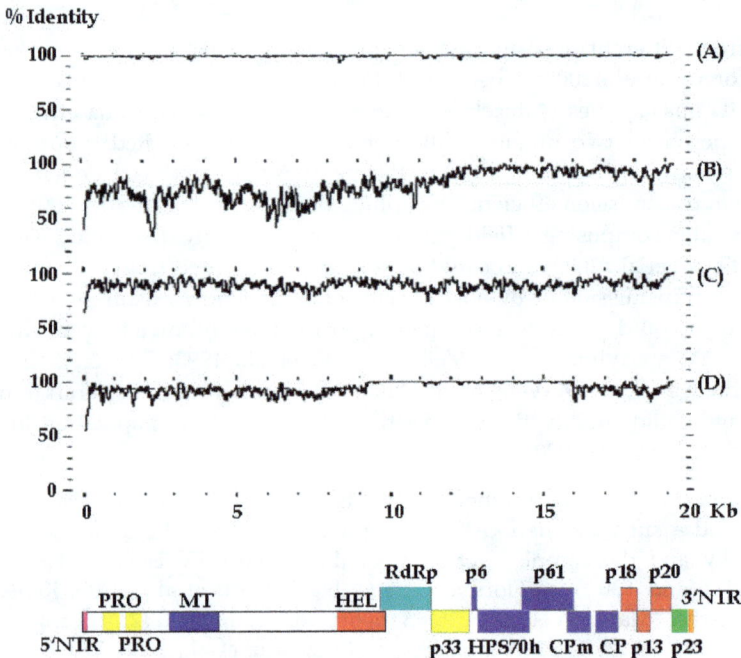

Fig. 3. Graphic of the sequence identities along the CTV genome between T30 genotype and (A) T385 (B) T36, (C) VT and (D) SY568 genotypes (Albiach-Marti et al., 2000b). PRO, MT, HEL and RdRp indicate protein domains of papain-like protease, methyltranferase, helicase and RNA-dependent RNA polymerase, respectively. HSP70h, CPm and CP indicate ORFs encoding a homologue of heat shock protein 70, the minor and the major capsid proteins, respectively. NTR indicates non-translated region.

In spite of this genetic variability, Albiach-Martí et al. (2000b) reported that the genomes of the symptomless isolates T30 from Florida and T385 from Spain, which where separated geographically and in time, were essentially identical (Figure 3). Moreover, these authors demonstrated that the T30/T385 genotype was distributed around the world and it could have been stable at least 500 years, which suggests that the T30/T385 genotype is well adapted to the citrus environment. Afterwards, this genetic similarity was found for other

CTV genotypes separated geographically and in time like T36 from Florida and Qaha from Egypt. Although, a great capacity for rapid evolution is a common feature of RNA viruses (Domingo & Holland, 1997), there are examples of long genetic stability in animal (Nakajima et al., 1978) and plant (Blok et al., 1987; Fraile et al., 1997) viral RNA populations, which were reported to be nearly identical after 22 years and from 100 to 13,000-14,000 years, respectively. This genetic stasis has been explained as a consequence of strong selection and competition between the mutants that arise in each replication cycle, which creates equilibrium in the viral quasispecies distribution (Blok et al., 1987; Fraile et al., 1997). Therefore, if some CTV sequences tend to remain relatively stable over periods of years, sequence-based control strategies like transgenic plants based in PTGS or cross-protection using recombinant mild CTV strains, have a higher probability of success (Albiach-Martí et al., 2000b, 2010; Folimonova et al., 2010).

A possible hypothesis to explain the actual high sequence variability found in the CTV isolates could be based in the fact that the main CTV genotypes evolved in different *Citrus* progenitors at its point of origin in Asia. Afterwards, the high viral diversity found *intra* and *inter* CTV isolates could has been generated by two processes acting in parallel. In one hand, the dispersal of the main CTV genotypes to different environments around the world by vegetative propagation of citrus, followed by the exposure during decades of the citrus-infected trees to repetitive inoculation by the natural aphid population, and by cultural practices like graft transmission, would create founder effects or bottlenecks. These could change the frequency of sequence variants within field isolates. On the other hand, RNA virus mutation due the error-prone nature of RNA-dependent RNA polymerases (Domingo & Holland, 1997), in addition to recombination events between diverged sequence variants, plus selection, genetic drift and gene flow, possibly might allow newly arising mutants to shift the sequence variants distribution of the isolates, becoming prevalent in CTV populations, and promoting rapid evolution (Moreno et al., 2008). In this context, identifying a specific genetic determinant that is responsible for a specific disease symptom under field or glasshouse conditions is, in the case of CTV, a real challenge.

3. Virus-host interactions in the *Citrus tristeza virus* pathosystem

Citrus tristeza virus natural plant hosts belong to the order *Geraniales*, family *Rutaceae*, subfamily *Aurantoidea*. Most of them are included in the genus *Citrus* L. except for kumquats (*Fortunella* spp) and other *Citrus* relatives (*Aegle*, *Aeglopsis*, *Afraegle*, *Atalantia*, *Citropsis*, *Clausena*, *Eremocitrus*, *Hesperthusa*, *Merrillia*, *Microcitrus*, *Pamburus*, *Pleiospermium*, and *Swinglea*). However, there are also non-rutaceous hosts that have been experimentally infected with CTV strains like *Passiflora gracilis* and *Passiflora coerulea* (Bar-Joseph et al, 1989; Moreno et al., 2008), and *N. benthamiana* protoplasts or agroinfiltrated leaves (Gowda et al., 2005; Navas-Castillo et al., 1997). When CTV interacts with the plant host there could be several plant responses. Depending on the CTV strain and the specific citrus host or variety/rootstock combination, CTV interactions with a particular citrus host might be pathogenesis or asymptomatic or from limited to complete plant resistance. In this way, Citron, Mexican lime, *C. macrophylla*, sour orange and lemon seedlings are usually susceptible to CTV infection. In addition, mandarins, clementine (*C. clementina* Hort. ex Tan.), satsuma [*C. unshiu* (Mak.) Marc.] and the citrus hybrids tangelos (mandarin × grapefruit or pummelo) and tangors (mandarin × sweet orange), as well as some pummelos and citrumelos (grapefruit × *P. trifoliata*), are among the commonly tolerant hosts to CTV.

Sweet orange [*C. sinensis* (L.) Osb.] and grapefruit could be susceptible or tolerant depending on the CTV pathotype (Bar-Joseph et al, 1989; Garnsey et al., 2005). Finally, pummelos, grapefruit, sour orange and the rootstock Swingle citrumelo exhibit a differential degree of resistance depending on the CTV strain (Bernet et al., 2008; Garnsey et al., 1996; Fang & Roose, 1999; Folimonova et al., 2008). Whereas, some *Citrus* relatives, within subfamily *Aurantioideae* like *Poncirus trifoliata* (L.) Raf., *Swinglea glutinosa* (Blanco) Merr., and *Severinia buxifolia* (Poir) Ten, as well as *P. trifoliata* intergenic hybrids like citranges (sweet orange × *P. trifoliata*), remain resistant or immune to most of the CTV strains (Garnsey et al., 1987; Yoshida, 1985, 1993). Therefore, there is an elevated complexity in the *Citrus*-CTV interactions that generate processes like pathogenesis or plant host resistance. Although there have been considerable advances in the study of the CTV genetics, the interaction between CTV and *Citrus* and in particular the mechanisms involved in the development of disease or plant resistance, are still poorly understood. Interactions between CTV and citrus could also affect the performance of the CTV arthropod vector, therefore affecting virus spread (Fereres & Moreno, 2009). Consequently, the different CTV-*Citrus* interactions as well the possible connection between CTV infected plant interactions and modulation of aphid transmission are reviewed in this section.

3.1 *Citrus tristeza virus* pathotypes

Phenotypically, CTV is a very complex virus, with three hallmark diseases, plus a myriad different symptom patterns in indexing plants. After the *Phytophthora* epidemics in 1836, commercial citrus varieties were mainly propagated on sour orange, a *Phytophthora* resistant rootstock exceptionally adaptable to all soil types that generates excellent fruit quality and elevated productivity. *Tristeza* disease or *Quick decline* (QD), the first known syndrome of CTV, appeared in 1930 as sour orange roostock resulted to be highly sensitive to CTV (Figure 4). The QD syndrome consists in the rapid death of the commercial varieties sweet orange, mandarin, grapefruit, Kumquats or limes on sour orange rootstock in field conditions (Bar-Joseph & Dawson, 2008; Moreno et al., 2008). During the development of QD the sieve tubes and companion cells close to the bud union between the scion and the rootstock collapse and necroses, producing an excessive amount of non-functional phloem (Schneider, 1959). This generates overgrowth of the scion at the bud union, loss of root mass, and therefore, drought sensitivity, stunting, leave chlorosis, reduced fruit size, poor growth, dieback, wilting and finally death of the tree. The QD symptomatology explains the disease name of *Tristeza*, which means "sadness" in Spanish and Portuguese. The CTV QD pathotype has been devastating for the commercial citrus industries around the world, since it has caused the death of hundred million trees worldwide. Moreover, the QD syndrome forced the transformation of a citrus industry based in sour orange rootstock to other established on *Tristeza*-tolerant rootstocks, which generate damages from soil salinity or alkalinity, water logging in heavy soils, or sensitivity to soil fungi and lower fruit yield than sour orange rootstock. The QD pathotypes are distributed in most of the citrus-growing areas, except a few places in the Mediterranean basin and Western USA. The second pathogenic interaction between citrus and CTV, *Stem pitting* (SP) (Figure 4) was first observed in orchards replanted with *Tristeza* resistant or tolerant rootstocks. The disease is produced by highly virulent strains that affect commercial lime, sweet orange, and grapefruit trees grafted on any rootstock. SP consists of deep pits in the wood under depressed areas of bark. Contrasting with QD, the SP pathotypes usually do not cause tree death, but chronically limit profitable growth of different varieties, significant reduction of plant vigor, severe stunting and low yield of unmarketable fruit, thus causing high economic losses (Bar-Joseph et al., 1989). The SP pathotypes are restricted to regions of Asia, Australia, South Africa and South America. However, these pathotypes have also been found, although at lower frequency, in California, Florida, and the Mediterranean area ((Bar-Joseph & Dawson, 2008; Moreno et al., 2008).

Fig. 4. CTV pathotypes. (Left panel): *Quick decline* symdrome. (Right panel): *Stem pitting* symdrome.

The third CTV-induced syndrome, *Seedling yellows* (SY) is observed in the greenhouse (Figure 5) but might also be found in the field in top–grafted plants. SY is characterized by stunting, leaf chlorosis and sometimes a complete cessation of growth of sour orange, grapefruit or lemon seedlings (Fraser, 1952). Although SY is not economically valuable, it can be examined in the greenhouse in a timely manner and has a substantial diagnostic value for CTV pathotype differentiation (Garnsey et al., 2005). In contrast, there are mild CTV strains (as the T30-like genotypes) that cause a complete lack of symptoms in almost all varieties of citrus, including those propagated on sour orange rootstocks, even though the virus multiplies to reach high titers (Albiach-Marti et al., 2000b). These mild isolates are common in almost all the citrus growing areas (Albiach-Marti et al., 2000b; Hilf et al., 2005; Roy et al., 2010), although their presence is frequently masked when they are present in mixed infections with severe isolates (Moreno et al., 2008).

The development of QD and SP extends over 10 to 40 years in the field, a period too long to screen the CTV pathotypes. Although the SP pathotype can be examined somehow in glasshouse conditions, there are no reliable methods to reproduce QD in those conditions (Moreno et al., 2008). Therefore, the degree of severity of a specific CTV isolate, strain or genotype usually is assessed by using indexing plants (seedlings from Mexican lime, sour orange, Madame vinous sweet orange and Duncan grapefruit), where the development of severe pathotypes could be determined in months (Garnsey et al., 2005). The majority of CTV isolates induce symptoms (vein clearing, leaf cupping, dwarfing and stem pitting), in Mexican Lime (Figure 5), the most sensitive host for CTV. In this case, the degree of CTV symptomology ranges from the mild phenotypes, which are almost asymptomatic, to the highly virulent CTV isolates that could kill the plant (Garnsey et al., 2005). However, there are exceptions, such as severe isolates that induce symptoms in sweet orange but not in Mexican lime (Harper et al., 2009).

Fig. 5. CTV symptomology in greenhouse conditions. (A) healthy sour orange plant. *Seedling yellows* syndrome in (B) sour orange and (C) grapefruit. Symptoms induced in Mexican lime: (D) leaf cupping, (E) vein clearing and (F) stem pitting.

3.2 Genetic determinants of the *Citrus tristeza virus* syndromes

Viruses possess the potential to disrupt host physiology by the interaction of specific viral components with the host components. During viral infection, the virus has to overcome the constitutive and/or inducible plant defences. The plant inducible defence could confine the virus and prevent systemic infection (Culver & Padmanabhan, 2007). The plant constitutive defense consist in the PTGS or the RNA interference (RNAi) pathway that implies the specific degradation of the viral dsRNA in small interfering RNAs (siRNAs), which guides a specific plant ribonuclease to degrade the viral genomes in the cytoplasm of the plant cell. Besides of the antiviral role, the gene silencing mechanism has important functions in regulating plant gene expression (miRNA metabolism) (Voinnet, 2005). Viruses contain RNAi suppressing genes or RNA silencing suppressors, allowing viral multiplication and interfering in host gene expression, thus inducing disease (Qu & Morris, 2005; Voinnet, 2005). The CTV genome contains three suppressors (CP, p20 and p23, see section 2) that block intracellular and/or intercellular RNA silencing mechanism in *N. benthamiana* and *N. tabacumm* plants (Lu et al., 2000). In spite of the presence of these three silencing suppressors, accumulation of siRNAs in CTV-infected susceptible hosts could be 50% of the total RNAs in the plant (Ruiz-Ruiz et al., 2011). Moreover, the siRNAs accumulation is directly proportional to the virus accumulation and varies depending on the CTV strain and the citrus host (Comellas, 2009). Deep sequencing analysis of siRNAs from CTV-infected plants indicated that they mainly consisted in small RNAs of 21-22 nt derived from essentially all the CTV genome (Ruiz-Ruiz et al., 2011). Although, CP, p20 and p23 have not been yet reported as RNAi suppressing genes in citrus, these three CTV genes could be candidates for symptom determinants.

Apart of CP, p20 and p23, CTV multiplication generates great quantities of other viral products (35 RNA species, 16 protein products and D-RNAs, see section 2.1 and 2.2) along with a complicated process of replication, gene expression, assembly and movement, where the interaction with host factors is essential. Consequently, during CTV-*Citrus* interactions there are multiple opportunities to generate disease. However, determination of which of the viral products induce a specific symptomology is a complicated task. Analysis of Mexican lime transcriptome after infection with a severe CTV isolate showed altered expression of 334 genes and about half of them without significant similarity with other known sequences, thus indicating elevated complexity in the citrus-CTV interaction during symptoms development (Gandia et al., 2007). Many attempts have been made to develop rapid diagnostics for specific CTV syndromes. Their application has led to the establishment of some correlations between various serological and molecular markers with CTV pathotypes (Hilf et al., 2005; López et al., 1998; Permar et al., 1990; Roy et al., 2010; Sambade et al., 2003). Although these molecular markers are a valuable tool for genotyping a particular CTV population, recombinants between genotypes affecting the disease determinants could be present, thus invalidating these methodologies for severe/mild strain differentiation. Additionally, direct linkage of these markers to symptom development has yet not been established. An important step through the identification of disease determinants was the sequencing of the nineteen CTV complete sequences. Their sequence comparison yielded an intriguing correspondence of the CTV phylogenetic clades (section 2.3 of this chapter) with CTV pathotypes that could point to a distribution of the symptom determinants along the CTV genome.

3.2.1 Genetic determinant of *Seedling yellows* syndrome

A distinctive phenotype of some isolates of CTV is the ability to induce *Seedling yellows* in sour orange, lemon and grapefruit seedlings (Fraser, 1952). The recombinant virus *T36-CTV9* and the original wild type T36 isolate produce identical SY symptoms in sour orange and grapefruit seedlings (Satyanarayana et al., 2001). T30, the type isolate of the widely distributed asymptomatic genotype T30 (T385), does not induce SY and consists of one genotype and its quasiespecies (Albiach-Marti et al., 2000b). To delimit the viral sequences associated with the SY syndrome, eleven T36/T30 hybrids were generated by substituting T36 sequences for homologous T30 sequences into the 3′ moiety of *T36-CTV9*, where both genome sequences (T30 and T36) are about 90% similar (Figure 3) (Albiach-Marti et al., 2010). However, hybrid constructs, which carried exchanges of T30 CP and CPm into the T36 genome, failed to passage through successive sets of *N. benthamiana* protoplasts (Albiach-Marti et al., 2010), probably due to deficient heteroencapsidation since incomplete virions do not withstand this procedure (Satyanarayana et al., 2000; 2004). Nevertheless, hybrid T30/T36 constructs [P23-3′NTR], [P13], [P61], [P18-3′NTR] and [HSP70h-P61] were sufficiently amplified to allow successful infection of the highly susceptible host *C.*

macrophylla by stem-slash or bark-flap inoculation (Robertson et al., 2005; Satyanarayana et al, 2001). Sour orange and Duncan grapefruit seedlings were graft-inoculated with tissues from the *C. macrophylla* plants infected with the five hybrid constructs as well as plants infected with T36 and T30 (as controls). Finally, analysis of the SY development demonstrated that the parental T36 and three of the T36/T30 hybrids induced SY symptoms while hybrid constructs [P23-3'NTR] and [P18-3'NTR] and the wild type T30 remained symptomless like the healthy controls (Figure 6, Left panel). Therefore, Albiach-Martí et al. (2010) demonstrated that SY is mapped to the region encompassing the p23 gene and the 3' NTR.

Fig. 6. (Left panel) Seedling yellows syndrome assay in sour orange seedlings inoculated with (A) T36/T30 hybrid [P23-3'NTR], (B) isolate T30 (C) healthy (D) construct *T36 CTV9* and (E) T36/T30 hybrid [HSP70h-P61]. (Right panel) T36/T30 hybrid [P23-3'NTR] protects against development of SY symptoms in sour orange seedlings. Sour orange plants inoculated with (A) T36/T30 hybrid [P23-3'UTR] (B) hybrid [P23-3'UTR] and then challenged with construct *T36-CTV9*, and (C) *T36 CTV9*. Pictures from Albiach-Marti et al. (2010).

Other methodology used to map disease determinants was the expression of CTV proteins in transgenic plants (Fagoaga et al., 2005; Ghorbel et al., 2001). When p23 is ectopically expressed in transgenic citrus induces virus-like symptoms. However, transgenic Mexican lime plants develop more intense vein clearing in the plant leaves and symptomatology like chlorotic pinpoints in leaves, stem necrosis and collapse (Ghorbel et al., 2001) that differs from those induced by natural virus infection (Figure 5). Additionally, transgenic sour orange plants expressing p23 develop vein clearing, leaf deformation, defoliation, and shoot necrosis (Fagoaga et al., 2005). However, these transgene-induced symptoms differ substantially from the virus-induced SY of uniform chlorosis and stunting of new shoot growth in sour orange (Figure 5). Transgenic limes differ from virus-infected limes in that symptom severity is proportional to the levels of p23 production, not to the source or sequence of the gene (Fagoaga et al., 2005; Ghorbel et al., 2001), whereas the symptom intensity in virus-infected limes is dramatically different between severe and mild isolates of virus. Yet, the different response in transgenic plants could be related to the fact that the p23 protein is produced constitutively in most cells, while in nature the expression of p23 ORF from the viral genome is limited to phloem-associated cells (Albiach-Marti et al., 2010).

If the symptoms induced by CTV in sour orange are determined by p23, they should be related to p23 sequence and not to protein expression levels, since there was no correlation between the amount of p23 and the intensity of the SY symptoms induced by T36 or by the T36/T30 hybrids, which did not induce SY in sour orange plants (Albiach-Marti et al., 2010). Since p23 is a suppressor of RNA-mediated gene silencing, it could potentially disrupt the miRNA metabolism thus inducing the SY syndrome. Several viral silencing suppressors have been identified as pathogenicity determinants (Qu & Morris, 2005) and p23 could be the obvious candidate for being the CTV determinant of SY syndrome development in sour orange and Duncan grapefruit seedlings. However, since a viral 3′ NTR has also been related to symptom development (Rodriguez-Cerezo et al., 1991), it cannot yet be concluded that the p23 protein directly induces SY. Additionally, the SY reaction is specific to only certain citrus hosts of CTV, such as lemons, sour orange and grapefruit, indicating that there are specific host factors involved in its expression in addition to the isolate-specific factors identified. Although Albiach-Marti et al. (2010) were able to map a determinant of the SY syndrome in T36, since this genotype is highly divergent from the other CTV genotypes (Harper et al., 2010; Mawassi et al., 1996), it is essential to assess whether this determinant is common to other CTV genotypes that also induce SY or if there are other possible SY determinants.

3.2.2 Genetic determinants of *Quick decline* and *Stem pitting* syndromes

From economic standpoint it would be highly valuable to map decline and stem pitting determinants, which could be developed into detection tools. It is possible, but not yet confirmed, that determinant(s) for the decline disease map similarly to that of SY, since a strong correlation between SY and decline has been observed in the biological evaluation of a wide range of CTV isolates (Garnsey et al., 2005). However, since some decline-inducing isolates do not produce obvious SY symptoms, the T36/T30 hybrids have to be directly evaluated in decline-susceptible grafted combinations of scion and rootstock. Unfortunately, clear decline assays need to be conducted during long periods in the field. In addition, since the hybrids are made by recombinant DNA technologies these assays require special permits from the plant protection and environmental safety authorities (Albiach-Marti et al., 2010).

In relation to the mapping of the stem pitting determinants, expression of p23 in transgenic plants of several citrus species, but not in tobacco plants, induced phenotypic aberrations resembling in some cases foliar symptoms induced by CTV, indicating that the stem pitting determinant could be also located in p23 (Fagoaga et al., 2005; Ghorbel et al., 2001). However, it seems that, in addition to p23, there are other genes related to the development of SP, at least in *C. macrophylla* plants infected with the four T36/T30 hybrids used to map the SY syndrome determinant, since the T36/T30 hybrid [p23-3′NTR] generate an attenuated phenotype for SP in this plant host (M.R. Albiach-Marti et al., unpublished data). Apart of p23, CTV genome codes for other two possible silencing suppressors in citrus plants (p20 and CP) that could be involved in the developing of QD and SP phenotypes. Consequently, there is no evidence that other CTV symptom determinants would map similarly to the SY determinant of the T36 isolate. Thus, it is necessary to promote the research of the mapping of the decline and stem pitting determinants and to discover the nature of these specific virus/host interactions.

3.2.3 The possible role of D-RNAs in *Citrus tristeza virus* pathogenicity modulation

Models for DI RNA-mediated reduction in helper virus levels and symptom modulation include the enhancement of the PTGS (Pathak & Nagy, 2009). At least in one case, the presence of CTV D-RNAs was suggested to modulate SY development either increasing or decreasing symptom expression (G. Yang et al., 1999). Most of the CTV D-RNAs contain a complete region p23 and the 3´NTR (G. Yang et al., 1997) that is associated with SY symptom development (Albiach-Martí et al., 2010), thus they could have a role in symptom modulation. Additionally, p23 could be a suppressor of PTGS in citrus (Lu et al., 2004), thus probably could act increasing symptom development. The isolate T30 usually generates elevated concentration of several small D-RNAs during replication in some species of citrus plants, while T36 generates sporadically small and large D-RNAs. Similarly, some of the T36/T30 hybrids infecting *C. macrophylla* also accumulated D-RNAs, which did not appear to affect the T36/T30 hybrid replication in *C. macrophylla* (Albiach-Martí et al., 2010). These D-RNAs, created during replication of the T36/T30 hybrids, were specific of the CTV *C. macrophylla* infection since the multiplication of the same T36/T30 hybrids in sour orange did not produce any D-RNA. These results suggest that the generation of the D-RNAs could depend in part on host factors. Further research would elucidate whether D-RNAs (or DI-RNAs) contribute in CTV disease modulation.

3.3 *Citrus* host resistance to *Citrus tristeza virus* infection

As mentioned above, while pummelos, grapefruit, sour orange and Swingle citrumelo exhibit a differential degree of resistance depending on the CTV strain, *P. trifoliata*, *Swinglea glutinosa*, *Severinia buxifolia*, and the citranges remain resistant or immune to most of the CTV strains (Bernet et al., 2008; Garnsey et al., 1996, 1987; Fang & Roose, 1999; Folimonova et al., 2008; Yoshida, 1985, 1993). The major component of CTV resistance in *P. trifoliata* appears to be a single-gene trait (*Ctv*) (Gmitter et al., 1996). There is little information concerning the nature of the resistance genes of *S. glutinosa* and *S. buxifolia*, but their resistance phenotypes seem to differ from that of *P. trifoliata* (Herrero et al., 1996; Mestre et al., 1997). Analysis of differential gene expression TAG libraries from CTV inoculated *P. trifoliata* tissues, yielded 289 sequences differentially expressed, mostly related with metabolism and defense responses indicating a complex resistance mechanism (Cristofani-Yaly et al., 2007). Additionally, resistance in Chandler pummelo [*C. maxima* (Burm.) Merrill] is controlled by a single dominant gene (*Ctv2*) different from the resistant gene of *P. trifoliata* (Fang & Roose, 1999). Resistance of plants to viruses results from blockage of some necessary step in the virus life cycle. This blockage can result from the lack of a factor(s) in the plant that is necessary for virus multiplication and movement (*passive resistance*) or activation of the plant defense mechanism (*active resistance*). One of the most effective methods of characterizing resistance mechanisms is to determine whether the resistance is expressed at the single-cell level. Albiach-Martí et al., (2004) studied these CTV resistance mechanisms and reported efficient multiplication of CTV in resistant *P. trifoliata* and its hybrids Carrizo citrange, US119 and Swingle citrumelo, and in *S. buxifolia* and *S. glutinosa* protoplasts. Thus, the resistance mechanism in these plant species affects a viral step subsequent to replication and assembly of viral particles, probably preventing CTV movement. Similar results were obtained from CTV-inoculated protoplasts from resistant pummelo and sour orange plants (Albiach-Martí et al., 2004; M.R. Albiach-Marti, unpublished data).

CTV resistance in Duncan grapefruit (a descent of pummelo) and in sour orange have been investigated recently (Bernet et al., 2008; Comellas, 2009; Folimonova et al., 2008). The systemic invasion of the stable virus-based vector CTV-BC5/GFP (descendent of the T36-CTV9 construct) in Duncan grapefruit and sour orange, compared to those of the susceptible hosts C. macrophylla and Mexican lime and the tolerant host Madam Vinous sweet orange, were examined (Folimonova et al., 2008). CTV infection sites, after cell to cell movement, consisted of clusters of 3 to 12 cells in the susceptible species, while in Duncan grapefruit and sour orange displayed fewer infection sites limited to single cells, indicating absence of viral movement in both cases (Folimonova et al., 2008). After the analysis of the sour orange resistance to mild, SP and T36-CTV9 CTV strains, Comellas (2009) found, similarly to Folimonova el al. (2008), a limitation of viral movement in this host. This limitation was more accentuated for T36 and mostly complete for the mild strain. However, after two years post inoculation, both, T36 and the mild strain, accumulated in sour orange similarly to in Mexican lime revealing a transitory viral resistance (Comellas, 2009), which was also noticed by Bernet et al. (2008) that using another CTV isolate and QTL-linked markers reported that CTV resistance in sour orange was distinct to that of P. trifoliata. Sour orange resistance to CTV infection could be due to the plant RNA silencing mechanism (Folimonova et al., 2008). However, the separate analysis of accumulation of RNA, concentration of siRNAs in plant, as well as changes in the transcriptome of sour orange during CTV-host resistance periode, indicated that the silencing mechanism was not activated as well as the known plant resistant genes (Comellas, 2009). Therefore, sour orange probably exhibits a passive resistance where an inefficient interaction between CTV and the host factors blocks viral movement. This plant-host interaction could be mediated by p33 gene (see Figure 1 and section 2.1), which is related with CTV systemic infection in sour orange (Tatineni et al., 2011).This resistance possibly is broken after the rising of movement competent CTV mutants. Similarly, the resistant-breaking (NZRB, see section 2) CTV genotype from New Zealand has been reported to overcome the resistance of the P. trifoliata and its intergenic hybrids and generate a SP syndrome in this host (Harper et al., 2010). The development of the NZRB genotype could be due to the extensive use of the P. trifoliata rootstock since the late 1920s, giving enough time to the adaptation of CTV to P. trifoliata host, followed by the rising of the NZRB genotypes able to overcome the resistance genes of this citrus host (Harper et al., 2010).

3.4 CTV-plant infected interactions and modulation of aphid transmission

One of the essential features of CTV, from the disease control standpoint, is that it is transmitted by aphids. In fact, without this feature, CTV would have been easy to eradicate by eliminating CTV-infected trees, and probably CTV strains would be less exposed to genetic variability, which could allow virulent genotypes to arise. Viruliferous aphids of Toxoptera citricida (Kirkaldy) and Aphis gossypii (Glover) are able to transmit CTV. However, A. spiraecola (Patch) and T. aurantii (Boyer de Fonscolombe) have also been reported as CTV vectors, although with less efficiency. The aphid T. citricida is able to transmit CTV 6 to 25 times more effectively than A. gossypii in greenhouse conditions, it enables experimental CTV transmission using single aphids and it is more efficient and fast in the spatial and temporal spreading of CTV in citrus orchards (Moreno et al., 2008). Citrus is the primary host of T. citricida, while A. gossypii populations build up in other crops. Probably T. citricida evolved with citrus and CTV, thus this could explain its high efficiency transmitting this virus. *T. citricida* is present in almost all the citrus producing areas except the Mediterranean basin and areas of North America, where *A. gossypii* is the main vector (Cambra et al., 2000; Hermoso de Mendoza et al., 1988; Yokomi & Garnsey, 1987). However, recently *T. citricida* became established in Florida (Halbert et al., 2004) and has been detected in Northern Spain and Portugal (Ilharco et al., 2005), representing a risk to these citrus production areas on the southern Iberian Peninsula. When *T. citricida* appears in a new citrus area, where mild or QD CTV phenotypes are endemic, existing minor virulent SP populations, which were masked by the predominant mild or QD genotypes, have become prevalent. Therefore, the interaction between CTV and *T. citricida* seems to shift a specific CTV population from mild or QD phenotypes to severe SP ones (Halbert et al., 2004; Rocha-Peña et al., 1995). This special ability of *T. citricida* is partially explained by its high efficiency in viral transmission.

CTV transmission efficiency depends on the aphid species, the viral strain, the host plant and the environmental conditions, however it is not reported to be dependent on the CTV pathotype (Moreno et al., 2008). Although relationships between viral pathogenicity and aphid transmission have been barely studied (Froissart et al., 2010), it was reported that in viral pathosystems involving transmission by aphids, trips or whiteflies, viruses transform infected-plants in host of superior quality for their vectors, promoting changes in attractiveness, settlement or feeding host plant preference, together with changes on vector performance (development, fecundity, rate of population increase and survival), therefore increasing vector fitness that promotes viral spread and alters disease epidemiology (Belliure et al., 2005, 2008; Bosque-Pérez & Eigenbrode, 2011; Fereres & Moreno, 2009; Froissart et al., 2010). In a recent study, it was shown that CTV affects the fitness of its vector *A. gossypii* developing on sweet orange and Mexican lime infected with four distinct CTV-isolates (mild, QD and SP strains). CTV affected the performance of *A. gossypii* from negative to positive depending on the host plant and the virus strain. Assuming equal transmission efficiency, the frequency in field of the CTV isolates neutral or beneficial for *A. gossypii* should be higher than the frequency of detrimental ones (B. Belliure-Ferrer & M.R. Albiach-Martí, unpublished results). Similarly, the capability of *T. citricida* of shifting the CTV population could be explained by the existence of a specific interaction between the virulent strain and the citrus host that alters the aphid performance, increasing viral spread of severe strains. The links between determinants of CTV aphid transmission and the aphid vector together with the interactions between the CTV-infected host, CTV pathogenicity and the aphid fitness seems to depend on numerous factors. The elucidation of these complex and specific interactions will promote the development of better biotechnological methods to manage viral epidemiology and control CTV diseases.

3.5 Application of the strategies based on plant-host interaction for viral control

The control of the CTV diseases constitutes a continuous challenge (Bar-Joseph et al., 1989). General strategies include quarantine and budwood certification programs, elimination of infected trees and, as mentioned above, the use of *Tristeza*-tolerant rootstocks. Mild strain cross protection has been widely applied for millions of citrus trees in Australia, Brazil and South Africa to protect against SP economic losses (Bar-Joseph et al., 1989; Broadbent et al., 1991; Costa and Müller, 1980; Van Vuuren et al., 1993). This technique consists of deliberate preinmunization of trees with a mild isolate of CTV that prevents or reduces the disease caused by a more virulent isolate (Fraser, 1998). However, mild strain cross-protection has not yet provided effective protection against QD isolates, and this remains an important goal since it would allow to recover the use of the sour orange, the rootstock with superior agronomic qualities (Bar-Joseph & Dawson, 2008). Additionally, incorporating resistance genes from *P. trifoliata* into commercial varieties as sour orange by conventional breeding is presently unfeasible and might need further research (Rai, 2006).

Advances in genetically engineered protection against viruses by the generation of transgenic plants have lately been remarkable. However, incorporation of pathogen-derived resistance by plant transformation of CP and p23 or the 3´NTR has yielded variable results (Cervera et al., 2010; Dominguez et al., 2002; Fagoaga et al., 2006; López et al., 2010). Another biotechnological approach to control the virus, and eventually turn it from a pathogen into a molecular tool for citrus improvement, is the custom engineering of a recombinant mild strain cross-protection (Albiach-Martí et al 2010). Wider application of natural mild strain cross-protection has been limited by difficulty in finding mild isolates of CTV that effectively protect against SP and QD pathotypes (Bar-Joseph et al., 1989). Another problem is that natural mild CTV isolates may contain minor severe stem pitting variants that, upon aphid transmission, could become prevalent (Moreno et al., 1993; Velazquez-Monreal et al., 2009). Since only isolates within a closely related sequence group will cross-protect (Folimonova et al. 2010), naturally occurring mild T30-like isolates (Albiach-Martí et al., 2000b), would not protect against disease inducing isolates from other genotypes.

A valuable outcome was that the recombinant mild hybrid virus [P23-3´NTR] developed by Albiach-Marti et al. (2010) is able to protect efficiently citrus trees from SY caused by the parental virus (T36) (Figure 6, Right panel) and their hybrid genomic sequences are highly stable in citrus plants. The use of these recombinant hybrid constructs could offer a mechanism to custom engineer isolates that are both protective and free of disease induction potential. The stability noted in the T30/T36 constructs is also important for its application. This means that if naturally occurring mild strains cannot be found for stem-pitting or decline diseases control, it would be possible to map the disease determinant, remove it by recombinant DNA technology, and use the recombinant mild virus as a cross-protecting strain. Therefore, the potential feasibility of using engineered constructs of CTV for mitigating disease has been demonstrated (Albiach-Marti et al., 2010).

4. Conclusions

Interactions between the different CTV strains and their citrus hosts assembled a complicated plant pathosystem. The large number of citrus species, cultivars, varieties and hybrids that could be infected with a virus with a large genome, complex genetics, as well as with an extreme diversity of viral populations, generates numerous possibilities of plant-host interactions. These factors complicate the study of the CTV pathogenicity and the development of reliable strategies for viral control. Although a remarkable advance in the knowledge of CTV genetics and the diversity of CTV viral populations have been achieved, the interaction between virus and host and particularly the mechanisms involved in the development of the disease are still mostly a mystery. Therefore, further attention needs the study of the interactions between viral products, the different citrus hosts and the vector transmission factors, which are the basis of pathogenicity, host resistance and viral epidemiology. The success of the citrus management strategies depends on a deep understanding of these interactions, as well as on the elucidation of the diversity, and evolutionary relationships of the CTV isolates present in a particular citrus area to protect. In addition to make available methods to rapidly discriminate virulent from mild isolates in order to reduce risks derived from introduction and dispersal of virulent isolates and to properly monitor mild cross-protection.

Recently, pushing the molecular virology methodology to further limits, molecular tools have been developed to clone each of the CTV pathotypes and examine them individually in *N. benthamiana* protoplasts or in a particular citrus host to study the genetics and biology of the virus and virus-host interactions like pathogenicity and host resistance. However, further efforts are needed for developing additional methodologies to map the QD and SP determinants and to study their pathogenicity mechanism, as well as to elucidate the possible role of CTV D-RNAs in symptom modulation, in addition to determine the viral factors related to sour orange and *P. trifoliata* resistance and the relationships between CTV pathogenicity, aphid fitness and virus dispersal. This knowledge must be applied to elaborate appropriate quarantine and eradication programs as well as to develop biotechnological approaches of viral control, which exploit virus plant-host interactions for viral control, such as sequence-based control strategies. Resistant transgenic plants based on PTGS and self-immunization by scFv expression mechanisms, against specific viral sequences, are already developed. In addition, engineered mild strain cross-protection demonstrated its potential in excluding superinfection by severe strains. Both biotechnological strategies retain high possibilities of success in the proper management of devastating CTV diseases.

5. Acknowledgments

The author is grateful to W.O. Dawson, S. Gowda, B. Belliure-Ferrer and Beatriz Sabater for their support, stimulating scientific discussions and critical review of the manuscript.

6. References

Albiach-Martí, M.R., Guerri, J., Hermoso de Mendoza, A., Laigret, F., Ballester-Olmos, J.F. & Moreno, P. (2000a). Aphid transmission alters the genomic and defective RNA populations of citrus tristeza virus. *Phytopathology*, Vol. 90, pp. (134-138).

Albiach-Martí, M.R., Mawassi, M., Gowda, S., Satyanarayana, T., Hilf, M.E., Shanker, S., Almira, E.C., Vives, M.C., López, C., Guerri, J., Flores, R., Moreno, P., Garnsey S.M. & Dawson W.O. (2000b). Sequences of *Citrus tristeza virus* separated in time and space are essentially identical. *Journal of Virology*, Vol. 74, pp. (6856-6865).

Albiach-Martí, M.R., Grosser, J.W., Gowda, S., Mawassi, M., Satyanarayana, T., Garnsey, S.M. & Dawson, W.O. (2004). Citrus tristeza virus replicates and forms infectious virions in protoplast of resistant citrus relatives. *Molecular Breeding*, Vol. 14, pp. (117-128).

Albiach-Marti, M.R., Robertson, C., Gowda, S., Tatineni, S., Belliure, B., Garnsey, S.M., Folimonova, S.Y., Moreno P. & Dawson.W.O. (2010). The pathogenicity determinant of *Citrus tristeza virus* causing the seedling yellows syndrome maps at the 3'-terminal region of the viral genome. *Molecular Plant Pathology*, Vol. 11, pp. (55–67).

Ayllón, M.A., López, C., Navas-Castillo, J., Mawassi, M. & Dawson W.O. (1999a). New defective RNAs from citrus tristeza virus: evidence for a replicase driven template switching mechanism in their generation. *Journal General Virology*, Vol. 80, pp. (871-82).

Ayllón, M.A., Rubio, L., Moya, A., Guerri, J. & Moreno, P. (1999b). The haplotype distribution of two genes of *Citrus tristeza virus* is altered after host change or aphid transmission. *Virology*, Vol. 255, pp. (32-39).

Ayllón, M.A., Satyanarayana, T., Gowda, S., & Dawson W.O. (2005). An atypical 3'-controller element mediates low-level transcription of the p6 subgenomic mRNA of *Citrus tristeza virus*. *Molecular Plant Pathology*, Vol. 6, No. 2, pp. (165–176).

Bar-Joseph, M., Marcus R. & Lee. R. F. (1989) The continuous challenge of citrus tristeza virus control. *Annual Review Phytopathology*, Vol. 27, pp. (291-316).

Bar-Joseph, M. & Dawson, W.O. (2008) *Citrus tristeza virus*. In *Encyclopedia of Virology*, Third edition evolutionary biology of viruses. Elsevier Ltd. Vol. 1, pp. (161–184)

Barzegar, A., Rahimian H., & Sohi H.H. (2009). Comparison of the minor coat protein gene sequences of aphid-transmissible and -nontransmissible isolates of *Citrus tristeza virus*. *Journal of General Plant Pathology*, Vol. 76, No.2, pp. (143-151).

Bernet, G.P., Gorris, M.T., Carbonell, E.A., Cambra, M. & Asins, M.J. (2008). Citrus tristeza virus resistance in a core collection of sour orange based on a diversity study of three germplasm collections using QTL-linked markers. *Plant Breeding*, Vol. 127, pp (398-406)

Belliure, B., Janssen, A., Maris, P.C., Peters, D. & Sabelis, M.W. (2005). Herbivore arthropods benefit from vectoring plant viruses. *Ecology Letters*, Vol. 8, pp. (70-79).

Belliure, B., Janssen, A., &Sabelis, M.W. (2008). Herbivore benefits from vectoring plant virus through reduction of period of vulnerability to predation. *Oecologia*, Vol. 156, pp. (797–806).

Blok, J., A. Mackenzie, P. Guy & A. Gibbs. 1987. Nucleotide sequence comparisons of turnip yellow mosaic virus from Australia and Europe. *Arch.Virol.*, Vol. 97, pp. (283-295).

Bosque-Pérez N.A. & Eigenbrode S.D. (2011). The influence of virus-induced changes in plants on aphid vectors: insights from luteovirus pathosystems. *Virus Research*, Vol.159, pp (201-205).

Broadbent, P., Bevington, K.B. & Coote, B.G. (1991) Control of stem pitting of grapefruit in Australia by mild strain cross protection. In *Proceedings of the 11th Conference of the International Organization of Citrus Virologists* (Brlansky, R.H., Lee, R.F. & Timmer, L.W., eds). Riverside, CA: IOCV, pp. (64-70).

Cambra, M., Gorris, M.T., Marroquín, C., Román, M.P., Olmos, A., Martínez, P.C., Hermoso de Mendoza, A.H., López, A. & Navarro, L. (2000) Incidence and epidemiology of citrus tristeza virus in the Valencian Community of Spain. *Virus Research*, Vol. 71, pp (85-95).

Cervera, M., Esteban, O., Gil, M., Gorris, M.T., Martínez, M.C., Peña, L. & Cambra, M. (2010). Transgenic expression in citrus of single-chain antibody fragments specific to *Citrus tristeza virus* confers long-term virus resistance. *Transgenic Research*, Vol. 19, pp. (1001-1015)

Che, X., Piestum, D., Mawassi, M., Satyanayanana, T., Gowda, S., Dawson W.O. & Bar-Joseph, M. (2001). 5'coterminal subgenonic RNAs in citrus tristeza virus-infected cells. *Virology*, Vol. 283, pp. (374-381).

Che, X., Dawson, W.O. & Bar-Joseph, M. (2003). Defective RNAs of Citrus tristeza virus analogous to Crinivirus genomic RNAs. *Virology*, Vol. 310, pp. (298-309).

Comellas, M. (2009). Estudio de la interacción entre naranjo amargo y el virus de la Tristeza de los Cítricos. *PhD*. Universidad Politécnica de Valencia. Valencia, Spain.

Costa, A.S. & Müller, G.W. (1980). Tristeza control by cross protection: a US-Brazil cooperative success. *Plant Disease*, Vol. 64, pp. (538–541).

Cristofani-Yaly, M., Berger, I.J., Targon, M.L.P.N., Takita, M.A., Dorta1, S.O., Freitas-Astúa, J., de Souza, A.A., Boscariol-Camargo, R.L., Reis, M.S. & Machado, M.A. (2007). Differential expression of genes identified from *Poncirus trifoliata* tissue inoculated with CTV through EST analysis and in silico hybridization. *Genetics and Molecular Biology*, Vol. 30, No. 3, pp. (972-979).

Culver, J.N. & Padmanabhan, M.S. (2007). Virus-Induced Disease: Altering Host Physiology One Interaction at a Time. *Ann. Rev. Phytopathol.*, Vol. 45, pp. (221-243).

Dolja, V.V., Kreuze, J.F. & Valkonen, J.P.T. (2006). Comparative & functional genomics of closteroviruses. *Virus Research*, Vol. 117, pp. (38-51).

Domingo, E. & Holland, J. J. (1997). RNA virus mutations & fitness for survival. *Ann. Rev. Microbiol.*, Vol. 51, pp. (151-178).

Domínguez, A., Hermoso de Mendoza, A., Guerri, J., Cambra, M., Navarro, L., Moreno, P. & Peña, L. (2002). Pathogen-derived resistance to *Citrus tristeza virus* (CTV) in transgenic Mexican lime (*Citrus aurantifolia* (Christm.) Swing.) plants expressing its p25 coat protein gene. *Molecular Breeding*, Vol. 10, pp. (1-10).

D'Urso, F., Ayllón M.A., Rubio, L., Sambade, A., Hermoso de Mendoza, A., Guerri, J. & Moreno, P. (2000). Contribution of uneven distribution of genomic RNA variants of citrus tristeza virus (CTV) within the plant to changes in the viral population following aphid transmission. *Plant Pathology*, Vol. 49, pp. (288–294).

Fagoaga, C., López, C., Moreno, P., Navarro, L., Flores R. & Peña L. (2005). Viral-like symptoms induced by the ectopic expresión of the p23 of *Citrus tristeza virus* are citrus specific & do not correlate with the patogenicity of the virus strain. *Molecular Plant Microbe Interaction*, Vol. 18, pp. (435-445).

Fagoaga, C., López, C., Hermoso de Mendoza, A.H., Moreno, P., Navarro, L., Flores R. & Peña L. (2006). Post-transcriptional gene silencing of the p23 silencing suppressor of *Citrus tristeza virus* confers resistance to the virus in transgenic Mexican lime. *Plant Molecular Biology*, Vol. 66, pp. (153-165).

Fang, D.Q. & Roose M.L. (1999). A novel gene conferring *Citrus tristeza virus* resistance in Citrus maxima (Burm.) Merrill. *Hort Science*, Vol. 34, pp. (334-335).

Febres, V.J., Ashoulin, L., Mawassi, M., Frank, A., Bar-Joseph, M., Manjunath, K.L., Lee R. F. & Niblett C.L. (1996) The p27 protein is present at one end of citrus tristeza virus particles. *Phytopathology*, Vol. 86, pp. (1331-1335).

Fereres, A. & Moreno, A. (2009). Behavioural aspects influencing plant virus transmission by homopteran insects. *Virus Research*, Vol. 141, No 2, pp. (158-168).

Folimonov, A.S., Folimonova, S.Y., Bar-Joseph, M. & Dawson, W.O. (2007). A stable RNA virus-based vector for citrus trees. *Virology*, Vol. 368, pp. (205-216).

Folimonova, S. Y., Folimonov, A. S., Tatineni, S. & Dawson, W. O. (2008). *Citrus tristeza virus*: survival at the edge of the movement continuum. *Journal of Virology*, Vol. 82, pp. (6546-6556).

Folimonova, S.Y., Robertson, C.J., Shilts, T., Folimonov, A.S., Hilf, M.E., Garnsey S.M., and Dawson W.O. (2010). Infection with strains of *Citrus tristeza virus* does not exclude superinfection by other strains of the virus. *Journal of Virology*, Vol. 84, No 3, pp. (1314-25)

Fraile, A., F. Escriu, M. A. Aranda, J. M. Malpica, A. J. Gibbs & Garcia-Arenal, F. (1997). A century of tobamovirus evolution in an Australian population of *Nicotiana glauca*. *Journal of Virology*, Vol. 71, pp. (8316-8320).

Fraser L. (1952) Seedling yellows, an unreported virus disease of citrus. *Agricultural Gazette N.S. Wales*, Vol. 63, pp. (125-131).

Fraser, R.S.S. (1998) Introduction to classical cross protection. In *Methods in Molecular Biology, Plant virus protocols* (Foster, D. & Taylor S.J., eds). Totowa, NJ, USA: Humana Press, Vol 81, pp. (13-24).

Froissart, R., Doumayrou, J.,Vuillaume, F., Alizon S.& Michalakis. Y. (2010). The virulence-transmission trade-off in vector-borne plant viruses: a review of (non-)existing studies. *Philosophycal Transactions of Royal Society B*, Vol. 365, pp. (1907–1918)

Gandia, M., Conesa, A., Ancillo, G., Gadea, J., Forment, J., Pallás, V., Flores, R., Duran-Vila, N., Moreno, P. & Guerri, J. (2007). Transcriptional response of Citrus aurantifolia to infection by *Citrus tristeza virus*. *Virology*, Vol. 367, pp. (298-306).

Garnsey, S.M., Barrett, H.C. & Hutchinson, D.J. (1987). Identification of citrus tristeza virus resistance in citrus relatives & its potential applications. *Phytophylactica*, Vol. 19, pp. (187-191).

Garnsey, S.M., Su, H.J. & Tsai, M.C. (1996). Differential susceptibility of pummelo & Swingle citrumelo to isolates of *Citrus tristeza virus*. In *Proceedings of the 13th Conference of the International Organization of Citrus Virologists* (da Graça, J.V., Moreno, P. & Yokomi., R.K., eds). Riverside, CA: IOCV, pp. (138-146).

Garnsey, S.M., Civerolo, E.L., Gumpf, D.J., Paul, C., Hilf, M.E., Lee, R.F., Brlansky, R.H., Yokomi, R.K. & Hartung, J.S (2005). Biological characterization of an international collection of *Citrus tristeza virus* (CTV) Isolates. In *Proceedings of the 16th Conference of the International Organization of Citrus Virologists* (Hilf, M.E., Duran-Vila, N. & Rocha-Peña, M.A., eds). Riverside, CA: IOCV, pp. (75-93).

Ghorbel, R., López, C., Moreno, P., Navarro, L., Flores, R. & Peña, L. (2001). Transgenic citrus plants expressing the Citrus tristeza virus p23 protein exhibit viral-like symptoms. *Molecular Plant Pathology*, Vol. 2, pp. (27-36).

Gmitter, F.G. Jr., Xiao, S.Y., Huang, S. & Hu, X.L. (1996). A localized linkage map of the citrus tristeza virus resistance gene region. *Theoretical Applied Genetics*. Vol.92, pp. (688-695).

Gowda, S., Satyanayanana, T., Davis, C.L., Navas-Castillo, J., Albiach-Martí, M.R., Mawassi, M., Valkov, N., Bar-Joseph, M., Moreno, P. & Dawson W.O. (2000). The p20 gene product of *Citrus tristeza virus* accumulates in the amorphous inclusion bodies. *Virology*, Vol. 274, pp. (246-254).

Gowda, S., Satyanayanana, T., Ayllón, M.A., Albiach-Martí, M.R., Mawassi, M., Rabindran S. & Dawson W.O. (2001). Characterization of the cis-acting elements controlling subgenomic mRNAs of *Citrus tristeza virus*: production of positive-and negative-stranded 3' -terminal and positive-stranded 5' terminal RNAs. Virology, Vol. 286, pp. (134-151).

Gowda, S., Ayllón, M.A., Satyanayanana, T., Bar-Joseph, M. & Dawson W.O. (2003a). Transcription strategy in a *Closterovirus*: a novel 5′-proximal controler element of *Citrus tristeza virus* produces 5′- & 3′-terminal subgenomic RNAs & differs from 3′open reading frame controler elements. *Journal of Virology*, Vol. 77, pp. (340-352).

Gowda, S., Satyanayanana, T., Ayllón, M.A., Moreno, P., Flores, R. & Dawson W.O. (2003b). The conserved structures of the 5′nontranslated region of *Citrus tristeza virus* are involved in replication & virion assembly. *Virology*, Vol. 317, pp. (50-64).

Gowda, S., Satyanarayana, T., Robertson, C.J., Garnsey, S.M. & Dawson, W.O. (2005). Infection of citrus plants with virions generated in *Nicotiana benthamiana* plants agroinfiltrated with binary vector based *Citrus tristeza virus*. In *Proceedings of the 16th Conference of the International Organization of Citrus Virologists* (Hilf, M.E., Duran-Vila, N. & Rocha-Peña, M.A., eds). Riverside, CA: IOCV, pp. (23-33).

Gowda, S., Tatineni S., Folimonova, S.Y., Hilf, M.E. &. Dawson, W.O. (2009). Accumulation of a 5′ proximal subgenomic RNA of *Citrus tristeza virus* is correlated with encapsidation by the minor coat protein. *Virology*, Vol. 389, pp. (122–131).

Halbert, S.E., Genc, H., Çevik, B., Brown, L.G., Rosales, I.M., Manjunath, K.L., Pomerinke, M., Davison, D.A., Lee, R.F. & Niblett, C.L. (2004) Distribution & characterization of *Citrus tristeza virus* in South Florida following establishment of *Toxoptera citricida*. *Plant Disease*, Vol. 88, pp. (935-941).

Harper, S. J., Dawson, T.E. & Pearson, M.N. (2010). Isolates of *Citrus tristeza virus* that overcome *Poncirus trifoliata* resistance comprise a novel strain. *Archives of Virolology*, Vol. 155, pp. (471-480.

Harper, S.J., Dawson, T.E.& Pearson, M.N. (2009). Complete genome sequences of two distinct and diverse Citrus tristeza virus isolates from New Zealand. *Archives of Virolology*, Vol. 154, pp. (1505–1510).

Herrero, R., Asins, M. J., Pina, J. A., Carbonell, E. A., & Navarro, L. (1996). Genetic diversity in the orange subfamily *Aurantioideae*. II. Genetic relationships among genera and species. *Theorethical Applied Genetics*, Vol. 93: pp. (1327-1334).

Hilf, M.E., Karasev, A.V., Pappu, H.R., Gumpf, D.J., Niblett, C.L. & Garnsey, S.M. (1995). Characterization of citrus tristeza virus subgenomic RNAs in infected tissue. *Virology*, Vol. 208, pp. (576-582).

Hilf, M.E., Karasev, A.V., Albiach-Martí, M.R., Dawson, W.O. & Garnsey, S.M. (1999). Two paths of sequence divergence in the citrus tristeza virus complex. *Phytopathology*, Vol. 89, pp. (336-342).

Hilf, M.E., Mavrodieva, V.A. & Garnsey, S.M. (2005) Genetic marker analysis of a global collection of isolates of *Citrus tristeza virus*: Characterization & distribution of CTV genotypes & association with symptoms. *Phytopathology*, Vol. 95, pp. (909-917).

Hermoso de Mendoza, A., Ballester-Olmos, J.F., Pina, J.A., Serra, J.A. & Fuertes, C. (1988). Difference in transmission efficiency of citrus tristeza virus by Aphis gossypii using sweet orange, mandarin or lemon trees as donor or receptor host plant. In *Proceedings of the 10th Conference of the International Organization of Citrus Virologists* (Timmer, L.W., Garnsey, S.M. & Navarro L., eds). Riverside, CA, IOCV, pp. (62-64).

Ilharco, F.A., Sousa-Silva, C.R. & Alvarez-Alvarez, A. (2005). First report on *Toxoptera citricidus* (Kirkaldy). *Spain and continental Portugal. Agron. Lusit.*, Vol. 51, pp. (19-21)

Kitajima, E.W., Silva, D.M., Oliveira, A.R., Muller, G.W. & Costa, A.S. (1964). Threadlike particles associated with tristeza disease of citrus. *Nature*, Vol. 201, No (101), pp. (1-112).

Karasev, A.V., Boyko, V.P., Gowda, S., Nikolaeva, O.V., Hilf, M.E., Koonin, E.V. Niblett, C.L. Cline, K., Gumpf, D.J., Lee, R.F., Garnsey, S.M., Lewandowski D.J. & Dawson W.O. (1995). Complete sequence of the citrus tristeza virus RNA genome. *Virology*, Vol. 208, pp. (511-520).

Karasev, A.V., Hilf, M.E., Garnsey, S.M. & Dawson, W.O. (1997). Transcriptional Strategy of *Closteroviruses*: Maping the 5′termini of the citrus tristeza virus subgenomic RNAs. *Journal of Virology*, Vol. 71, pp. (6233-6236).

López, C., Ayllón, M.A., Navas-Castillo, J., Guerri, J., Moreno P. & Flores R. (1998). Molecular variability of the 5′ & 3′ terminal regions of citrus tristeza virus RNA. *Phytopathology*, Vol. 88, pp. (685-691).

López, C., Navas-Castillo, J., Gowda, S., Moreno P. & Flores R. (2000). The 23-kDa protein coded by the 3′-terminal gene of citrus tristeza virus is an RNA-binding protein. *Virology*, Vol. 269, pp. (462-470).

Lopez, C., Cervera, M., Fagoaga, C., Moreno, P., Navarro, L., Flores R. & Peña L. (2010). Accumulation of transgene-derived siRNAs is not sufficient for RNAi-mediated protection against *Citrus tristeza virus* in transgenic Mexican lime. *Molecular Plant Pathology*, Vol. 11, pp (33–41).

Lu, R., Folimonov, A., Shintaku, M., Li, W.X., Falk, B.W., Dawson, W.O. & Ding, S.W. (2004). Three distinct suppressors of RNA silencing encoded by a 20-Kb viral RNA genome. *Proc. Natl. Acad. Sci. USA*, Vol. 101, pp. (15742-15747).

Mawassi, M., Karasev, A.V., Mietkiewska, E., Gafny, R., Lee, R.F., Dawson, W.O. & Bar-Joseph M. (1995a). Defective RNA molecules associated with citrus tristeza virus. *Virology*, Vol. 208, pp. (383-387).

Mawassi, M., Mietkiewska, E., Hilf, M.E., Ashoulin, L., Karasev, A.V., Gafny, A.V., Lee, R.F., Garnsey, S.M., Dawson W.O. & Bar-Joseph M. (1995b) Multiple species of defective RNAs in plants infected with citrus tristeza virus. *Virology*, Vol. 214, pp. (264-268).

Mawassi, M., Mietkiewska, E., Gofman, R., Yang G. & Bar-Joseph M. (1996). Unusual sequence relationships between two isolates of citrus tristeza virus. *Journal General Virology*, Vol. 77, pp. (2359-2364).

Mawassi, M., Satyanayanana, T., Albiach-Martí, M.R., Gowda, S., Ayllón, M.A, Robertson, C. & Dawson W.O. (2000). The fitness of *Citrus tristeza virus* defective RNA is affected by the length of their 5′ and 3′ termini and by coding capacity. *Virology*, Vol. 275, pp. (42-56).

Melzer, M.J., Borth, W.B., Sether, D.M., Ferreira, S., Gonsalves D. & Hu J.S. (2010). Genetic diversity and evidence for recent modular recombination in Hawaiian *Citrus tristeza virus*. *Virus Genes*, Vol. 40, No. 1, pp. (111-118).

Mestre, P.F., Asins, M.J., Pina, J.A. & Navarro, L. (1997b). Efficient search for new resistant genotypes to the citrus tristeza closterovirus in the orange subfamily *Aurantioideae*. *Theorethical Applied Genetics*, Vol. 95, pp. (1282-1288).

Moreno, P., Guerri, J., Ballester-Olmos, J.F., Albiach, R. & Martínez, M.E. (1993). Separation and interference of strains from a citrus tristeza virus isolate evidenced by biological activity and double-stranded RNA (dsRNA) analysis. *Plant Patholology*, Vol. 42, pp. (35–41).

Moreno P., Ambrós S., Albiach-Martí M.R., Guerri J. & Peña, L. (2008). *Citrus tristeza virus*: a pathogen that changed the course of the citrus industry *Molecular Plant Patholology*, Vol. 9, pp. (251- 268).

Nakajima, K., Desselberger, U. & Palese, P. (1978). Recent human influenza A (H1N1) viruses are closely related genetically to strains isolated in 1950. *Nature* Vol. 247, pp. (334-339.

Navas-Castillo J., Albiach-Martí, M.R., Gowda, S., Hilf, M.E., Garnsey S.M & Dawson W.O. (1997) Kinetics of accumulation of *Citrus tristeza virus* RNAs. *Virology*, Vol. 228, pp. (92-97).

Ng, J.C.K., & Falk, B.W. (2006). Virus-vector interactions mediating nonpersistent and semipersistent plant virus transmission. *Ann. Rev. Phytopathol.* Vol. 44, pp. (183-212).

Nicolosi, E., Deng, Z. N., Gentile, A., La Malfa, S., Continella, G., & Tribulato, E. (2000). Citrus phylogeny & genetic origin of important species as investigated by molecular markers. Theoretical Applied Genetics, Vol. 100, pp. (1155-1166).

Pathak K.B. & Nagy P.D. (2009). Defective Interfering RNAs: Foes of Viruses & Friends of Virologists *Viruses*, Vol. 1, pp. (895-919).

Permar, T.A., Garnsey, S.M., Gumpf, D.J. & Lee, R.F. (1990), A monoclonal antibody which discriminates strains of citrus tristeza virus. *Phytopathology*, Vol. 80, pp. (224–228).

Qu, F. & Morris, J. (2005). Suppressors of RNA silencing encoded by plant viruses and their role in viral infections. *FEBS Letters*. Vol. 579, pp. (5958-5964).

Rai, M. (2006) Refinement of the *Citrus tristeza virus* resistance gene (*Ctv*) positional map in *Poncirus trifoliata* and generation of transgenic grapefruit (*Citrus paradisi*) plant lines with candidate resistance genes in this region. *Plant Molecular Biology*, Vol. 61, pp. (399-414).

Robertson, C.J., Garnsey, S.M., Satyanarayana, T., Folimonova, S., & Dawson, W.O. (2005). Efficient infection of citrus plants with different cloned constructs of *Citrus tristeza virus* amplified in *Nicotiana benthamiana* protoplasts. In *Proceedings of the 16th Conference of the International Organization of Citrus Virologists* (Hilf, M.E., Duran-Vila, N. & Rocha-Peña, M.A., eds). Riverside, CA: IOCV, pp. (.187-195).

Rocha-Peña, M.A., Lee, R.F., Lastra, R., Niblet, C.L., Ochoa-Corona, F.M., Garnsey, S.M. & Yokomi, R.K. (1995) *Citrus tristeza virus* and its vector *Toxoptera citricida*. *Plant Disease*, Vol. 79, pp. (437-445).

Rodríguez- Cerezo, E., Gamble Klein, P. & Shaw, J.G (1991). A determinant of disease symptom severity is located in the 3`-terminal noncoding regions of the RNA of a plant virus. *Proc. Natl. Acad. Sci. USA*. Vol. 88, pp. (9863-9867).

Roy, A. & Brlansky, R. H. (2009). Population dynamics of a Florida *Citrus tristeza virus* isolate and aphid-transmitted subisolates: identification of three genotypic groups and recombinants after aphid transmission. *Phytopathology*, Vol. 99, No (11), pp. (1297-1306).

Roy, A. & Brlansky, R. H. (2010).Genome analysis of an orange stem pitting *Citrus Tristeza Virus* isolate reveals a novel recombinant genotype. *Virus Research*,. Vol. 151(2), pp. (118-30).

Roy, A., Ananthakrishnan, G., Hartung, J.S. & Brlansky, R. H. (2010).Development and application of a multiplex reverse-transcription polymerase chain reaction assay for screening a global collection of *Citrus tristeza virus* isolates. *Phytopathology*, Vol. 100, pp (1077-1088).

Ruiz-Ruiz, S., Moreno, P., Guerri, J. & Ambrós, S. (2006). The complete nucleotide sequence of a severe stem pitting isolate of *Citrus tristeza virus* from Spain: comparison with isolates from different origins. *Archives of Virology*, Vol. 151, pp. (387-398).

Ruiz-Ruiz, S., Navarro, B., Gisel, A., Peña, L., Navarro, L., Moreno, P., Di Serio, F. & Flores, R. (2011). *Citrus tristeza virus* infection induces the accumulation of viral small RNAs (21-24-nt) mapping preferentially at the 3'-terminal region of the genomic RNA and affects the host small RNA profile. *Plant molecular biology*, Vol. 75, pp. (607–619)

Sambade, A., López, C., Rubio, L., Flores, R., Guerri, J. & Moreno, P. (2003). Polymorphism of a specific region in the gene p23 of *Citrus tristeza virus* allows differentiation between mild & severe isolates. *Archives of Virology*, Vol. 148, pp. (2281–2291.

Satyanayanana, T., Gowda, S., Boyko, V.P., Albiach-Martí, M.R., Mawassi, M., Navas-Castillo, J., Karasev, A.V., Dolja, V., Hilf, M.E., Lew&owsky, D.J., Moreno, P., Bar-Joseph, M., Garnsey S. M. & Dawson W.O. (1999). An engineered closterovirus RNA replicon & analysis of heterologous terminal sequences for replication. *Proc. Natl. Acad. Sci. USA*, Vol. 96, pp. (7433-7438).

Satyanayanana, T., Gowda, S., Mawassi, M., Albiach-Martí, M.R., Ayllón, M.A., Robertson, C., Garnsey S. M. & Dawson W.O. (2000). Closterovirus encoded HSP70 homolog & p61 in addition to both coat proteins function in efficient virion assembly. *Virology*, Vol. 278, pp. (253-265).

Satyanayanana, T., Bar-Joseph, M., Mawassi, M., Albiach-Martí, M.R., Ayllón, M.A., Gowda, S., Hilf, M.E., Moreno, P., Garnsey S. M. & Dawson W.O. (2001). Amplification of *Citrus tristeza virus* from a cDNA clone & infection of citrus trees. *Virology*, Vol. 280, pp. (87-96).

Satyanarayana, T., Gowda, S., Ayllón, M.A., Albiach-Martí, M.R. & Dawson W.O. (2002a). Mutational analysis of the replication signals in the 3´-non translated region of *Citrus tristeza virus*. *Virology*, Vol. 300, pp. (140-152).

Satyanarayana, T., Gowda, S., Ayllón, M.A., Albiach-Martí, M.R., Rabindram, R. & Dawson W.O. (2002b). The p23 protein of *Citrus tristeza virus* controls asymmetrical RNA accumulation. *Journal Virology*, Vol. 76, pp. (473-483).

Satyanayanana, T., Gowda, S., Ayllón, M.A. & Dawson W. O. (2004). Closterovirus bipolar virion: evidence for initiation of assembly by minor coat protein and its restriction to the genomic RNA 5´region. *Proc. Natl. Acad. Sci. USA*, Vol. 101, pp. (799-804).

Scora, R.W. (1988). Biochemistry, taxonomy & evolution of modern cultivated citrus. In *Proceedings of the 6th International Citrus Congress*. (Goren, R. & Mendel, K., eds). Margraf Scientific Books, Weikersheim, pp. (277-289).

Schneider, H. (1959) The anatomy of tristeza-virus-infected citrus. In *Citrus virus diseases* (Wallace, J.M., eds). Berkeley, CA: Univ. Calif. Div. Agr. Sci. pp. (73-84).

Stewart, L.R., Medina, V., Tian, T., Turina, M., Falk, B.W., Ng, J.C.K. (2010). A mutation in the *Lettuce infectious yellows virus* minor coat protein disrupts whitefly transmission but not in planta systemic movement. *Journal of Virology*, Vol. 84: pp. (12165-12173).

Suastika, G., Natsuaki, T., Terui, H., Kano, T. Ieki, H., & Okuda, S. (2001). Nucleotide Sequence of *Citrus tristeza virus* seeding yellows isolate. *Journal General Plant Pathology*, Vol. 67, pp. (73-77).

Tatineni, S., Robertson, C., Garnsey, S. M. Bar-Joseph, M., Gowda, S., & Dawson, W.O. (2008). Three genes of *Citrus tristeza virus* are dispensable for infection and movement throughout some varieties of citrus trees. *Virology*. Vol. 376, No. 2, pp. (297-307).

Tatineni, S. , Gowda, S. & Dawson, W. (2010). Heterologous minor coat proteins of Citrus tristeza virus strains affect encapsidation, but the coexpression of HSP70h and p61 restores encapsidation to wild-type levels. *Virology*, Vol. 402, pp. (262-270).

Van Vuuren, S.P., Collins, R.P. & da Graça, J.V. (1993) Evaluation of citrus tristeza virus isolates for cross protection of grapefruit in South Africa. *Plant Disease*, Vol. 77, pp. (24-28).

Velazquez-Monreal, J.J., Mathews, D.M., Dodds, J.A. (2009). Segregation of Distinct Variants from *Citrus tristeza virus* Isolate SY568 Using Aphid Transmisión. *Phytopathology*, Vol. 99, pp. (1168-1176).

Vives, M.C., Rubio, L., López, C., Navas-Castillo, J., Albiach-Martí, M.R., Dawson, W.O., Guerri, J., Flores R. & Moreno P. (1999). The complete genome sequence of the major component of a mild citrus tristeza virus isolate. *Journal General Virology*, Vol. 80, pp. (811-816).

Vives, M.C., Rubio, L., Sambade, A., Mirkov, Moreno, P. & Guerri, J. (2005). Evidence of multiple recombination events between two RNA sequence variants within a *Citrus tristeza virus* isolate. *Virology*, Vol. 331, pp. (232-237).

Voinnet, O. (2005) Induction & suppression of RNA silencing: insights from viral infections. *Nature Gen. Rev.* Vol. 6, pp. (206-220).

Weng, Z., Barthelson, R., Gowda, S., Hilf, M. E., Dawson, W. O., Galbraith, D. W., & Xiong, Z. (2007). Persistent infection & promiscuous recombination of multiple genotypes of an RNA virus within a single host generate extensive diversity. *PLoS ONE*, Vol. 2, No (9), e917.

Yang, G., Mawassi, M., Gofman, R., Gafny, R. & Bar-Joseph, M. (1997) Involvement of a subgenomic mRNA in the generation of a variable population of defective citrus tristeza virus molecules. *Journal of Virology*, Vol. 71, pp. (9800–9802).

Yang, G., Che, X., Gofman, R., Ben Shalom, Y., Piestun, D., Gafny, R., Mawassi, M., Bar-Joseph, M. (1999). D-RNA molecules associated with subisolates of the VT strain of citrus tristeza virus which induce different seedling-yellows reactions. *Virus Genes*, Vol. 19, pp. (5-13).

Yang Z.N., Mathews, D.M., Dodds, J.A. & Mirkov T.E. (1999). Molecular characterization of an isolate of citrus tristeza virus that causes severe symptoms in sweet orange. *Virus Genes*, Vol. 19, pp. (11-142).

Yokomi, R.K. & Garnsey, S.M. (1987) Transmision of citrus tristeza virus by A. gossypii & A. citricola in Florida. *Phytophylactica*, Vol. 19, pp. (169-172).

Yoshida, T. (1985) Inheritance of susceptibility to citrus tristeza virus in trifoliate orange (Poncirus trifoliata Raf.). *Bull. Fruit Tree Res. Sta.*, Vol.12, pp. (17-25).

Yoshida, T. (1993) Inheritance of immunity to citrus tristeza virus of trifoliate orange in some citrus intergeneric hybrids. *Bull. Fruit Tree Res. Sta.*, Vol. 25, pp. (33-43).

Permissions

The contributors of this book come from diverse backgrounds, making this book a truly international effort. This book will bring forth new frontiers with its revolutionizing research information and detailed analysis of the nascent developments around the world.

We would like to thank Profs. María Laura García and Víctor Romanowski, for lending their expertise to make the book truly unique. They have played a crucial role in the development of this book. Without their invaluable contribution this book wouldn't have been possible. They have made vital efforts to compile up to date information on the varied aspects of this subject to make this book a valuable addition to the collection of many professionals and students.

This book was conceptualized with the vision of imparting up-to-date information and advanced data in this field. To ensure the same, a matchless editorial board was set up. Every individual on the board went through rigorous rounds of assessment to prove their worth. After which they invested a large part of their time researching and compiling the most relevant data for our readers. Conferences and sessions were held from time to time between the editorial board and the contributing authors to present the data in the most comprehensible form. The editorial team has worked tirelessly to provide valuable and valid information to help people across the globe.

Every chapter published in this book has been scrutinized by our experts. Their significance has been extensively debated. The topics covered herein carry significant findings which will fuel the growth of the discipline. They may even be implemented as practical applications or may be referred to as a beginning point for another development. Chapters in this book were first published by InTech; hereby published with permission under the Creative Commons Attribution License or equivalent.

The editorial board has been involved in producing this book since its inception. They have spent rigorous hours researching and exploring the diverse topics which have resulted in the successful publishing of this book. They have passed on their knowledge of decades through this book. To expedite this challenging task, the publisher supported the team at every step. A small team of assistant editors was also appointed to further simplify the editing procedure and attain best results for the readers.

Our editorial team has been hand-picked from every corner of the world. Their multi-ethnicity adds dynamic inputs to the discussions which result in innovative outcomes. These outcomes are then further discussed with the researchers and contributors who give their valuable feedback and opinion regarding the same. The feedback is then collaborated with the researches and they are edited in a comprehensive manner to aid the understanding of the subject.

Apart from the editorial board, the designing team has also invested a significant amount of their time in understanding the subject and creating the most relevant covers. They scrutinized every image to scout for the most suitable representation of the subject and create an appropriate cover for the book.

The publishing team has been involved in this book since its early stages. They were actively engaged in every process, be it collecting the data, connecting with the contributors or procuring relevant information. The team has been an ardent support to the editorial, designing and production team. Their endless efforts to recruit the best for this project, has resulted in the accomplishment of this book. They are a veteran in the field of academics and their pool of knowledge is as vast as their experience in printing. Their expertise and guidance has proved useful at every step. Their uncompromising quality standards have made this book an exceptional effort. Their encouragement from time to time has been an inspiration for everyone.

The publisher and the editorial board hope that this book will prove to be a valuable piece of knowledge for researchers, students, practitioners and scholars across the globe.

List of Contributors

Yongjie Wang
Laboratory of Marine and Food Microbiology, College of Food Science and Technology, Shanghai Ocean University, Shanghai, China

Olaf R.P. Bininda-Emonds
Institute for Biology and Environmental Sciences (IBU), Carl von Ossietzky University Oldenburg, Oldenburg, Germany

Johannes A. Jehle
Institute for Biological Control, Federal Research Centre for Cultivated Plants, Julius Kühn-Institut, Darmstadt, Germany

M. Leticia Ferrelli and Víctor Romanowski
Instituto de Biotecnología y Biología Molecular, Facultad de Ciencias Exactas, Universidad Nacional de La Plata, CONICET, Argentina

Marcelo F. Berretta and Alicia Sciocco-Cap
Laboratorio de Ingeniería Genética y Biología Celular y Molecular - Area Virosis de Insectos, Departamento de Ciencia y Tecnología, Universidad Nacional de Quilmes, Argentina

Mariano N. Belaich and P. Daniel Ghiringhelli
Instituto de Microbiología y Zoología Agrícola, INTA Castelar, Argentina

Consuelo Carrillo
APHIS-NVSL-FADDL, USA

Maria Laura Garcia
Instituto de Biotecnología y Biología Molecular, Facultad de Ciencias Exactas, Universidad, Nacional de La Plata, CONICET, Argentina

Mario P.S. Chin
Department of Microbiology and Immunology, Center for Substance Abuse Research, Temple University School of Medicine, USA

Ewan P. Plant
Food and Drug Administration, USA

Julie Lucifora and Ulrike Protzer
Institute of Virology, Technische Universität, München / Helmholtz Zentrum München, Germany

Majid Laassri and Konstantin Chumakov
Center for Biologics Evaluation and Research, U.S. Food and Drug Administration, Rockville, USA

Elena Cherkasova
National Heart, Lung, and Blood Institute, National Institutes of Health, Bethesda, USA

Mones S. Abu-Asab
National Cancer Institute, National Institutes of Health, Bethesda, USA

Andriyan Grinev, Zhong Lu, Vladimir Chizhikov and Maria Rios
Center for Biologics Evaluation and Research, US Food and Drug Administration, USA

Day-Yu Chao
National Chung Hsing University, Taiwan

Limin Chen, Ning Li and Cheng Luo
Drug Discovery and Design Center, Shanghai Institute of Materia Medica, Chinese Academy of Sciences, Shanghai, China

Jaime E. Castellanos
Grupo de Virología, Universidad El Bosque, Colombia
Grupo Patogénesis Viral, Universidad Nacional de Colombia Bogotá, Colombia

Myriam Lucia Velandia
Grupo de Virología, Universidad El Bosque, Colombia

Jiann-Ruey Hong
Laboratory of Molecular Virology and Biotechnology, Institute of Biotechnology, National Cheng Kung University, Taiwan

Jen-Leih Wu
Laboratory of Marine Molecular Biology and Biotechnology, Institute of Cellular and Organismic Biology, Academia Sinica, Nankang, Taiwan

Maria R. Albiach-Marti
Instituto Valenciano de Investigaciones Agrarias, Valencia, Spain

www.ingramcontent.com/pod-product-compliance
Lightning Source LLC
Chambersburg PA
CBHW070734190326
41458CB00004B/1164